SUPER-RESOLUTION
IMAGING

SUPER-RESOLUTION IMAGING

Edited by

Peyman Milanfar

CRC Press
Taylor & Francis Group
Boca Raton   London   New York

CRC Press is an imprint of the
Taylor & Francis Group, an **informa** business

CRC Press
Taylor & Francis Group
6000 Broken Sound Parkway NW, Suite 300
Boca Raton, FL 33487-2742

© 2011 by Taylor and Francis Group, LLC
CRC Press is an imprint of Taylor & Francis Group, an Informa business

No claim to original U.S. Government works

International Standard Book Number: 978-1-4398-1930-2 (Hardback)

Visit the Taylor & Francis Web site at
http://www.taylorandfrancis.com

and the CRC Press Web site at
http://www.crcpress.com

To my wife Sheila, and
our children Leila and Sara

Contents

3 Locally Adaptive Kernel Regression for Space-Time Super-Resolution   63

Hiroyuki Takeda and Peyman Milanfar

4 Super-Resolution with Probabilistic Motion Estimation   97

Matan Protter and Michael Elad

Preface

"Enhance 34 to 36. Pan right and pull back. Stop. Enhance 34 to 46. Pull back. Wait a minute, go right, stop. Enhance 57 to 19. Track 45 left. Stop. Enhance 15 to 23. Give me a hard copy right there."

– Harrison Ford in *Blade Runner*, 1982

If you have never seen the movie *Blade Runner*, you should. Aside from being one of the greatest science fiction films of all time, it is uniquely relevant to the subject of this book: almost 30 years ago, the opening scene of this movie foresaw the idea of super-resolution. In the intervening years, a great deal has transpired: computing power has increased by orders of magnitude, digital cameras are everywhere, and of course digital displays have become magnificently detailed. Along with these advances, the public's expectations for high-quality imagery has naturally intensified, often out of proportion with the state-of-the-art technology. In fact, in the last few years, the visual quality of captured images and video has not kept pace with these lofty expectations. By packing increasingly larger number of pixels into ever smaller spaces, and using less sophisticated optical elements, public, commercial, and official users alike have seen an overall decline in the visual quality of their recorded content. So despite what might at first seem like a losing battle against better and cheaper sensors, super-resolution technology (and image enhancement more generally) has really become more relevant than ever. Given that almost all recorded visual content is now enhanced in one form or another by just about every digital camera sold today, it is not entirely outrageous then to believe that before long, super-resolution will become the "killer application" for imaging.

Ironically, only two years after the release of *Blade Runner*, the seminal paper by Tsai and Huang kick-started the modern idea of computational super-resolution. While results in sampling theory dating as far back as the '50s (Yan) and the '70s (Papoulis) had hinted at the idea, it was Tsai and Huang who explicitly showed that, at least in theory, it was possible to improve resolution by registering and fusing multiple images. The rest, as they say, is history. We are fortunate to be able to write a little bit of that history in this book. In the last five years or so, the field of super-resolution imaging has truly flourished both academically and commercially. The growing importance of super-resolution imaging has manifested itself in an explosive growth in the number of papers in this area and citations to these papers (a few dozen in 1994, to more than 500 in 2004, and to more than 2000 in 2008). What has been missing, however, is a book to not only gather key recent contributions in one place, but also to serve as a starting point for those interested in this field to begin learning about and exploring the state of the art. This is what this book hopes to accomplish.

As is probably well-known by now, every super-resolution algorithm ever developed is sabotaged by at least one spoke of our triumvirate "axis of evil": the need for (1) (subpixel) accurate motion estimation, (2) (spatially varying) deblurring, and (3) robustness to modeling and stochastic errors. To be sure, these are not independent problems and should ideally be treated in unison (ambitious graduate students take note!). But realistically, each is sufficiently complex as to merit its own section in the library, or at least a couple of nice chapters in this book. This books gathers contributions that will present the reader with a snapshot of where the field stands, a reasonable idea of where the field is heading—and perhaps where it should be heading. Chapter 1 provides an introduction to the history of the subject that should be of broad interest. Indeed, the collection of citations summarized in this chapter is an excellent wellspring for continued research on super-resolution.

One of the most active areas of work in image and video enhancement in recent years has been the subject of locally adaptive processing methods, which are discussed in Chapters 2, 3, 4, and 5. In contrast to globally optimal methods (treated later in the book), these methods are built on the notion that processing should be strongly tailored to the local behavior of the given data. An explicit goal in some cases, and a happy consequence in others, local processing enables us to largely avoid direct and detailed estimation of motion. Readers interested in methods for explicit motion estimation will find an excellent overview of modern techniques in Chapter 6.

While motion estimation is typically the first step in many super-resolution algorithms, deblurring is typically the last step. Unfortunately, having been relegated to the last position has meant that this important aspect of enhancement has not received the respect and attention it deserves. Despite heavy recent activity in both the image processing and machine vision community, and some notable successes, deblurring even in its simplest (space-invariant, known point-spread-function) form is still largely an unsolved problem. Inasmuch as we would like to hope, blur almost never manifests itself in a spatially uniform fashion. In Chapter 7, the reader will find a well-motivated and direct attack at this challenging problem. Despite our best efforts, a sequential approach to super-resolution consisting of motion estimation, fusion, and deblurring will always be subject to the vagaries of the data, the models, and noise. As such, building robustness into the reconstruction process, as treated in Chapter 8, is vital if the algorithm is to be practically useful.

As with most inverse problems, super-resolution is highly ill-posed. In the most general case, the motion between the frames, the blur kernel(s), and the high-resolution image of interest are three interwoven unknowns that should ideally be estimated together (rather than sequentially), and whose effect is directly felt in the three points of weakness to which I referred earlier. Principled Bayesian statistical approaches addressing these issues are presented in Chapters 9 and 10, where the ever important prior information is brought to bear on the super-resolution problem. Of course, prior information can be brought to the table either in "bulk" form as a statistical distribution, or in more specific "piecemeal" form as examples. Naively speaking, this distinction is indeed what leads us to learning-based methods described in Chapter 11.

In the final three chapters of the book, we concentrate on applications. Among the many areas of science to which super-resolution has been successfully applied in recent years, medicine and remote sensing have perhaps seen the most direct impact. In Chapter 12, a novel application of super-resolution to massive multispectral remote-sensing data sets is detailed. Medical imaging applications of super-resolution in Chapter 13 discuss two important problems. In what should be good news to everyone, high resolution X-ray imaging is made possible at lower radiation dosages thanks to super-resolution. In another application, detailed imaging of the retina is made possible for diagnostic purposes. Finally, in Chapter 14, a successful commercial application of super-resolution (in which I was fortunate to have a hand) is discussed. This chapter is quite informative not only because of the interesting perspective it provides, but also because of the valuable practical nuggets it imparts to the reader. It is an interesting glimpse into what it really takes to make super-resolution tick.

Perhaps it is worth saying a few words about how this book can be used. As with any edited volume, it is intended to provide a snapshot of the field, which is sure to evolve over time. Yet, I have endeavored to organize the chapters to be used as a teaching tool as well. Indeed, the first four chapters can easily be incorporated into the latter part of a graduate-level course on image processing. Other selected chapters of the book can be used to offer short courses on the subject to a wide audience of engineers and scientists. The book as a whole can also be used as a text for a semester-long focused seminar course on the topic, with one or two lectures dedicated to each chapter. It is hoped that over time the authors may provide supplementary material for each chapter, including slides, code, or data, which will be archived at the book Website – so the interested reader is encouraged to revisit the site.

This book is the collective effort of a kind group of friends and colleagues. I am grateful to each of the authors for giving generously of their time and contributing to this book. I am also thankful to my students past and present for their contributions to this topic, and to this book in particular. Specifically, I acknowledge (soon to be Dr.) Hiroyuki Takeda for his assistance with myriad LaTeX issues.

It is my hope that this book will help to promote this field of endeavor for many years to come.

Peyman Milanfar
Menlo Park, March 2010

1

Image Super-Resolution: Historical Overview and Future Challenges

Jianchao Yang

University of Illinois at Urbana-Champaign

Thomas Huang

University of Illinois at Urbana-Champaign

CONTENTS

1.1 Introduction to Super-Resolution

In most digital imaging applications, high-resolution images or videos are usually desired for later image processing and analysis. The desire for high-resolution stems from two principal application areas: improvement of pictorial

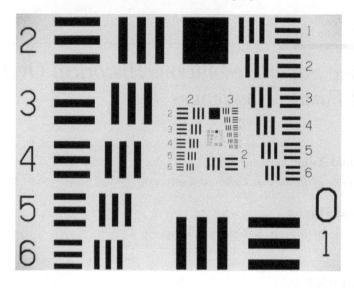

FIGURE 1.1: The 1951 USAF resolution test target, a classic test target used to determine spatial resolution of imaging sensors and imaging systems.

information for human interpretation; and helping representation for automatic machine perception. Image resolution describes the details contained in an image, the higher the resolution, the more image details. The resolution of a digital image can be classified in many different ways: pixel resolution, spatial resolution, spectral resolution, temporal resolution, and radiometric resolution. In this context, we are mainly interested in spatial resolution.

Spatial resolution: *a digital image is made up of small picture elements called pixels. Spatial resolution refers to the pixel density in an image and measures in pixels per unit area. Figure 1.1 shows a classic test target to determine the spatial resolution of an imaging system.*

The image spatial resolution is first limited by the imaging sensors or the imaging acquisition device. A modern image sensor is typically a charge-coupled device (CCD) or a complementary metal-oxide-semiconductor (CMOS) active-pixel sensor. These sensors are typically arranged in a two-dimensional array to capture two-dimensional image signals. The sensor size or equivalently the number of sensor elements per unit area in the first place determines the spatial resolution of the image to capture. The higher density of the sensors, the higher spatial resolution possible of the imaging system. An imaging system with inadequate detectors will generate low-resolution images with blocky effects, due to the aliasing from low spatial sampling frequency. In order to increase the spatial resolution of an imaging system, one straight-forward way is to increase the sensor density by reducing the sensor size. However, as the sensor size decreases, the amount of light incident on each sensor also decreases, causing the so-called shot noise. Also, the hardware cost of a sensor increases with the increase of sensor density or corresponding image pixel density. Therefore, the hardware limitation on the size of the sensor restricts the spatial resolution of an image that can be captured.

While the image sensors limit the spatial resolution of the image, the image details (high-frequency bands) are also limited by the optics, due to lens blurs (associated with the sensor point spread function (PSF)), lens aberration effects, aperture diffractions, and optical blurring due to motion. Constructing imaging chips and optical components to capture very high-resolution images is prohibitively expensive and not practical in most real applications, e.g., widely used surveillance cameras and cell phone built-in cameras. Besides the cost, the resolution of a surveillance camera is also limited in the camera speed and hardware storage. In some other scenarios such as satellite imagery, it is difficult to use high resolution sensors due to physical constraints. Another way to address this problem is to accept the image degradations and use signal processing to post-process the captured images, to trade off computational cost with the hardware cost. These techniques are specifically referred to as Super-Resolution (SR) reconstruction.

Super-Resolution (SR) are techniques that construct high-resolution (HR) images from several observed low-resolution (LR) images, thereby increasing the high-frequency components and removing the degradations caused by the imaging process of the low-resolution camera. The basic idea behind SR is to combine the non-redundant information contained in multiple low-resolution frames to generate a high-resolution image. A closely related technique with SR is the single-image interpolation approach, which can be also used to increase the image size. However, since there is no additional information provided, the quality of the single-image interpolation is very much limited due to the ill-posed nature of the problem, and the lost frequency components cannot be recovered. In the SR setting, however, multiple low-resolution observations are available for reconstruction, making the problem better constrained. The nonredundant information contained in the these LR images is typically introduced by subpixel shifts between them. These subpixel shifts may occur due to uncontrolled motions between the imaging system and scene, e.g., movements of objects, or due to controlled motions, e.g., the satellite imaging system orbits the earth with predefined speed and path.

Each low-resolution frame is a decimated, aliased observation of the true scene. SR is possible only if there exists subpixel motions between these low-resolution frames,[1] and thus the ill-posed upsampling problem can be better conditioned. Figure 1.2 shows a simplified diagram describing the basic idea of SR reconstruction. In the imaging process, the camera captures several LR frames, which are downsampled from the HR scene with subpixel shifts between each other. SR construction reverses this process by aligning the LR

---

[1]The mainstream SR techniques rely on motions, although there are some works using defocus as a cue.

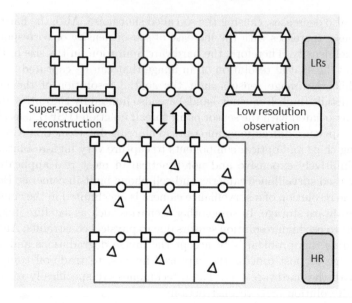

FIGURE 1.2: The basic idea for super-resolution reconstruction from multiple low-resolution frames. Subpixel motion provides the complementary information among the low-resolution frames that makes SR reconstruction possible.

observations to subpixel accuracy and combining them into a HR image grid (interpolation), thereby overcoming the imaging limitation of the camera. SR (some of which described in this book), arises in many areas such as:

1. Surveillance video [20, 55]: frame freeze and zoom region of interest (ROI) in video for human perception (e.g., look at the license plate in the video), resolution enhancement for automatic target recognition (e.g., try to recognize a criminal's face).

2. Remote sensing [29]: several images of the same area are provided, and an improved resolution image can be sought.

3. Medical imaging (CT, MRI, Ultrasound, etc.) [59, 70, 47, 60]: several images limited in resolution quality can be acquired, and SR technique can be applied to enhance the resolution.

4. Video standard conversion, e.g., from NTSC video signal to HDTV signal.

This chapter targets an introduction to the SR research area, by explaining some basic techniques of SR, an overview of the literature, and discussions about some challenging issues for future research.

1.2 Notations

Before talking about SR techniques, we introduce the notations we use in this chapter. Uppercase bold letters X and Y denote the vector form in lexicographical order for HR and LR images respectively. Lowercase bold letters x and y denote the vector form in lexicographical order for HR and LR image *patches* respectively. Underlined uppercase bold letters are used to denote a vector concatenation of multiple vectors, e.g., $\underline{Y}$ is a vector concatenation of Y_k $(k = 1, 2, ..., K)$. We use plain uppercase symbols to denote matrices, and plain lowercase symbols to denote scalars.

1.3 Techniques for Super-Resolution

SR reconstruction has been one of the most active research areas since the seminal work by Tsai and Huang [99] in 1984. Many techniques have been proposed over the last two decades [4, 65] representing approaches from frequency domain to spatial domain, and from the signal processing perspective to the machine learning perspective. Early works on super-resolution mainly followed the theory of [99] by exploring the shift and aliasing properties of the Fourier transform. However, these frequency domain approaches are very restricted in the image observation model they can handle, and real problems are much more complicated. Researchers nowadays most commonly address the problem mainly in the spatial domain, for its flexibility to model all kinds of image degradations. This section talks about these techniques, starting from the image observation model.

1.3.1 Image Observation Model

The digital imaging system is not perfect due to hardware limitations, acquiring images with various kinds of degradations. For example, the finite aperture size causes optical blur, modeled by Point Spread Function (PSF). The finite aperture time results in motion blur, which is very common in videos. The finite sensor size leads to sensor blur; the image pixel is generated by integration over the sensor area instead of impulse sampling. The limited sensor density leads to aliasing effects, limiting the spatial resolution of the achieved image. These degradations are modeled fully or partially in different SR techniques.

Figure 1.3 shows a typical observation model relating the HR image with LR video frames, as introduced in the literature [65, 82]. The input of the imaging system is continuous natural scenes, well approximated as band-limited signals. These signals may be contaminated by atmospheric turbulence be-

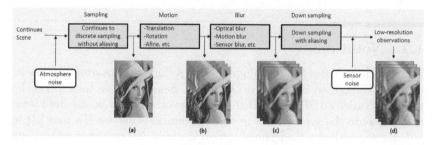

FIGURE 1.3: The observation model of a real imaging system relating a high resolution image to the low-resolution observation frames with motion between the scene and the camera.

fore reaching the imaging system. Sampling the continues signal beyond the Nyquist rate generates the high resolution digital image (a) we desire. In our SR setting, usually there exists some kind of motion between the camera and scene to capture. The inputs to the camera are multiple frames of the scene, connected by possibly local or global shifts, leading to image (b). Going through the camera, these motion related high-resolution frames will incur different kinds of blurring effects, such as optical blur and motion blur. These blurred images (c) are then downsampled at the image sensors (e.g., CCD detectors) into pixels, by an integral of the image falling into each sensor area. These downsampled images are further affected by the sensor noise and color filtering noise. Finally, the frames captured by the low-resolution imaging system are blurred, decimated, and noisy versions of the underlying true scene.

Let X denote the HR image desired, i.e., the digital image sampled above Nyquist sampling rate from the band-limited continuous scene, and Y_k be the k-th LR observation from the camera. X and $Y_k's$ are represented in lexicographical order. Assume the camera captures K LR frames of X, where the LR observations are related with the HR scene X by

$$Y_k = D_k H_k F_k X + V_k, k = 1, 2, ..., K, \tag{1.1}$$

where F_k encodes the motion information for the k-th frame, H_k models the blurring effects, D_k is the down-sampling operator, and V_k is the noise term. These linear equations can be rearranged into a large linear system

$$\begin{bmatrix} Y_1 \\ Y_2 \\ \cdot \\ \cdot \\ Y_K \end{bmatrix} = \begin{bmatrix} D_1 H_1 F_1 \\ D_2 H_2 F_2 \\ \cdot \\ \cdot \\ D_K H_K F_K \end{bmatrix} X + \underline{V} \tag{1.2}$$

or equivalently

$$\underline{Y} = MX + \underline{V} \tag{1.3}$$

The involved matrices D_k, H_k, F_k, or M are very sparse, and this linear system is typically ill-posed. Furthermore, in real imaging systems, these matrices are unknown and need to be estimated from the available LR observations, leaving the problem even more ill-conditioned. Thus, proper prior regularization for the high resolution image is always desirable and often even crucial. In the following, we will introduce some basic super-resolution techniques proposed in the literature and give an overview of the recent developments.

1.3.2   Super-Resolution in the Frequency Domain

The pioneering work for super-resolution traces back to Tsai and Huang [99], in which the authors related the high resolution image with multiple shifted low-resolution images by a frequency domain formulation based on the shift and aliasing properties of the Continuous and Discrete Fourier Transforms. Let $x(t_1, t_2)$ denote a continuous high-resolution scene. The global translations yield K shifted images, $x_k(t_1, t_2) = x(t_1 + \triangle_{k_1}, t_2 + \triangle_{k_2})$, with $k = 1, 2, ..., K$, where $\triangle_{k_1}$ and $\triangle_{k_2}$ are arbitrary but known shifts. The continuous Fourier transform (CFT) of the scene is given by $\mathcal{X}(u_1, u_2)$ and those of the translated scenes by $\mathcal{X}_k(u_1, u_2)$. Then by the shifting properties of the CFT, the CFT of the shifted images can be written as

$$\mathcal{X}_k(u_1, u_2) = \exp\left[j2\pi(\triangle_{k_1} u_1 + \triangle_{k_2} u_2)\right] \mathcal{X}(u_1, u_2). \qquad (1.4)$$

The shifted images are impulse sampled with the sampling period T_1 and T_2 to yield observed low-resolution image $y_k[n_1, n_2] = x_k(n_1 T_1 + \triangle_{k_1}, n_2 T_2 + \triangle_{k_2})$ with $n_1 = 0, 1, 2, ..., N_1 - 1$ and $n_2 = 0, 1, 2, ..., N_2 - 1$. Denote the discrete Fourier transforms (DFTs) of these low-resolution images by $\mathcal{Y}_k[r_1, r_2]$. The CFTs of the shifted images are related with their DFTs by the aliasing property:

$$\mathcal{Y}_k[r_1, r_2] = \frac{1}{T_1 T_2} \sum_{m_1=-\infty}^{\infty} \sum_{m_2=-\infty}^{\infty} \mathcal{X}_k\left(\frac{2\pi}{T_1}\left(\frac{r_1}{N_1} - m_1\right), \frac{2\pi}{T_2}\left(\frac{r_2}{N_2} - m_2\right)\right).$$
$$(1.5)$$

Assuming $\mathcal{X}(u_1, u_2)$ is band-limited, $|\mathcal{X}(u_1, u_2)| = 0$ for $|u_1| \geq (N_1\pi)/T_1$, $|u_2| \geq (N_2\pi)/T_2$, combining Eqn. 1.4 and Eqn. 1.5 we relate the DFT coefficients of $\mathcal{Y}_k[r_1, r_2]$ with the samples of the unknown CFT of $x(t_1, t_2)$ in matrix form as [2]

$$\underline{y} = \Phi \underline{x}, \qquad (1.6)$$

where $\underline{y}$ is a $K \times 1$ column vector with the kth element being the DFT coefficient $\mathcal{Y}_k[r_1, r_2]$, $\underline{x}$ is a $N_1 N_2 \times 1$ column vector containing the samples of the unknown CFT coefficients of $x(t_1, t_2)$, and Φ is a $K \times N_1 N_2$ matrix relating

---

[2]Strictly, subscripts $\{r_1, r_2\}$ should be used in the following equation. We omit those for an uncluttered presentation.

$\mathcal{Y}$ and $\mathcal{X}$. Eqn. 1.6 defines a set of linear equations from which we intend to solve $\mathcal{X}$ and then use the inverse DFT to obtain the reconstructed image.

The above formulation for SR reconstruction assumes a noise-free and global translation model with known parameters. The downsampling process is assumed to be impulse sampling, with no sensor blurring effects modeled. Along this line of work, many extensions have been proposed to handle more complicated observation models. Kim et al. [49] extended [99] by taking into account the observation noise as well as spatial blurring. Their later work in [5] extend the work further by introducing Tikohonov regularization [95]. In [89], a local motion model is considered by dividing the images into overlapping blocks and estimating motions for each local block individually. In [98], the restoration and motion estimation are done simultaneously using an EM algorithm. However, the frequency domain SR theory of these works did not go beyond as what was initially proposed. These approaches are computationally efficient, but limited in their abilities to handle more complicated image degradation models and include various image priors as proper regularization. Later works on super-resolution reconstruction have been almost exclusively in the spatial domain.

1.3.3   Interpolation-Restoration: Non-Iterative Approaches

Many spatial domain approaches [4, 82, 65, 2] have been proposed over the years to overcome the difficulties of the frequency domain methods. As the HR image and the LR frames are related in a sparse linear system (1.3), similar to the traditional single image restoration problem [26], many flexible estimators can be applied to the SR reconstruction. These include Maximum Likelihood (ML), Maximum *a Posteriori* (MAP)[84, 35], and Projection Onto Convex Sets (POCS)[88]. In this section, we start with the simplest and a noniterative forward model for SR reconstruction in the spatial domain, in analogy to the frequency domain approach.

Assume H_k is Linearly Spatial Invariant (LSI) and is the same for all K frames, and we denote it as H. Suppose F_k considers only simple motion models such as translation and rotation, then H and F_k commute [27, 30] and we get

$$Y_k = D_k F_k H X + V_k = D_k F_k Z, k = 1, 2, ..., K, \qquad (1.7)$$

which motivates a forward noniterative approach based on interpolation and restoration. There are three stages for this approach 1) low-resolution image registration; 2) nonuniform interpolation to get Z, and 3) deblurring and noise removal to get X. Figure 1.4 shows the procedure of such an approach. The low-resolution frames are first aligned by some image registration algorithm [77] to subpixel accuracy. These aligned low-resolution frames are then put on a high-resolution image grid, where nonuniform interpolation methods are used to fill in those missing pixels on the HR image grid to get Z. At last, Z is deblurred by any classical deconvolutional algorithm with noise removal to achieve $\boldsymbol{X}$. Keren et al. [48] proposed an early two-step approach to SR reconstruction based on a global translation and rotation model. Gross et al. [101] proposed a nonuniform interpolation of a set of spatially shifted low-resolution images by utilizing the generalized multi-channel sampling theorem by Yen [109] and later Papulis [64], followed by deblurring. Nguyen and Milanfar [62] proposed an efficient wavelet-based interpolation SR reconstruction algorithm by exploiting the interlacing sampling structure in the low-resolution data. Alam et al. [1] presented an efficient interpolation scheme based on weighted nearest neighbors, followed by Wiener filtering for deblurring. Focusing on the special case of SR reconstruction where the observation is composed of pure translation, space invariant blur, and additive Gaussian noise, Elad and Hel-Or [27] presented a very computationally efficient algorithm. [52] proposed a triangulation-based method for interpolating irregularly sampled data. The triangulation method, however, is not robust to noise commonly present in real applications. Based on the normalized convolution [50], Pham et al. [71] proposed a robust certainty and a structure-adaptive applicability function to the polynomial facet model and applied it to fusion of irregularly sampled data. Recently, Takeda et al. [91] proposed an adaptive steering kernel regression for interpolation on the high-resolution image grid where the low-resolution images are registered and mapped on.

These interpolation-restoration forward approaches are intuitive, simple, and computationally efficient [30], [18], assuming simple observation models. However, the step-by-step forward approach does not guarantee optimality of the estimation. The registration error can easily propagate to the later processing. Also, the interpolation step is suboptimal without considering the noise and blurring effects. Moreover, without the HR image prior as proper regularization, the interpolation based approaches need special treatment of limited observations in order to reduce aliasing.

1.3.4 Statistical Approaches

Unlike the interpolation-restoration approaches, statistical approaches relate the SR reconstruction steps stochastically toward optimal reconstruction. The HR image and motions among low-resolution inputs can be both regarded as stochastic variables. Let $M(\nu, h)$ denote the degradation matrix defined by the motion vector ν and blurring kernel h, the SR reconstruction can be cast

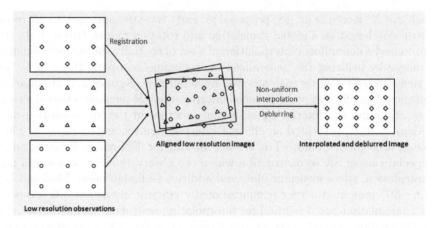

FIGURE 1.4: The interpolation SR approach based on alignment and post processing of deblurring.

into a full Bayesian framework:

$$
\begin{aligned}
\boldsymbol{X} &= \arg\max_{\boldsymbol{X}} Pr(\boldsymbol{X}|\underline{\boldsymbol{Y}}) \\
&= \arg\max_{\boldsymbol{X}} \int_{\nu,h} Pr(\boldsymbol{X}, M(\nu,h)|\underline{\boldsymbol{Y}})d\nu \\
&= \arg\max_{\boldsymbol{X}} \int_{\nu,h} \frac{Pr(\underline{\boldsymbol{Y}}|\boldsymbol{X}, M(\nu,h))Pr(\boldsymbol{X}, M(\nu,h))}{Pr(\underline{\boldsymbol{Y}})}d\nu \\
&= \arg\max_{\boldsymbol{X}} \int_{\nu,h} Pr(\underline{\boldsymbol{Y}}|\boldsymbol{X}, M(\nu,h))Pr(\boldsymbol{X})Pr(M(\nu,h))d\nu.
\end{aligned}
\tag{1.8}
$$

Note that $\boldsymbol{X}$ and $M(\nu,h)$ are statistically independent [35]. Here $Pr(\underline{\boldsymbol{Y}}|\boldsymbol{X}, M(\nu,h))$ is the data likelihood, $Pr(\boldsymbol{X})$ is the prior term on the desired high-resolution image and $Pr(M(\nu,h))$ is a prior term on the motion estimation. $\underline{V}$ in Eqn. 1.3 usually stands for additive noise, assumed to be a zero-mean and white Gaussian random vector. Therefore,

$$
Pr(\underline{\boldsymbol{Y}}|\boldsymbol{X}, M(\nu,h)) \propto \exp\left\{-\frac{1}{2\sigma^2}\|\underline{\boldsymbol{Y}} - M(\nu,h)\boldsymbol{X}\|^2\right\}.
\tag{1.9}
$$

$Pr(\boldsymbol{X})$ is typically defined using the Gibbs distribution in an exponential form

$$
Pr(\boldsymbol{X}) = \frac{1}{Z}\exp\{-\alpha A(\boldsymbol{X})\},
\tag{1.10}
$$

where $A(\boldsymbol{X})$ is a non-negative potential function, and Z is just a normalization factor. The Bayesian formulation in Equation 1.8 is complicated and difficult to evaluate due to the integration over motion estimates. If $M(\nu,h)$ is given or estimated beforehand (denote as M), Eqn. 1.8 can be simplified as

$$\begin{aligned} \boldsymbol{X} &= \arg\max_{\boldsymbol{X}} Pr(\underline{\boldsymbol{Y}}|\boldsymbol{X}, M)Pr(\boldsymbol{X}) \\ &= arg\min_{\boldsymbol{X}}\{\|\underline{\boldsymbol{Y}} - M\boldsymbol{X}\|^2 + \lambda A(\boldsymbol{X})\}, \end{aligned} \tag{1.11}$$

where λ absorbs the variance of the noise and α in Eqn. 1.10, balancing the data consistence and the HR image prior strength. Eqn. 1.11 is the popular *Maximum a Posteriori* (MAP) formulation for SR, where M is assumed to be known. The statistical approaches discussed below vary in the ways of treating degradation matrix $M(\nu, h)$, prior term $Pr(\boldsymbol{X})$, and statistical inference methods toward Equation 1.8.

1.3.4.1   Maximum Likelihood

If we assume uniform prior over $\boldsymbol{X}$, Eqn. 1.11 reduces to the simplest maximum likelihood (ML) estimator (motion estimation is assumed as a prior). The ML estimator relies on the observations only, seeking the most likely solution for the observations to take place by maximizing $p(\underline{\boldsymbol{Y}}|\boldsymbol{X})$, giving

$$\hat{\boldsymbol{X}}_{ML} = \arg\min_{\boldsymbol{X}} \|\underline{\boldsymbol{Y}} - M\boldsymbol{X}\|^2. \tag{1.12}$$

Differentiating Eqn. 1.12 with respect to $\boldsymbol{X}$ and setting the derivative to be zero gives the classical pseudo-inverse result

$$\hat{\boldsymbol{X}}_{ML} = (M^T M)^{-1} M^T \underline{\boldsymbol{Y}}. \tag{1.13}$$

If $M^T M$ is singular, the problem is ill-posed and there are infinite many possible solutions due to the null space of M. This naturally leads to the term of regularization for the sake of a unique solution from purely the algebraic point of view, which although can be interpreted in the MAP framework. For computation, direct inverse of matrix as $M^T M$ is usually prohibitive in practice due to the high-dimensionality problem. For example, if the low-resolution images are of size 100×100 and are to be zoomed to a single high-resolution frame $\boldsymbol{X}$ of 300×300, M is of the size 90000×90000, requiring inverse of a matrix of size 90000×90000. Therefore, many iterative methods for practical ways to solve this large set of sparse linear equations have been suggested in the literature [111].

Irani and Peleg proposed a simple but very popular method, based on an error back-projection scheme inspired by computer-aided tomography, in [39, 40, 41]. The algorithm iteratively updates the current estimation by adding back the warped simulation error convolved with a back-projection function (BPF):

$$\boldsymbol{X}^{i+1} = \boldsymbol{X}^i + c\sum_k F_k^{-1}[h_{bpf} * S \uparrow (\widehat{\boldsymbol{Y}}_k - \boldsymbol{Y}_k)], \tag{1.14}$$

where c is a constant, h_{bpf} is the back-projection kernel, $S \uparrow$ is the upsampling operator, and $\widehat{\boldsymbol{Y}}_k$ is the simulated k-th LR frame from the current HR estimation. In [41], the authors applied this idea to real applications by incorporating a multiple motion tracking algorithm to deal with partially occluded objects, transparent objects and some objects of interest. The back-projection algorithm is simple and flexible in handing many observations with different degradation procedures. However, the solution of back-projection is not unique, depending on the initialization and the choice of back-projection kernel. As shown in [26] and [10], the back-projection algorithm is none other than an ML estimator. The choice of BPF implies some underlying assumption about the noise covariance of the observed low-resolution pixels [10]. Treating the motion estimates $M(\nu)$ as unknown, Tom et al. [98] proposed an ML SR image estimation algorithm to estimate the subpixel shifts, noise of the image, and the HR image simultaneously. The proposed ML estimation is treated by the Expectation-Maximization (EM) algorithm.

As in the image denoising and single image expansion case, direct ML estimator without regularization in SR where the number of observations is limited can be severely ill-posed, especially when the zoom factor is large (e.g. greater than 2). The ML estimator is usually very sensitive to noise, registration estimation errors, and PSF estimation errors [10], and therefore proper regularization on the feasible solution space is always desirable. This leads to the mainstream SR reconstruction approaches based on MAP.

1.3.4.2 Maximum a Posteriori

Many works [46, 84, 15] in SR reconstruction have followed the MAP approach in Eqn. 1.11, where the techniques vary in the observation model assumptions and the prior term $Pr(\boldsymbol{X})$ for the desired solution. Different kinds of priors for natural images have been proposed in the literature, but none of them stands out as the lead. In the following, we list three commonly used image priors for the SR reconstruction techniques.

1. **Gaussian MRF**. The Gaussian Markov Random Field (GMRF) [37, 33] takes the form

$$A(\boldsymbol{X}) = X^T Q X, \qquad (1.15)$$

   where Q is a symmetric positive matrix, capturing spatial relations between adjacent pixels in the image by its off-diagonal elements. Q is often defined as $\Gamma^T \Gamma$, where Γ acts as some first or second derivative operator on the image $\boldsymbol{X}$. In such a case, the log likelihood of the prior is

$$\log p(\boldsymbol{X}) \propto \|\Gamma \boldsymbol{X}\|^2, \qquad (1.16)$$

   which is well-known as the *Tikhonov regularization* [95, 26, 63], the most commonly used method for regularization of ill-posed problems. Γ is usually referred as Tikhonov matrix. Hardie et al. [35] proposed a joint MAP framework for simultaneous estimation of the high-resolution image and motion parameters with Gaussian MRF prior for the HR image. Bishop and Tipping [96] proposed a simple Gaussian process prior where the covariance matrix Q is constructed by spatial correlations of the image pixels. The nice analytical property of Gaussian process prior allows a Bayesian treatment of the SR reconstruction problem, where the unknown high-resolution image is integrated out for robust estimation of the observation model parameters (unknown PSFs and registration parameters). Although the GMRF prior has many analytical advantages, a common criticism for it associated with super-resolution reconstruction is that the results tend to be overly smooth, penalizing sharp edges that we desire to recover.

2. **Huber MRF**. The problem with GMRF can be ameliorated by modeling the image gradients with a distribution with heavier tails than Gaussian, leading to the popular Huber MRF (HMRF) where the Gibbs potentials are determined by the Huber function,

$$\rho(a) = \begin{cases} a^2 & |a| \leq \alpha \\ 2\alpha|a| - \alpha^2 & otherwise, \end{cases} \tag{1.17}$$

where a is the first derivative of the image. Such a prior encourages piecewise smoothness, and can preserve edges well. Schultz and Stevenson [83] applied this Huber MRF to the single image expansion problem, and later to the SR reconstruction problem in [84]. Many later works on super-resolution employed the Huber MRF as the regularization prior, such as [11, 12, 15, 13, 73, 74] and [3].

3. **Total Variation**. The Total Variation (TV) norm as a gradient penalty function is very popular in the image denoising and deblurring literature [81, 54, 16]. The TV criterion penalizes the total amount of change in the image as measured by the ℓ_1 norm of the magnitude of the gradient

$$A(\boldsymbol{X}) = \|\nabla \boldsymbol{X}\|_1 \tag{1.18}$$

where ∇ is a gradient operator that can be approximated by Laplacian operators [81]. The ℓ_1 norm in the TV criterion favors sparse gradients, preserving steep local gradients while encouraging local smoothness[13]. Farsiu et al. [30] generalized the notation of TV and proposed the so called bilateral TV (BTV) for robust regularization.

For more comparisons of these generic image priors on effecting the solution of super-resolution, one can further refer to [10] and [25].

1.3.4.3 Joint MAP Restoration

Multiple frame SR reconstruction can be divided into two subproblems: LR registration and HR estimation. Many previous algorithms treat these two processes as two distinct processes: first do registration and then estimation by MAP, which is suboptimal as registration and estimation are interdependent. Motion estimation and HR estimation can benefit each other if interactions between them are allowed. In joint MAP restoration, Eqn. 1.11 is extended to include the motion and PSF estimates as unknowns for inference:

$$
\begin{aligned}
\{\boldsymbol{X}, \nu, h\} &= \arg \max_{\boldsymbol{X}, \nu, h} Pr(\underline{\boldsymbol{Y}}|\boldsymbol{X}, M(\nu, h)) Pr(\boldsymbol{X}) Pr(M(\nu, h)) \\
&= \arg \min_{\boldsymbol{X}, \nu, h} -\log\left[Pr(\underline{\boldsymbol{Y}}|\boldsymbol{X}, M(\nu, h))\right] - \log\left[Pr(\boldsymbol{X})\right] \\
&\quad - \log\left[Pr(M(\nu, h))\right].
\end{aligned} \tag{1.19}
$$

Tom et al. [98] divided the SR problem into three subproblems, namely registration, restoration and interpolation. Instead of solving them independently they simultaneously estimated registration and restoration by maximizing likelihood using Expectation-Maximization (EM). Later they included interpolation into the framework and estimated all of the unknowns using EM in [97]. [35] applied the MAP framework for simultaneous estimation of the high-resolution image and translation motion parameters (PSF is taken as a known prior). The high-resolution image and motion parameters are estimated using a cyclic coordinate-descent optimization procedure. The algorithm converges slowly but improves the estimation a lot. Segall *et al.* [86, 85] presented an approach of joint estimation of dense motion vectors and HR images applied to compressed video. Woods *et al.*[105] treated the noise variance, regularization and registration parameters all as unknowns and estimated them jointly in a Bayesian framework based on the available observations. Chung *et al.* [19] proposed a joint optimization framework and showed superior performance to the coordinate descent approach [46]. The motion model they handled is affine transformations. To handle more complex multiple moving objects problems in the SR setting, Shen et al. [87] addressed the problem by MAP formulation combining motion estimation, segmentation and SR reconstruction together. The optimization is done in a cyclic coordinate descent process similar to [46].

1.3.4.4   Bayesian Treatments

Due to limited low-resolution observations, the SR reconstruction problem is ill-posed in nature. Joint MAP estimation of motion parameters, PSF, and the HR image may face the problem of overfitting [96]. While motion and blurring is difficult to model in general, simple models spanned by few parameters are sufficient for SR applications in many scenarios. Given the low-resolution observations, however, estimating these parameters by integrating over the unknown high-resolution image is a useful approach. Bishop and Tipping [96] proposed such a Bayesian approach for SR where the unknown high-resolution image is integrated out and the marginal is used to estimate the PSF and motion parameters. To make the problem analytically tractable, a Gaussian Markov Random Field (GMRF) prior is used to model the high-resolution image. Even though an unfavorable GMRF is used for the high-resolution image, the PSF and motion parameters can still be estimated quite accurately. Then the estimated parameters are fixed, and a MAP estimation of the HR image is performed. An in-depth analysis similar to this for blind deconvolution is discussed in [53]. Such a Bayesian approach outperforms the joint MAP approaches in Subsection 1.3.4.3, which will easily get overfitting with the PSF parameters. However, the integration over the high-resolution image is computationally heavy and the Gaussian prior over the image leads the final reconstruction toward excess smoothness. Instead of marginalization over the unknown high-resolution image, Pickup et al. proposed in their recent works [73, 74, 72] to integrate over the unknown PSF and motion parameters as in Eqn. 1.8, which is motivated to overcome the uncertainty of the registration parameters [79]. The registration parameters are estimated beforehand and then treated as Gaussian variables with the pre-estimated values as the means to model their uncertainty. The HR image estimation can be combined with any favorable image prior for MAP estimation after integrating the observation model parameters. Sharper results can be obtained with such an approach compared with [96] as reported in [73, 74, 72].

Such Bayesian treatments, by marginalizing the unknowns, demonstrate promising power for SR recovery. However, in order for integration to be tractable, image priors or registration parameters have to take simple parametric forms, limiting these models in dealing with more complex cases that may happen in real videos. Computation could also be a concern for such algorithms in realistic applications.

1.3.5 Example-Based Approaches

Previous super-resolution approaches rely on aggregating multiple frames that contain complementary spatial information. Generic image priors are usually deployed to regularize the solution properly. The regularization becomes especially crucial when an insufficient number of measurements is supplied, as in the extreme case, only one single low-resolution frame is observed. In such cases, generic image priors do not suffice as an effective regularization for SR [2]. A recently emerging methodology for regularizing the ill-posed super-resolution reconstruction is to use examples, in order to break the super-resolution limit caused by inadequate measurements. Different from previous approaches where the prior is in a parametric form regularizing on the whole image, the example-based methods develop the prior by sampling from other images, similar to [24],[38] in a local way.

One family of example-based approaches is to use the examples directly, with the representative work proposed by Freeman et al. [31]. Such approaches usually work by maintaining two sets of training patches, $\{x_i\}_{i=1}^n$ sampled from the high-resolution images, and $\{y_i\}_{i=1}^n$ sampled from the low-resolution images correspondingly. Each patch pair (x_i, y_i) is connected by the observation model $y_i = DHx_i + v$. This high- and low-resolution co-occurrence model is then applied to the target image for predicting the high-resolution image in a patch-based fashion, with a Markov Random Field (MRF) model as shown in Figure 1.5. The observation model parameters have to be known as a prior, and the training sets are tightly coupled with the image targeted.

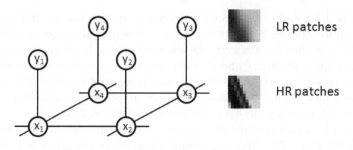

FIGURE 1.5: The MRF model for single frame super-resolution.

Patch size should also be chosen properly. If the patch size is very small, the co-occurrence prior is too weak to make the prediction meaningful. On the other hand, if the patch size is too large, one may need a huge training set to find proximity patches for the current observations.

A naive way to do super-resolution with such a coupled training sets is, for each low-resolution patch in the low-resolution image, find its nearest neighbor $\tilde{\boldsymbol{y}}$ in $\{\boldsymbol{y}_i\}_{i=1}^n$, and then put the corresponding $\tilde{\boldsymbol{x}}$ from $\{\boldsymbol{x}_i\}_{i=1}^n$ to the high-resolution image grid. Unfortunately, this simple approach will produce disturbing artifacts due to noise and the ill-posed nature of super-resolution [25]. Relaxing the nearest neighbor search to k-nearest neighbors can ensure that the proximity patch we desire will be included. Freeman et al. [31] proposed a belief propagation [108] algorithm based on the above MRF model to select the best high-resolution patch found by k-nearest neighbors that has best compatibility with adjacent patches. Sun et al.[90] extended this idea using the sketch prior to enhance only the edges in the image, aiming to speed up the algorithm. The IBP [39] algorithm is then applied as a post processing step to ensure data consistency on the whole image. Wang et al. [103] further followed this line of work and proposed a statistical model that can handle unknown PSF.

The above methods are based on image patches directly, requiring large training sets to include any patterns possibly encountered in testing. Chang et al. [17] proposed another simple but effective method based on neighbor embedding [93], with the assumption of correspondence between the two manifolds formed by the low- and high-resolution image patches. For each low-resolution image patch $\boldsymbol{y}_k^t$ from the test image (superscript "t" distinguishes the test patch from the training patches), the algorithm finds its k-nearest neighbors $\mathcal{N}_t$ from $\{\boldsymbol{y}_i\}_{i=1}^n$, and computes the reconstruction weights by neighbor embedding

$$\hat{w}_s = \arg\min_{w_s} \left\| \boldsymbol{y}_k^t - \sum_{\boldsymbol{y}_s \in \mathcal{N}_t} w_s \boldsymbol{y}_s \right\|^2,$$

$$s.t. \sum_{\boldsymbol{y}_s \in \mathcal{N}_t} w_s = 1. \tag{1.20}$$

The reconstruction weights are then applied to generate the corresponding high-resolution patch $\hat{\boldsymbol{x}}_k^t = \sum_{\boldsymbol{y}_s \in \mathcal{N}_t} \hat{w}_s \boldsymbol{x}_s$. To handle the compatibility problem between adjacent patches, simple averaging in the overlapping regions is performed. The algorithm works nicely even with smaller patch databases than [108]. However, fixing k for each low-resolution patch may result in overfitting or underfitting. Yang et al. [107] proposed another patch-based single frame super-resolution method. The method is derived from the compressive sensing theory, which ensures that linear relationships among high-resolution signals can be precisely recovered from their low-dimensional projections [9], [22]. The algorithm models the training sets as two dictionaries: $D_h = [\boldsymbol{x}_1, \boldsymbol{x}_2, ..., \boldsymbol{x}_n]$ and $D_l = [\boldsymbol{y}_1, \boldsymbol{y}_2, ..., \boldsymbol{y}_n]$. Given a test low-resolution image patch $\boldsymbol{y}_k^t$, the approach basically seeks the supports by an ℓ_1 minimization [23]

$$\hat{w} = \arg\min_{w} \|w\|_1$$

$$s.t. \quad \|\boldsymbol{y}_t - D_l w\|^2 \le \sigma^2, \tag{1.21}$$

which can be rewritten with Lagrange multiplier as an unconstrained optimization problem known as Lasso in the statistics literature [94]. The corresponding high-resolution patch is recovered by $\boldsymbol{x}_k^t = D_h \hat{w}$. Compared to the neighbor embedding method with fixed k neighbors, Yang's method adaptively chooses the fewest necessary supports for reconstruction, avoiding overfitting. Moreover, the ℓ_1 minimization formulation is more robust to noise than the previous mentioned patch-based methods. In a later version [42], this approach is further extended by learning a coupled dictionary instead of using the raw patches, allowing the algorithm to be much more efficient.

One criticism with the aforementioned methods with direct examples is that operating on local patches cannot guarantee global optimality of the estimation. Another kind of example-based approach seeks to perform MAP estimation with local priors on the image space sampled from examples. The pioneering work by Baker and Kanade [2] formulated an explicit regularization that demands proximity between the spatial derivatives of the unknown image to those of the found examples. The examples are formed by a pyramid derivative set of features, instead of raw data directly. Similar method is applied to text super-resolution in [75]. Datsenko and Elad [21] presented a global MAP estimation where the example-based regularization is given by a binary weighted average instead of the nearest neighbor, bypassing outliers due to noise. This work is further extended and elaborated in [21], where the binary weighting scheme is relaxed. Another noteworthy approach for example-based approach is by Protter et al. [78], generalized from the nonlocal means denoising algorithm [8]. Instead of sampling examples from other training images, the algorithm explores self-similarities within the image (or sequence) and extract the example patches from the target image (or sequence) itself. A recent work by Glasner et al. further explored self-similarities in images for SR by combining the classical algorithm based on subpixel displacements and the example-based method based on patch pairs extracted from the target image.

The use of examples can be much more effective when dealing with narrow families of images, such as text and face images. A group of algorithms have emerged targeting face super-resolution in recent years due to its importance in surveillance scenarios. Face super-resolution is usually referred to as face hallucination, following the early work by Baker and Kanade [2]. Capel and Zisserman [14] proposed an algorithm where PCA [45] subspace models are used to learn parts of the faces. Liu et al. [58], [57] proposed a two-step approach toward super-resolution of faces, where the first step uses the eigenface [100] to generate a medium resolution face, followed by the nonparametric patch-based approach [31] in the second step. Such an Eigenface-based approach has been explored in several later works [32],[104] too. Yang et al. [106] proposed a similar two-step approach. Instead of using the holistic PCA subspace, [106] uses local Nonnegative Matrix Factorization (NMF)[51] to model faces and the patch-based model in the second step is adopted from [107]. Jia and Gong [43], [44] proposed the tensor approach to deal with more face variations, such as illuminations and expressions. Although these face hallucination algorithms work surprisingly well, they only apply to frontal faces, and only few works have been devoted on evaluating face hallucination for recognition [32], [36].

Example-based regularization is effective in our SR problem when insufficient observations are available. There are still a number of questions we need to answer regarding this kind of approaches. First, how to choose the optimal patch size given the target image. Perhaps a multiresolution treatment is needed. Second, how to choose the database. Different images have different statistics, and thereby need different databases. An efficient method for dictionary adaptation to the current target image may suggest a way out. Third, how to use the example-based prior more efficiently. The computation issue could be a difficulty for practical applications. Readers are suggested to refer to [25] for more detailed analysis on example-based regularization for inverse problems.

1.3.6 Set Theoretic Restoration

Besides the optimization approaches derived from stochastic view as discussed above, another stream of methods is through the well-known Projection onto Convex Sets (POCS) [110]. The POCS methods approach the SR problem by formulating multiple constraining convex sets containing the desired image as a point within the sets. Defining such convex sets is flexible and can incorporate different kinds of constraints or priors, even nonlinear and nonparametric constraints. As an example, we introduce several commonly used convex sets in the POCS methods. The data consistency or reconstruction constraints can be modeled as K convex sets:

$$\mathcal{C}_k = \left\{ \boldsymbol{X} \middle| \|D_k H_k F_k \boldsymbol{X} - \boldsymbol{Y}_k\|^2 \leq \sigma^2, 1 \leq k \leq K \right\}. \tag{1.22}$$

Smoothness constraints can be defined as

$$\mathcal{C}_\Gamma = \left\{ \boldsymbol{X} \middle| \|\Gamma \boldsymbol{X}\|_p < \sigma_2 \right\}. \tag{1.23}$$

where $p = 1, 2, \infty$ denotes different norms. Amplitude constraints can also be modeled:

$$\mathcal{C}_A = \left\{ \boldsymbol{X} \middle| A_1 \leq \boldsymbol{X}[m, n] \leq A_2 \right\}. \tag{1.24}$$

With a group of $\mathcal{M}$ convex sets, the desired solution lies in the intersection of these sets $\boldsymbol{X} \in \mathcal{C}_s = \bigcap_{i=1}^{\mathcal{M}} \mathcal{C}_i$. The POCS technique suggests the following recursive algorithm for finding a point within the intersection set given an initial guess:

$$\boldsymbol{X}_{k+1} = P_\mathcal{M} P_{\mathcal{M}-1} \cdots P_2 P_1 \boldsymbol{X}_k, \tag{1.25}$$

where $\boldsymbol{X}_0$ is an initial guess, and P_i is the projection operator that projects a point onto a closed, convex set $\mathcal{C}_i$.

Early POCS techniques for SR reconstruction were proposed by Stark and Oskoui [88]. Extensions were proposed to handle space-varying PSF, motion blur, sensor blur, and aliasing sampling effects in [68], [67], [69]. Many super-resolution works only consider nonzero aperture size (the lens blur, PSF), but not finite aperture time (motion blur) which is quite common in real low-resolution videos. [69] is the early work to take into account the motion blur in SR reconstruction of videos based on POCS technique. As the motion blurring caused by a finite aperture time will in general be space- and perhaps time-varying, it cannot be factored out of the SR restoration problem and performed as a separate post-processing step. The POCS technique can conveniently handle such problems. Extending this method, Eren et al. [28] proposed a POCS-based approach for robust, object-based SR reconstruction. The proposed method employs a validity map to disable projections based on observations with inaccurate motion estimation, and a segmentation map for object-based processing. Elad and Feuer [26] analyzed and compared the ML, MAP and POCS methods for super-resolution, and proposed a hybrid approach. Patti and Altunbasak [66] extended their earlier work in the image observation model to allow high order interpolation and modified constraint sets to reduce the edge ringing artifacts.

The advantage of the POCS technique lies in its simplicity to incorporate any kinds of constraints and priors that may present as impossible for those stochastic approaches. However, POCS is notorious for its heavy computation and slow convergence. The solution is not unique, depending on the initial guess. The POCS methods also assume priors on the motion parameters and system blurs. They cannot estimate those registration parameters and the high-resolution image as in the stochastic approaches simultaneously. The hybrid approach combining a stochastic view and the POCS philosophy suggests a promising way to pursue.

1.4 Challenge Issues for Super-Resolution

In the previous sections, we have discussed several basic techniques for SR reconstruction. Although many different approaches have been proposed since the SR concept was introduced, most approaches work well on toy data rather than in real problems. In building a practical SR system, many challenging issues still lay ahead preventing the SR techniques from wide applications. In the following, we list several challenges that we think are important for the future development and application of SR techniques.

1.4.1 Image Registration

Image registration is critical for the success of multiframe SR reconstruction, where complementary spatial samplings of the HR image are fused. The image registration is a fundamental image processing problem that is well known as ill-posed. The problem is even more difficult in the SR setting, where the observations are low-resolution images with heavy aliasing artifacts. The performance of the standard image registration algorithms decreases as the resolution of the observations goes down, resulting in more registration errors. Artifacts caused by these registration errors are visually more annoying than the blurring effect resulting from the interpolation of a single image. Traditional SR reconstruction usually treats image registration as a distinct process from the HR image estimation. Therefore, the recovered HR image quality depends largely on the image registration accuracy from the previous step. Many image registration techniques derived from different principles have been proposed in the literature [7, 114]. However, Robinson and Milanfar [79] showed that the registration performance is bounded even for the simplest case of global translation.

LR image registration and the HR image estimation are actually interdependent [80]. On one hand, accurate subpixel motion estimation benefits HR image estimation. On the other hand, high-quality HR image can facilitate accurate motion estimation. Therefore, tailored to the SR reconstruction problem, the LR image registration can be addressed together with the HR image reconstruction, leading to the joint ML [97] or MAP [35, 87, 76] framework for simultaneous estimation. These joint estimation algorithms capture the dependence between LR image registration and HR image estimation, and performance improvements are observed. However, with limited observations, the joint estimation for registration parameters and HR image may result in overfitting. To overcome this overfitting problem, Tipping and Bishop [96] employed a Bayesian approach for estimating both registration and blur parameters by marginalizing the unknown high-resolution image. The algorithms shows noteworthy estimation accuracy both for registration and blur parameters, however the computation cost is very high. Pickup et al. [73, 74, 72] instead cast the Bayesian approach in another way by marginalizing the unknown registration parameters, to address the uncertainty inherent with the image registration [79].

The stochastic approaches associating the HR image estimation toward image registration do demonstrate promising results, however such parametric methods are limited in the motion models they can effectively handle. Usually, some simple global motion models are assumed. Real videos are complicated comprising arbitrary local motions, where parametrization of the motion models may be intractable. Optical flow motion estimation can be applied to such scenarios. However, the insufficient measurements for local motion estimations make these algorithms vulnerable to errors, which may cause disasters for SR reconstruction [112]. Another promising approach toward SR reconstruction is the nonparametric methods, which try to bypass the explicit motion estimation. Protter et al. [78] extended the non-local means denoising algorithm to SR reconstruction, where fuzzy motion estimation based on block matching is used. Later they proposed a probabilistic motion model in [77], which is a nonparametric model analogy to [72]. Both [78] and [77] can handle complex motion patterns in real videos. Compared to the classical SR methods based on optical flow motion estimation, Protter's methods reduce the errors caused by misalignment by a weighting strategy over multiple possible candidates. Takeda et al. [92] on the other hand applied an 3-D steering kernel proposed in their early work [91] to video, which also avoids explicit motion estimation, for denoising and SR reconstruction. The 3-D steering kernel captures both spatial and temporal structure, encoding implicit motion information, and thus can be applied to both spatial and temporal SR for video with complex motion activities. While methods without explicit motion estimation indeed produce promising results toward the practical applicability of SR techniques, further improvements may include computation efficiency, combining adaptive interpolation or regression together with deblurring, and generalizing observation models to 3-D motions in video, e.g., out-of-plane rotation.

1.4.2   Computation Efficiency

Another difficulty limiting practical application of SR reconstruction is its intensive computation due to a large number of unknowns, which require expensive matrix manipulations. Real applications always demand efficiency of the SR reconstruction to be of practical utility, e.g., in the surveillance video scenarios, it is desired for the SR reconstruction to occur in real time. Efficiency is also desirable for SR systems with users in the loop for tuning parameters. Many SR algorithms targeting efficiency fall into the previously discussed interpolation-restoration approach, such as [27], [1], [61], and [34]. In [34], Hardie showed the computation superiority of his algorithm over previous efficient algorithms proposed in [1] and [61], and claimed that the algorithm can be applied in real time with global translation model. However, the computation goes up significant when nontranslation model occurs, which can be ameliorated by massive parallel computing. Others tried to examine particular modeling scenarios to speed up the optimization problem. Zomet and Peleg [115] and Farsiu et al. [30] studied the application of D_k, H_k, and F_k directly as the corresponding image operations of downsampling, blurring and shifting, bypassing the need to explicitly construct the matrices, bringing significant speed ups. [6] combined a slightly modified version of [27] and [30] and implemented a real-time SR system using FPGA, a nice attempt to the practical use of SR.

However, such algorithms require precise image registration, which is computation intensive in the first place. Moreover, these algorithm can only handle simple motion models efficiently up to now, far from application in real complex video scenarios. For videos with arbitrary motions, [92] suggests promising directions for seeking efficient algorithms. It is also interesting to see how parallel computing, e.g., GPU, and hardware implementations affect the future applications of SR techniques.

1.4.3   Robustness Aspects

Traditional SR techniques are vulnerable to the presence of outliers due to motion errors, inaccurate blur models, noise, moving objects, motion blur, etc. These inaccurate model errors cannot be treated as Gaussian noise as the usual assumption with ℓ_2 reconstruction residue. Robustness of SR is of interest because the image degradation model parameters cannot be estimated perfectly, and sensitivity to outliers may result in visually disturbing artifacts, which are intolerable in many applications, e.g., video standard conversion. However, not enough work has been devoted to such an important aspect. Chiang and Boulte [18] used median estimation to combine the upsampled images to cope with outliers from non-stationary noise. Zomet et al. [116] cast the problem in a different way, where a robust median-based gradient is used for the optimization to bypass the influence of outliers. Farsiu et al. [30],[82] changed the commonly used ℓ_2 norm into ℓ_1 norm for robust estimation similar to [18] and robust regularization. [113] introduced a simultaneous super-resolution with Huber norm as the prior for robust regularization. Pham et al. [71] proposed a robust certainty to each neighboring sample for interpolating unknown data, with the same photometric-based weighting scheme used in bilateral filtering. A similar uncertainty scheme is also used in the probabilistic motion model [77] for taking care of optical flow motion estimation errors based on block matching. Many of these algorithms showed improvements for outliers assumed on the toy data, where more experimental evaluations are needed to see how much the robustness efforts can benefit real SR performance.

1.4.4 Performance Limits

The SR reconstruction has become a hot research topic since it was introduced, and thousands of SR papers have bloomed into publications. However, not much work has been devoted to the fundamental understanding of the performance limits of these SR reconstruction algorithms. Such a performance limit understanding is important. For example, it will shed light on SR camera design, helping to analyze factors such as model errors, zooming factors and number of frames, etc. In general, an ambitious analysis of the performance limits for all SR techniques could be intractable. First, SR reconstruction is a complex task that consists of many interdependent components. Second, it is still unknown what is the most informative prior given the SR task, especially for the example-based approaches. Last, a good measure instead of simple MSE is still needed for performance evaluation. It has been recognized that an estimation with higher MSE does not have to be visually more appealing. For example, bicubic interpolation usually achieves smaller MSE compared with those recovered by some example-based approaches [107].

Several works attempting at the performance understanding have been proposed over the last several years. [2] analyzed the numerical conditions of the SR linear systems, and concluded that as the zoom factor increases the general image prior is of less and less help for SR. [56] derived the performance limits based on matrix perturbation, but with the assumption that image registration is known as a prior. With simple translation model, Robinson and Milanfar in [79], use the Cramér-Rao (CR) bounds to analyze the registration performance limit. They extend this work in [80] to give a thorough analysis of SR performance with factors such as motion estimation, decimation factor, number of frames, and prior information. The analysis is based on the MSE criterion and the motion model is again assumed to be simple global translational. Eekeren et al. [102] evaluated several SR algorithms on real-world data exploring several influential factors empirically. Even though these efforts at understanding performance bounds are far from enough about SR, they indeed suggest ways for people to follow.

While it is hard to draw consistent conclusions for different SR techniques, in terms of performance evaluation, some benchmark and realistic datasets are needed for fair comparison and algorithm understanding. Future research should pursue more theoretical analysis and performance evaluation for directing SR technique developments.

Bibliography

[1] M. S. Alam, J. G. Bognar, R. C. Hardie, and B. J. Yasuda. Infrared image registration and high-resolution reconstruction using multiple translationally shifted aliased video frames. *IEEE Transactions on Instrumentation and Measurement*, 49(5):915–923, 2000.

[2] S. Baker and T. Kanade. Limits on super-resolution and how to break them. *IEEE Transactions on Pattern Analysis and Machine Intelligence*, 24(9):1167–1183, 2002.

[3] S. Borman and R. L. Stevenson. Simultaneous multi-frame MAP super-resolution video enhancement using spatio-temporal priors. In *Proceedings of IEEE International Conference on Image Processing*, volume 3, pages 469–473, 1999.

[4] Sean Borman and Robert L. Stevenson. Super-resolution from image sequences - A review. In *Proceedings of the 1998 Midwest Symposium on Circuits and Systems*, pages 374–378, 1998.

[5] N. K. Bose, H. C. Kim, and H. M. Valenzuela. Recursive implementation of total least squares algorithm for image reconstruction from noisy, undersampled multiframes. In *Proceedings of the IEEE Conference on Acoustics, Speech and Signal Processing*, volume 5, pages 269–272, 1993.

[6] O. Bowen and C. S. Bouganis. Real-time image super resolution using an FPGA. In *International Conference on Field Programmable Logic and Applications*, pages 89–94, 2008.

[7] L. Brown. A survey of image registration techniques. *ACM Computing Surveys*, 24(4):325–376, 1992.

[8] A. Buades, B. Coll, and J. M. More. A non-local algorithm for image denoising. In *Proceedings of IEEE Computer Society Conference on Computer Vision and Pattern Recognition*, pages 60–65, 2005.

[9] E. Candes. Compressive sensing. In *Proceedings of International Congress of Mathematicians*, volume 3, pages 1433–1452, 2006.

[10] D. Capel. *Image Mosaicing and Super-resolution*. Springer, 2004.

[11] D. Capel and A. Zisserman. Automated mosaicing with super-resolution zoom. In *Proceedings of IEEE Computer Society Conference on Computer Vision and Pattern Recognition*, pages 885–891, 1998.

[12] D. Capel and A. Zisserman. Super-resolution enhancement of text image sequences. In *Proceedings of the International Conference on Pattern Recognition*, volume 1, pages 1600–1605, 2000.

[13] D. Capel and A. Zisserman. Super-resolution enhancement of text image sequences. In *Proceedings of 15th International Conference on Pattern Recognition*, pages 600–605, 2000.

[14] D. Capel and A. Zisserman. Super-resolution from multiple views using learnt image models. In *Proceedings of IEEE Computer Society Conference on Computer Vision and Pattern Recognition*, volume 2, pages 627–634, 2001.

[15] D. Capel and A. Zisserman. Computer vision applied to super-resolution. *IEEE Signal Processing Magazine*, 20(3):75–86, 2003.

[16] T. F. Chan, S. Osher, and J. Shen. The digital TV filter and nonlinear denosing. *IEEE Transaction on Image Processing*, 10(2):231–241, 2001.

[17] H. Chang, D. Y. Yeung, and Y. Xiong. Super-resolution through neighbor embedding. In *Proceedings of IEEE Computer Society Conference on Computer Vision and Pattern Recognition*, volume 1, pages 275–282, 2004.

[18] M. C. Chiang and T. E. Boulte. Efficient super-resolution via image warping. *Image and Vision Computing*, 18(10):761–771, 2000.

[19] J. Chung, E. Haber, and J. Nagy. Numerical methods for coupled super-resolution. *Inverse Problems*, 22(4):1261–1272, 2006.

[20] Marco Crisani, Dong Seon Cheng, Vittorio Murino, and Donato Pannullo. Distilling information with super-resolution for video surveillance. In *Proceedings of the ACM 2nd International Workshop on Video Surveillance and Sensor Networks*, pages 2–11, 2004.

[21] D. Datsenko and M. Elad. Example-based single document image super-resolution: a global MAP approach with outlier rejection. *Multidimensional System and Signal Processing*, 18(2-3):103–121, 2007.

[22] D. L. Donoho. Compressed sensing. *IEEE Transactions on Information Theory*, 52(4):1289–1306, 2006.

[23] D. L. Donoho. For most large underdetermined systems of linear equations, the minimal ℓ^1-norm near-solution approximates the sparsest near-solution. *Communications on Pure and Applied Mathematics*, 59(7):907–934, 2006.

[24] A. A. Efros and T. K. Leung. Texture synthesis by non-parametric sampling. In *Proceedings of IEEE International Conference on Computer Vision*, pages 1033–1038, 1999.

[25] M. Elad and D. Datsenko. Example-based regularization deployed to super-resolution reconstruction of a single image. *The Computer Journal*, 52(1):15–30, 2007.

[26] M. Elad and A. Feuer. Restoration of single super-resolution image from several blurred, noisy and down-sampled measured images. *IEEE Transaction on Image Processing*, 6(12):1646–1658, 1997.

[27] M. Elad and Y. Hel-Or. A fast super-resolution reconstruction algorithm for pure translational motion and common space invariant blur. *IEEE Transactions on Image Processing*, 10(8):1187–1193, 2001.

[28] P. E. Eren, M. I. Sezan, and A. M. Tekalp. Robust, object-based high resolution image reconstruction from low-resolution video. *IEEE Transactions on Image Processing*, 6(10):1446–1451, 1997.

[29] X. Jia F. Li and D. Fraser. Universal HMT based super resolution for remote sensing images. In *IEEE International Conference on Image Processing*, pages 333–336, 2008.

[30] S. Farsiu, D. Robinson, M. Elad, and P. Milanfar. Fast and robust multi-frame super-resolution. *IEEE Transaction on Image Processing*, 13(10):1327–1344, 2004.

[31] W. T. Freeman, T. R. Jones, and E. C. Pasztor. Example-based super-resolution. *IEEE Computer Graphics and Applications*, 22(2):56–65, 2002.

[32] B. K. Gunturk, A. U. Batur, Y. Altunbasak, M. H. Hayes, and R. M. Mersereau. Eigenface-domain super-resolution for face recognition. *IEEE Transactions on Image Processing*, 12(5):597–606, 2003.

[33] K. M. Hanson and G. W. Wecksung. Bayesian approach to limited-angle reconstruction in computed tomography. *Journal of Optical Society of America*, 73(11):1501–1509, 1983.

[34] R. Hardie. A fast image super-resolution algorithm using an adaptive Wiener filter. *IEEE Transactions on Image Processing*, 16(12):2953–2964, 2007.

[35] R. C. Hardie, K. J. Barnard, and E. E. Armstrong. Join MAP registration and high resolution image estimation using a sequence of undersampled images. *IEEE Transactions on Image Processing*, 6(12):1621–1633, 1997.

[36] P. H. Hennings-Yeomans, S. Baker, and B. V. K. V. Kumar. Simultaneous super-resolution and feature extraction for recognition of low-resolution faces. In *Proceedings of IEEE Computer Society Conference on Computer Vision and Pattern Recognition*, pages 1–8, 2008.

[37] G. T. Herman, H. Hurwitz, A. Lent, and H-P. Lung. On the Bayesian approach to image reconstruction. *Information and Control*, 42(1):60–71, 1979.

[38] A. Hertzmann, C. E. Jacobs, N. Oliver, B. Curless, and D. H. Salesin. Image analogies. In *Proceedings of the 28th annual conference Computer Graphics and Interactive Techniques*, pages 327–340, 2001.

[39] M. Irani and S. Peleg. Super resolution from image sequences. In *Proceedings of 10th International Conference on Pattern Recognition*, volume 2, pages 115–120, 1990.

[40] M. Irani and S. Peleg. Improving resolution by image registration. *CVGIP: Graphical Models and Imaging Processing*, 53(3):231–239, 1991.

[41] M. Irani and S. Peleg. Motion analysis for image enhancement: resolution, occlusion and tranparency. *Journal of Visual Communications and Image Representation*, 4(4):324–335, 1993.

[42] T. S. Huang J. Yang, J. Wright and Y. Ma. Super-resolution via sparse representation. *Submitted to IEEE Transactions on Image Processing*, 2009.

[43] K. Jia and S. Gong. Multi-model tensor face for simultaneous super-resolution and recognition. In *Proceedings of IEEE International Conference on Computer Vision*, volume 2, pages 1683–1690, 2005.

[44] K. Jia and S. Gong. Generalized face super-resolution. *IEEE Transactions on Image Processing*, 17(6):873–886, 2008.

[45] I. T. Jolliffe. *Principal Component Analysis*. Series: Springer Series in Statistics, 2002.

[46] E. Kaltenbacher and R. C. Hardie. High-resolution infrared image reconstruction using multiple low resolution aliased frames. In *Proceedings of the IEEE National Aerospace Electronics Conference*, volume 2, pages 702–709, 1996.

[47] J. A. Kennedy, O. Israel, A. Frenkel, R. Bar-Shalom, and A. Haim. Super-resolution in PET imaging. *IEEE Transactions on Medical Imaging*, 25(2):137–147, 2006.

[48] D. Keren, S. Peleg, and R. Brada. Image sequence enhancement using subpixel displacements. In *Proceedings of the IEEE Conference on Computer Vision and Pattern Recognition*, pages 742–746, 1988.

[49] S. P. Kim, N. K. Bose, and H. M. Valenzuela. Recursive reconstruction of high resolution image from noisy undersampled multiframes. *IEEE Transactions on Acoustics, Speech and Signal Processing*, 38(6):1013–1027, 1990.

[50] H. Knutsson and C.-F. Westin. Normalized and differential convolution. In *Proceedings of IEEE Computer Society Conference on Computer Vision and Pattern Recognition*, pages 515–523, 1993.

[51] D. D. Lee and H. S. Seung. Learning the parts of objects by non-negative matrix factorizaiton. *Nature*, 401(6755):788–791, 1999.

[52] S. Lerttrattanapanich and N. K. Bost. High resolution image formation from low resolution frames using delaunay triangulation. *IEEE Transaction on Image Processing*, 11(12):1427–1441, 2002.

[53] A. Levin, Y. Weiss, F. Durand, and W. Freeman. Understanding and evaluating blind deconvolution algorithms. In *Proceedings of IEEE Computer Society Conference on Computer Vision and Pattern Recognition*, pages 1964–1971, 2009.

[54] Y. Li and F. Santosa. A computational algorithm for minimizing total variation in image restoration. *IEEE Transactions on Image Processing*, 5(6):987–995, 1996.

[55] Frank Lin, Clinton B. Fookes, Vinod Chandran, and Sridha Sridharan. Investigation into optical flow super-resolution for surveillance applications. In *The Austrilian Pattern Recognition Society Worshop on Digital Image Computing*, 2005.

[56] Z. Lin and H.-Y. Shum. Fundamental limits on reconstruction-based superresolution algorithms under local translation. *IEEE Transactions on Pattern Analysis and Machine Intelligence*, 26(1):83–97, 2004.

[57] C. Liu, H. Y. Shum, and W. T. Freeman. Face hallucination: theory and practice. *International Journal of Computer Vision*, 75(1):115–134, 2007.

[58] C. Liu, H. Y. Shum, and C. S. Zhang. Two-step approach to hallucinating faces: global parametric model and local nonparametric model. In *Proceedings of IEEE Computer Society Conference on Computer Vision and Pattern Recognition*, volume 1, pages 192–198, 2001.

[59] J. Maintz and M. Viergever. A survey of medical image registration. *Medical Image Analysis*, 2(1):1–36, 1998.

[60] K. Malczewski and R. Stasinski. Toeplitz-based iterative image fusion scheme for MRI. In *IEEE International Conference on Image Processing*, pages 341–344, 2008.

[61] B. Narayanan, R. C. Hardie, K. E. Barner, and M. Shao. A computationally efficient super-resolution algorithm for video processing using partition filters. *IEEE Transactions on Circuits and Systems for Video Technology*, 17(5):621–634, 2007.

[62] N. Nguyen and P. Milanfar. An efficient wavelet-based algorithm for image super-resolution. In *Proceedings of International Conference on Image Processing*, volume 2, pages 351–354, 2000.

[63] N. Nguyen, P. Milanfar, and G. H. Golub. A computationally efficient image superresolution algorithm. *IEEE Transactions on Image Processing*, 10(5):573–583, 2001.

[64] A. Papulis. Generalized sampling expansion. *IEEE Transactions on Circuits and Systems*, 24(11):652–654, 1977.

[65] Sung C. Park, Min K. Park, and Moon G. Kang. Super-resolution image reconstruction: a technical overview. *IEEE Signal Processing Magazine*, 20(3):21–36, 2003.

[66] A. J. Patti and Y. Altunbasak. Aritifact reduction for set theoretic super resolution image reconstruction with edge adaptive constraints and higher-order interpolants. *IEEE Transaction on Image Processing*, 10(1):179–186, 2001.

[67] A. J. Patti, M. Sezan, and A. M. Tekalp. Robust methods for high quality stills from interlaced video in the presence of dominant motion. *IEEE Transactions on Circuits and Systems for Video Technology*, 7(2):328–342, 1997.

[68] A. J. Patti, M. I. Sezan, and A. M. Tekalp. High-resolution image reconstruction from a low-resolution image sequence in the presence of time-varing motion blur. In *Proceedings of the IEEE International Conference on Image Processing*, volume 1, pages 343–347, 1994.

[69] A. J. Patti, M. I. Sezan, and A. M. Tekalp. Superresolution video reconstruction with arbitrary sampling lattices and nonzero aperture time. *IEEE Transactions on Image Processing*, 6(8):1064–1076, 1997.

[70] S. Peleg and Y. Yeshurun. Superresolution in MRI: application to human white matter fiber tract visualization by diffusion tensor imaging. *Magnetic Resonance in Medicine*, 45(1):29–35, 2001.

[71] T. Q. Pham, L. J. Vliet, and K. Schutte. Robust fusion of irregularly sampled data using adaptive normalized convolution. *EURASIP Journal on Applied Signal Processing*, 2006.

[72] L. C. Pickup, D. P. Capel, S. J. Robert, and A. Zisserman. Overcoming registration uncertainty in image super-resolution: maximize or marginalize? *EURASIP Journal on Advances in Signal Processing*, 2007.

[73] L. C. Pickup, D. P. Capel, S. J. Roberts, and A. Zisserman. Bayesian image super-resolution, continued. In *Proceedings of Advances in Neural Information and Proceedings Systems*, pages 1089–1096, 2006.

[74] L. C. Pickup, D. P. Capel, S. J. Roberts, and A. Zisserman. Bayesian methods for image super-resolution. *The Computer Journal*, 52(1):101–113, 2009.

[75] L. C. Pickup, S. J. Robert, and A. Zisserman. A sampled texture prior for image super-resolution. In *Proceedings of Advances in Neural Information and Processing System*, pages 1587–1594, 2003.

[76] L. C. Pickup, S. J. Roberts, and A. Zisserman. Optimizing and learning for super-resolution. In *British Machine Vision Conference*, volume 2, pages 439–448, 2006.

[77] M. Protter and M. Elad. Super resolution with probabilistic motion estimation. *IEEE Transactions on Image Processing*, 18(8):1899–1904, 2009.

[78] M. Protter, M. Elad, H. Takeda, and P. Milanfar. Generalizing the nonlocal-means to super-resolution reconstruction. *IEEE Transactions on Image Processing*, 18(1):36–51, 2009.

[79] D. Robinson and P. Milanfar. Fundamental performance limits in image registration. *IEEE Transactions on Image Processing*, 13(9):1185–1199, 2004.

[80] D. Robinson and P. Milanfar. Statistical performance analysis of super-resolution. *IEEE Transactions on Image Processing*, 15(6):1413–1428, 2006.

[81] L. Rudin, S. Osher, and E. Fatemi. Nonlinear total variation based noise removal algorithms. *Physica D: Nonlinear Phenomena*, 60(1-4):259–268, 1992.

[82] M. Elad S. Farsiu, D. Robinson and P. Milanfar. Advances and challenges in super-resolution. *International Journal of Imaing Systems and Technology*, 14(2):47–57, 2004.

[83] R. R. Schultz and R. L. Stevenson. A Bayesian approach to image expansion for improved definition. *IEEE Transactions on Image Processing*, 3(3):233–242, 1994.

[84] R. R. Schultz and R. L. Stevenson. Extraction of high-resolution frames from video sequences. *IEEE Transactions on Image Processing*, 5(6):996–1011, 1996.

[85] C. A. Segall, A. K. Katsaggelos, R. Molina, and J. Mateos. Bayesian resolution enhancement of compressed video. *IEEE Transactions on Image Processing*, 13(7):898–910, 2004.

[86] C. A. Segall, R. Molina, and A. K. Katsaggelos. High resolution images from low-resolution compressed video. *IEEE Signal Processing Magazine*, 20(3):37–38, 2003.

[87] H. Shen, L. Zhang, B. Huang, and P. Li. A MAP approach for joint motion estimation, segmentation and super-resolution. *IEEE Transactions on Image Processing*, 16(2):479–490, 2007.

[88] H. Stark and P. Oskoui. High-resolution image recovery from image-plane arrays, using convex projections. *Journal of Optical Society of America A*, 6(11):1715–1726, 1989.

[89] W. Su and S. P. Kim. High-resolution restoration of dynamic image sequences. *International Journal of Imaging Systems and Technology*, 5(4):330–339, 1994.

[90] J. Sun, N. N. Zheng, H. Tao, and H. Shum. Image hallucination with primal sketch priors. In *Proceedings of IEEE Computer Society Conference on Computer Vision and Pattern Recognition*, volume 2, pages 729–736, 2003.

[91] H. Takeda, S. Farsiu, and P. Milanfar. Kernel regression for image processing and reconstruction. *IEEE Transactions on Image Processing*, 16(2):349–366, 2007.

[92] H. Takeda, P. Milanfar, M. Protter, and M. Elad. Super-resolution without explicit subpixel motion estimation. *IEEE Transaction on Image Processing*, 18(9):1958–1975, 2009.

[93] J. B. Tenenbaum, V. Silva, and J. C. Langford. A global geometric framework for nonlinear dimensionality reduction. *Science*, 290(5500):2319–2323, 2000.

[94] R. Tibshirani. Regression shrinkge and selection via the Lasso. *Journal of Royal Statistical Society: Series B (Statistical Methodology)*, 59(1):267–288, 1996.

[95] A. N. Tikhonov and V. A. Arsenin. *Solution of ill-posed problems*. Winston & Sons, Washington, 1997.

[96] Michael E. Tipping and Christopher M. Bishop. Bayesian image super-resolution. In *Proceedings of Advances in Neural Information Proceeding Systems*, pages 1279–1286, 2003.

[97] B. C. Tom and A. K. Katsaggelos. Reconstuction of a high-resolution image by simultaneous registration, restoration and interpolation of low-resolution images. In *Proceedings of the IEEE International Conference on Image Processing*, volume 2, page 2539, 1995.

[98] B. C. Tom, A. K. Katsaggelos, and N. P. Galatsanos. Reconstruction of a high resolution image from registration and restoration of low resolution images. In *Proceedings of IEEE International Conference on Image Processing*, pages 553–557, 1994.

[99] R. Y. Tsai and T. S. Huang. Multipleframe image restoration and registration. In *Advances in Computer Vision and Image Processing*, pages 317–339. Greenwich, CT: JAI Press Inc., 1984.

[100] M. Turk and A. Pentland. Face recognition using eigenfaces. In *Proceedings of IEEE Computer Society Conference on Computer Vision and Pattern Recognition*, pages 586–591, 1991.

[101] H. Ur and D. Gross. Improved resolution from subpixel shifted pictures. *CVGIP: Graphical Models and Image Processing*, 54(2):181–186, 1992.

[102] A. W. M. van Eekeren, K. Schutte, O. R. Oudegeest, and L. J. van Vilet. Performance evaluation of super-resolution reconstruction methods on real-world data. *EURASIP Journal on Advances in Signal Processing*, 2007.

[103] Q. Wang, X. Tang, and H. Shum. Patch based blind image super-resolution. In *Proceedings of IEEE International Conference on Computer Vision*, volume 1, pages 709–716, 2005.

[104] X. Wang and X. Tang. Hallucinating face by eigentransformation. *IEEE Transactions on Systems, Man, and Cybernetics*, 35(3):425–434, 2003.

[105] N. A. Woods, N. P. Galatsanos, and A. K. Katsaggelos. Stochastic methods for joint registration, restoration and interpolation of multiple undersampled images. *IEEE Transactions on Image Processing*, 15(1):210–213, 2006.

[106] Jianchao Yang, Hao Tang, Yi Ma, and Thomas Huang. Face hallucination via sparse coding. In *Proceedings of IEEE International Conference on Image Processing*, pages 1264–1267, 2008.

[107] Jianchao Yang, John Wright, Thomas Huang, and Yi Ma. Image super-resolution as sparse representation of raw image patches. In *Proceedings of IEEE Computer Society Conference on Computer Vision and Pattern Recognition*, pages 1–8, 2008.

[108] J. S. Yedidia, W. T. Freeman, and Y. Weiss. Generalized belief propagation. In *Proceedings of Advances in Neural Information Processing Systems*, pages 689–695, 2001.

[109] L. J. Yen. On non-uniform sampling of bandwidth limited signals. *IRE Transactions on Circuits Theory*, 3(4):251–257, 1956.

[110] D. C. Youla and H. Webb. Image registration by the method of convex projections: Part 1-thoery. *IEEE Transactions on Medical Imaging*, 1(2):81–94, 1982.

[111] D. M. Young. *Iterative solution of large linear systems*. New York: Academic, 1971.

[112] W. Zhao and H. S. Sawhney. Is super-resolution with optical flow feasible? In *Proceedings of the 7th European Conference on Computer Vision*, pages 599–613, 2002.

[113] M. V. W. Zibetti and J. Mayer. Outlier robust and edge-preserving simultaneous super-resolution. In *Proceedings of IEEE International Conference on Image Processing*, pages 1741–1744, 2006.

[114] B. Zitová and J. Flusser. Image registration methods: a survey. *Image and Vision Computing*, 21(11):977–1000, 2003.

[115] A. Zomet and S. Peleg. Efficient super-resolution and applications to mosaics. In *Proceedings of International Conference on Pattern Recognition*, pages 579–583, 2000.

[116] A. Zomet, A. Rav-Acha, and S. Peleg. Robust super-resolution. In *Proceedings of the IEEE Workshop on Applications of Computer Vision*, pages 645–650, 2001.

2

Super-Resolution Using Adaptive Wiener Filters

Russell C. Hardie

University of Dayton

CONTENTS

Abstract — This chapter will describe an approach to super-resolution (SR) based on adaptive Wiener filters. In this approach, multiple frames are registered relative to a common grid. Output SR pixel values are estimated using a type of adaptive Wiener filter that forms a weighted sum of neighboring observed pixels on the registered high-resolution grid. The filter weights are selected to minimize the mean squared error based on statistical correlation models or empirical training data. This estimation step simultaneously serves both a nonuniform interpolation function and a restoration function. This SR approach is most appropriate when the point spread function model and the motion model commute. If the filter weights can be precomputed, this approach can lead to very fast implementations. Note that the weights adapt to different spatial positions of the observed samples and can also be designed to adapt to local structure in the intensity data (e.g., edges, lines, and flat regions). We consider adaption based on vector quantization and local variance.

We also consider fast algorithms with precomputed weights and algorithms with weights computed on the fly.

2.1   Introduction

The spatial sampling rate of an imaging system is determined by the spacing of the detectors in the focal plane array (FPA). The spatial frequencies present in the image on the focal plane are band-limited by the optics. This is due to diffraction through a finite aperture. To guarantee that there will be no aliasing during image acquisition, the Nyquist criterion dictates that the sampling rate must be greater than twice the cut-off frequency of the optics. However, optical designs involve a number of trade-offs and typical imaging systems are designed with some level of aliasing. We will refer to such systems as detector limited, as opposed to optically limited. Furthermore, with or without aliasing, imaging systems invariably suffer from diffraction blur, optical abberations, and noise.

Multiframe super-resolution (SR) processing has proven to be successful in reducing aliasing and enhancing the resolution of images from detector limited imaging systems [25]. If relative motion between the scene and camera is present, sampling diversity is provided by the multiple looks at the scene that can be exploited to combat undersampling. Such processing can be viewed as trading temporal resolution for spatial resolution. This allows us to reduce or eliminate aliasing artifacts. Furthermore, if aliasing can be reduced to a minimal level, linear restoration techniques can be successfully applied to deconvolve the blurring effects of the system point spread function (PSF). Note that if little or no aliasing is present in the uncompensated imaging system, single frame restoration may be a more appropriate choice for many applications. If one does employ multiframe SR, it is critical that the motion include a subpixel component and be estimated accurately.

A class of computationally simple multiframe SR methods are those based on nonuniform interpolation [12, 2, 19, 30, 1, 36, 26, 34, 23, 10, 11, 33, 28, 35]. Such methods are of particular interest for implementing SR in real-time or soft real-time. These nonuniform interpolation based methods typically begin by using image registration to position the observed pixel values from each frame onto a common high-resolution (HR) grid. However, the extra samples are generally distributed nonuniformly, unless the motion is very carefully controlled. The nonuniformly sampled HR grid is of little practical use. Therefore, a nonuniform interpolation operation is used to create a uniformly sampled high-resolution (HR) image with reduced aliasing. The nonuniform interpolation based SR methods then typically employ a restoration step to reduce the blurring effects of the system PSF.

Most nonuniform interpolation based SR methods use independent interpolation and restoration steps. One potential downside of this is that an independent restoration step may aggravate artifacts from an imperfect nonuniform interpolation step. For example, when the distribution of low resolution pixels on the HR grid is poor, any nonuniform interpolation step will suffer. An independent restoration step can easily exaggerate any resulting artifacts. A new approach, using an adaptive Wiener filter (AWF), combines the nonuniform interpolation and restoration into a single step [22, 14, 15]. This provides potential computational savings as well as robustness to the spatial distribution of low-resolution pixels.

The AWF SR method forms a nonuniformly sampled HR grid in the traditional way. However, the AWF method produces the final output pixels with a single spatially-adaptive weighted sum operation using a finite moving window. Here each output HR pixel is formed as a weighted sum of neighboring observed pixels from the nonuniform HR grid. By designing appropriate weights, the output from this single weighted sum operation is not only on a uniform grid, but also restored from the system PSF. Note that the weights in the AWF SR approach are optimized for the specific local spatial arrangement of the neighboring pixels on the HR grid. In contrast, when independent nonuniform interpolation and restoration steps are used, the restoration step does not exploit knowledge of the original distribution of the observed pixels. Rather, it only sees a uniform HR produced the nonuniform interpolation.

Some variations of the AWF SR have been explored. In [22], the HR grid is discrete and the spatial locations of the observed pixels are quantized to fit on the discrete grid. If more than one observed pixel falls at a particular location, those multitemporal pixel values are averaged to fill the HR grid location. Furthermore, the work in [22] uses vector quantization on the pixel intensities to select the weights used for each observation window, in addition to the specific spatial distribution of samples. Empirical models for the required correlations are used in [22] based on training data. In contrast, the work in [14] uses an unquantized HR grid and theoretical parametric correlation models. This eliminates the need for training data and reduces error due to sample position quantization. Local estimates of the signal variance are employed in [14] to allow the weights to adapt to both the spatial distribution of samples and the local signal-to-noise ratio. The vector quantization in [22] and locally adaptive SNR in [14] have been observed to be beneficial in moderate to low SNR application. In high SNRs, these extra steps may not be necessary and can be avoided to reduce computational complexity.

The central challenge with the AWF SR method lies in determining the weights. The approach for designing the filter weights for the AWF SR method is based on a finite impulse response Wiener filter. Using correlation models, weights that minimize the mean squared error (MSE) are found. The approach is adaptive in the sense that as the spatial distribution of samples in the HR grid vary with the position of the observation window, so do the weights. Furthermore, spatially varying local statistics may be used to modify the correlation model for each observation window position. Note that for pure global translational motion, the spatial distribution of samples on the HR grid is periodic. This means that the number of filter weights required is relatively small. Since computing the weights is the bulk of the computational load, such imagery can be processed very fast with the AWF SR algorithm, even with a nonquantized HR grid [14]. For nontranslational motion, a potentially unique spatial pattern can be seen for each observation window. While it is possible to calculate all of these on-the-fly, this is a high computational burden. The work in [15] has proposed a modified version of the AWF SR algorithm that uses a specially selected partial observation window applied to a quantized HR grid. The partial observation window and quantized HR grid reduces the number of distinct spatial patterns observed, reducing the number of weights to be computed. With a reduced number of weight vectors, these can all be precomputed prior to processing video. This allows the AWF SR algorithm to process frames with very little computational load (given the precomputed weights), even for nontranslational motion.

In this chapter, we review the AWF SR methods. We discuss the observation models used including motion models and system PSF models. Also, a number of experimental results are presented to demonstrate the performance of the AWF SR methods. The organization of the remainder of the chapter is as follows. Section 2.2 presents the relevant observation models for the AWF SR methods. The AWF SR algorithms are presented in Section 2.3. Experimental results are provided in Section 2.4 and conclusions are presented in Section 2.5.

2.2 Observation Model

In this section we begin with the overall image formation model. We then discuss the motion model, registration, and finally the system PSF model.

2.2.1 Image Formation Model

The low-resolution (LR) image formation model is shown in Figure 2.1. The model is used for many SR algorithms including those in [22, 14]. The model begins with a desired 2-D continuous scene, $d(x, y)$. Here this desired image is assumed to be geometrically aligned with one of the observed frames, referred to as the reference frame. A geometric transformation is used to account for any motion between acquired frames, $d_k(x, y) = T_{p_k}\{d(x, y)\}$, for $k = 1, 2, ..., P$, and the reference frame. Note that the transformation depends on the motion model parameters in p_k. Details of the motion model are presented in Section 2.2.2. Blurring from the system PSF [17, 14] occurs next in the observation model, yielding $f_k(x, y) = d_k(x, y) * h(x, y)$, where $h(x, y)$ is the system PSF. More will be said about the PSF in Section 2.2.4. The

FPA in the camera serves to sample the scene for each frame yielding the vector of samples denoted $f(k)$ for frame k. Here we shall assume that the detector pitch is not sufficiently small to meet the Nyquist criterion, hence the need for multiframe SR. Finally, additive noise corrupts the samples yielding $g(k) = f(k) + n(k)$, where $n(k)$ contains the additive noise samples. Note that ideal sampling of the scene would give rise to the ideal image, represented here in lexicographical form as the vector $z = [z_1, z_2, \ldots, z_K]^T$. An equivalent and entirely discrete observation model can be found to relate z to the observed frames $g(k)$ using an impulse invariant PSF and downsampling [17].

Let is now consider undersampling in the observation model. First note that the optics serve to bandlimit the image in the focal plane. For example, the radial cut-off frequency associated with PSF from diffraction-limited optics with a circular exit pupil is [13] $\rho_c = \frac{1}{\lambda N}$, where N is the f-number of the optics and λ is the wavelength of light. To avoid aliasing, the FPA must sample at more than twice this cut-off frequency. This dictates that to avoid aliasing, the detector pitch must be less than $\frac{\lambda N}{2}$. Consider two imaging systems in Table 2.1 that will be used for experimental results in this chapter. System 1 uses a 256×256 Amber FPA with detector pitch of $50\,\mu$m and the optics have an f-number of $N = 3$. System 2 uses a $20\,\mu$m pitch 640×512 FPA from L-3 Cincinnati Electronics also equipped with $N = 3$ optics. Both systems produce 14 bit data. As can be seen in Table 2.1, both systems allow for the acquisition of aliased imagery with the selected optics. This is very common in imaging system design, since the desire for wide field-of-view small f-number optics often outweighs concerns over aliasing. Thus, such systems may be considered detector limited. Multiframe SR methods are a good choice for resolution enhancement for such systems, given that the pixel motion can be reliably estimated with subpixel accuracy.

Given this observation model in Figure 2.1, one approach to SR is to treat this as an inverse problem. This generally leads to iterative image re-

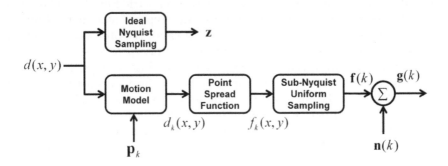

FIGURE 2.1: Observation model relating the desired 2-D continuous scene, $d(x, y)$, and the observed LR frames.

TABLE 2.1: Undersamping factors for two imaging systems considered here.

Imaging System	Optics	Cutoff Frequency (ρ_c)	Detector Pitch	Sampling Frequency	Under-sampling
1 Amber	F/3	83.3 cyc/mm	50 μm	20 cyc/mm	8.33×
2 L-3	F/3	83.3 cyc/mm	20 μm	50 cyc/mm	3.33×

construction SR algorithms that seek to find a z that would give rise to the observed $g(k)$ when put through a discrete observation model and be consistent with prior statistical models [18, 21, 4, 9, 7, 17, 39, 24, 8, 31, 16]. However, these iterative approaches can be computationally costly. A simpler and often faster class of SR algorithms is based on nonuniform interpolation [12, 2, 19, 30, 1, 36, 26, 34, 23, 10, 11, 33, 28, 35]. To understand these, consider the case where the motion model and the PSF degradation processes commute [11]. For such cases, consider switching the order of the motion model and PSF in the block diagram in Figure 2.1. Now the motion model and uniform sampling blocks are back-to-back. These can equivalently be combined into a single nonuniform sampling block. That is, motion followed by sampling is equivalent to simply altering the sampling locations according to the motion. Such an alternative observation model is shown in Figure 2.2. This figure includes a block diagram along with a representation of the imagery at various stages in the observation model.

Using registration, the LR pixels can be placed on a common HR grid represented by g in Figure 2.2. Nonuniform interpolation can be used to estimate a uniform grid of samples of $f(x, y)$ at or above the Nyquist rate. Finally, image restoration can be applied to reduce noise and reduce the blur caused by the system PSF, yielding an estimate of z. Note that the motion model and PSF operations commute for translational motion due to the shift invariance property of convolution. For a circularly symmetric PSF, it can be shown that they also commute for rotation. With other types of motion, the PSF and motion do not necessarily commute. In such cases, the observation model in Figure 2.2 does not strictly apply and hence the interpolation-restoration SR approaches that are based on Figure 2.2 may be less effective. However, we have observed that useful results can often be obtained for other types of motion models using the interpolation-restoration approaches. In this chapter, we compare the performance of several interpolation-restoration SR approaches applied to affine motion. Note that for high levels of affine zoom or skew, one might expect some degradation in performance because such motion does not commute with the PSF blur. With a highly noncircularly symmetric PSF, we might also expect to see problems with rotational motion.

2.2.2   Image Motion Model

Consider a static 3-D scene and a moving camera acquiring video for SR. The resulting 2-D optical flow models for various scenarios are summarized in Table 2.2 [38]. Note that for a planar scene and orthographic projection, the 2-D flow is affine. However, for the more realistic prospective projection case, but still with a planar scene, the flow is nonlinear and can be approximated with a quadratic flow model. For nonplanar scenes and arbitrary camera motion, the 2-D motion is dependent on the specific 3-D scene geometry. In most cases the scene 3-D geometry is not known and estimating it is an extremely demanding problem in its own right. Furthermore, occlusion effects and motion parallax that can lead to discontinuities in the 2-D flow for general camera motion and nonplanar scenes. These factors make accurate subpixel registration of multiple frames to a common grid for SR a very difficult task. It is interesting to note that when no translational camera motion is present (i.e., stationary camera with angular camera pointing motion only), a simple quadratic model is a good approximation, even for a nonplanar scene. In this camera angle variation only scenario, there are no motion parallax or occlusion effects to contend with and the 2-D optical flow is not scene dependent. From Table 2.2, it can be seen that for relatively small image regions, the resulting 2-D flow can be approximated by a affine model for the camera angle variation only case. This is particularly true for longer focal length optics (as the nonlinear terms are divided by the focal length). Thus, perhaps one of the most favorable

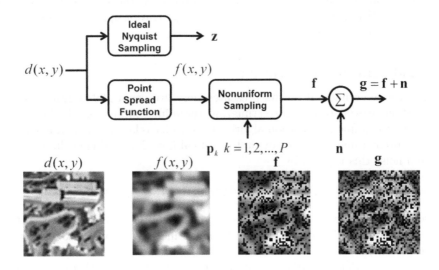

FIGURE 2.2: Alternative observation model relating a desired 2-D continuous scene, $d(x, y)$, with a set of corresponding LR frames. This model is appropriate when the motion model and PSF commute.

TABLE 2.2: 2-D optical flow models based on relative motion between a 3-D rigid scene and camera [38].

Flow Type	Model	When Applicable
Affine	$v_x(x,y) = p_1 x + p_2 y + p_3$ $v_y(x,y) = p_4 x + p_5 y + p_6$	Planar scene with orthographic projection
Quadratic (8)	$v_x(x,y) \approx p_1 y + p_2 xy + p_3 x^2 + p_4$ $= \omega_Z y + \frac{\omega_X xy}{l} - \frac{\omega_Y x^2}{l} - \omega_Y l$ $v_y(x,y) \approx p_5 x + p_6 xy + p_7 y^2 + p_8$ $= -\omega_Z x - \frac{\omega_Y xy}{l} + \frac{\omega_X y^2}{l} + \omega_X l$	Approximate for prospective projection projection with camera angle variation only. ω_X, ω_Y, ω_Z are camera angles in radians, l is focal length
Quadratic (10)	$v_x(x,y) \approx$ $p_1 x + p_2 y + p_3 x^2 + p_4 xy + p_5$ $v_y(x,y) \approx$ $p_6 x + p_7 y + p_9 xy + p_{10}$	Approximate for planar scene with full prospective projection
Planar Projective	$v_x(x,y) = \frac{p_1 + p_2 x + p_3 y}{p_7 + p_8 x + p_9 y} - x$ $v_y(x,y) = \frac{p_4 + p_5 x + p_6 y}{p_7 + p_8 x + p_9 y} - y$	Exact for planar scene with full prospective projection

acquisition scenarios for video for SR involves a camera at a fixed location panning and/or rotating from that fixed position relative to a static scene.

In this chapter, we focus on an affine motion model. Although it does not fully capture all of types of motion in Table 2.2, it can often serve as a useful approximation and has a number of useful properties. In particular, it only has 6 parameters to estimate and multiple sequential affine transformations is still an affine transformation. Thus, we can register each frame to the previous frame and then accumulate these to reference all the frames to a common frame or HR grid. Iterative and multiscale registration approaches also benefit from the accumulation property of affine flow.

2.2.3 Image Registration

For the experimental results presented here, we use a global gradient-based least-squares algorithm for estimating the affine parameters [5, 17, 27]. To define the affine registration algorithm, consider relating a new frame $d(x,y)$ to a prior frame $\tilde{d}(x,y)$ through 2-D optical flow. Neglecting occlusion effects and noise this is given by

$$d(x,y) = \tilde{d}(\tilde{x}, \tilde{y}) = \tilde{d}\left(x + v_x(x,y), y + v_y(x,y)\right), \quad (2.1)$$

where $v_x(x,y)$ and $v_y(x,y)$ are the polynomial optical flow functions. A truncated Taylor series approximation for small motions allows us to express this as

$$d(x,y) \approx \tilde{d}(x,y) + v_x(x,y)g_x(x,y) + v_y(x,y)g_y(x,y). \quad (2.2)$$

Now that we have removed the polynomial functions $v_x(x,y)$ and $v_y(x,y)$ from the argument of $\tilde{d}(\cdot)$, we get a set of linear equations (one per pixel) that can be solved using least squares. If we assume affine flow, then $v_x(x,y) =$

$p_1x + p_2y + p_3$ and $v_y(x, y) = p_4x + p_5y + p_6$. Because of the truncated Taylor series approximation, the least squares estimate is accurate only for small motions. To address this, an iterative approach is recommended. Here the initial registration parameters are found between two images using the method described above. Then one of the images is repositioned using interpolation according to the registration estimate and a new estimate is formed. This process repeats until the final incremental estimate is judged to be sufficiently small. The final registration estimate is formed by accumulating all of the incremental estimates. The repositioning at each iteration is always done directly from the original image (using the currently accumulated transform estimate) so as to avoid accumulating interpolation errors.

While this iterative method extends the useful range of the registration technique, it may still fail if the initial estimate moves the image in the wrong direction. This can happen with very large motions between frames. So to deal with very large motions, a multiscale approach is recommended. Here the registration begins using low-resolution versions of the two images and the iterative registration technique is applied. These registration parameters are used to initialize the registration at the next resolution level. This continues until registration at the full resolution is complete. Finally, for improved numerical stability, it is recommended that the x, y coordinates of the center of the image be represented as $0, 0$ when setting up the least squares equations.

The affine registration method described can be applied to the entire image. However, to deal with more complex motion, it might be beneficial to use a piecewise affine model. That is, break the image up into subimages and estimate affine parameters for each subimage. A practical way to do this and deal with large motions is to first do a global affine registration with the entire image and then refine these estimates in subimages with local affine estimates. Note that as the size of the subimages gets smaller, fewer equations are used in the least squares estimate. Hence, the accuracy of the subpixel flow estimate can be expected to suffer. Thus, a trade-off must be found to balance the the accuracy of the model parameter estimate (favoring larger subimages) with the accuracy of the flow model itself (favoring smaller subimages).

Note that deformable scene motion, or simply a nonstatic scene, greatly complicates the registration process [39]. Numerous additional registration parameters need to be estimated. Changing the pose of objects within the scene as well as occlusion further complicate matters in the general motion case. This makes subpixel registration accuracy across the full image very difficult, if not impossible. Furthermore, because the image data being used for registration is aliased and noisy, highly overdetermined linear equations are generally needed to get an accurate estimate. Joint SR and registration approaches with global motion models have been proposed to improve registration in the presence of high levels of aliasing [6, 16, 29]. Handling various types of more complex scene motion is an ongoing area of research in multiframe SR.

2.2.4   System Point-Spread Function

In this section, we address the modeling of the system PSF. We begin by modeling the optical transfer function (OTF) with three components as follows

$$H(u, v) = H_{\text{dif}}(u, v)H_{\text{abr}}(u, v)H_{\text{det}}(u, v), \tag{2.3}$$

where u and v are the horizontal and vertical spatial frequencies in cycles per milimeter. Diffraction limited optics contributes $H_{\text{dif}}(u, v)$, optical aberrations contribute $H_{\text{abr}}(u, v)$, and detector integration contributes $H_{\text{det}}(u, v)$. Other factors such as defocus, motion blur, and atmospheric effects are not considered here, but could be included if they are deemed to be significant in a particular application. The blurring effects from diffraction limited optics with a circular pupil function are described by the following OTF [13]

$$H_{dif}(u, v) = \begin{cases} \frac{2}{\pi}\left[\cos^{-1}(\rho/\rho_c) - (\rho/\rho_c)\sqrt{1 - (\rho/\rho_c)^2}\right] & \text{for } \rho < \rho_c \\ 0 & \text{else} \end{cases},$$
$$\tag{2.4}$$

where $\rho = \sqrt{u^2 + v^2}$ and $\rho_c = \frac{1}{\lambda N}$. It is this function that provides the band-limiting of the continuous scene, and consequently determines the necessary detector spacing to prevent aliasing. Even very well designed optical systems are likely to have aberrations that alter this diffraction limited model. One such aberration model is given by the following OTF [32]

$$H_{abr}(u, v) = \begin{cases} 1 - (W_{RMS}/0.18)^2(1 - 4(\rho/\rho_c - 0.5))^2 & \text{for } \rho < \rho_c \\ 0 & \text{else} \end{cases}. \tag{2.5}$$

Note that for a well-tuned system, $W_{RMS} = 1/14$ may be a good choice [32]. Finally, the detector component of the frequency response, $H_{\text{det}}(u, v)$, is obtained from the Fourier transform of the function describing the active area of an individual detector (assuming all detectors in the FPA have the same active area shape). Cross sections of the overall 2-D modulation transfer function (MTF) and its components are shown in Figure 2.3 for the imaging systems in Table 2.1. Note that the MTF is simply the magnitude of the OTF. Here, we assume $\lambda = 4$ μm and $W_{RMS} = 1/14$. It can been seen that the detector MTF dominates the Amber system because of the relatively large detectors in the FPA. For the L-3 system, the optics dominates the overall MTF. Note that the Nyquist frequency in both cases is below the cut-off frequency of the optics. Any frequency content above the Nyquist frequency is "folded" into lower frequencies, creating aliasing artifacts. In addition to aliasing, another important thing to note from Figure 2.3 is that, like most any imaging system, the MTF is not flat and high frequencies are attenuated (reducing image detail). The problem is that one cannot simply apply a high-boost filter to the imagery to restore the attenuated frequencies due to aliasing. However, if the effective sampling rate is increased by a multiframe nonuniform interpolation process, such restoration is then possible. The system PSF can be found by

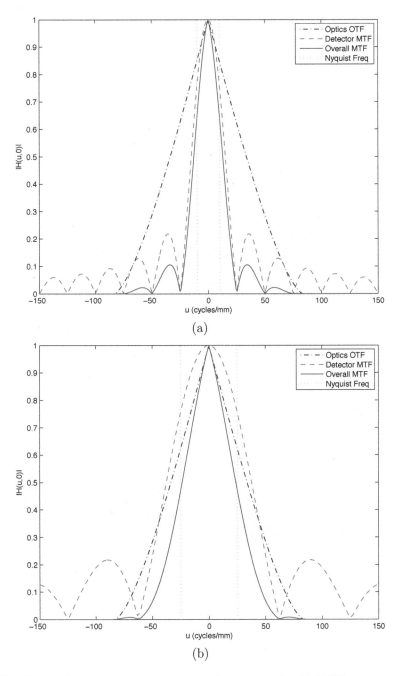

FIGURE 2.3: Cross sections of the overall theoretical 2-D MTF and its components with $\lambda = 4$ μm and $W_{RMS} = 1/14$ for (a) Imaging System 1 (Amber FPA) in Table 2.1 and (b) Imaging System 2 (L-3 FPA) in Table 2.1.

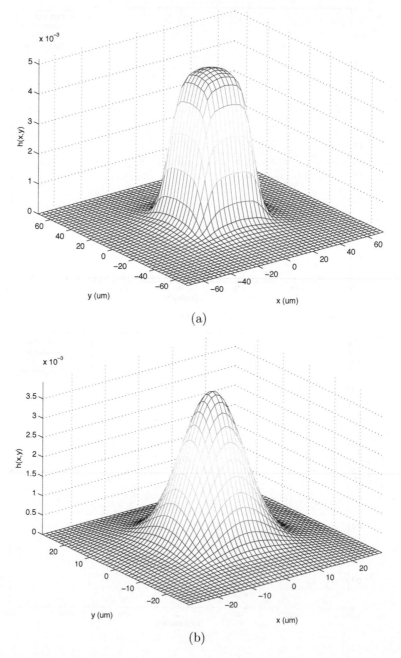

FIGURE 2.4: Theoretical PSF with $\lambda = 4$ μm and $W_{RMS} = 1/14$ for (a) Imaging System 1 (Amber FPA) in Table 2.1 and (b) Imaging System 2 (L-3 FPA) in Table 2.1.

taking the inverse Fourier transform of the optical transfer function. These are shown in Figure 2.4. Note that the discrete impulse invariant PSF can be found by sampling this PSF at spacings of the detector pitch divided by the SR upsampling factor.

2.3 AWF SR Algorithms

The AWF SR algorithms use $g(k)$, for $k = 1, 2, \ldots, P$ to form an estimate of z, denoted $\hat{z}$. The SR algorithms can be applied to video using a temporal sliding window of frames, or it can be used to generate a single output from an input sequence. The effective sampling rate for the estimated image is defined to be L times greater than that of the observed imagery. Ideally, L would selected to meet the Nyquist criterion. However, often a lower value of L may be sufficient to provide a useful result with minimal aliasing. Inspection of the system MTF can be useful in making this selection.

The basic AWF SR algorithm is illustrated in the block diagram in Figure 2.5. As mentioned above, registration allows us to create a nonuniform HR grid image denoted g in Figure 2.5. A moving window filter spanning $W_x \times W_y$ HR pixels processes this HR grid to produce the final output. Let the pixel values spanned by the moving window at position i be denoted $g_i = [g_{i,1}, g_{i,2}, \ldots, g_{i,K_i}]^T$, where K_i is the number of LR pixels within the i'th observation window.

For each observation window, an estimate of the pixel at the center of the

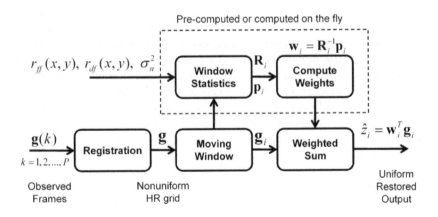

FIGURE 2.5: Overview of the proposed SR algorithm.

observation window is formed as a weighted sum. This is expressed as

$$\hat{z}_i = w_i^T g_i, \tag{2.6}$$

where $\hat{z}_i$ is the estimate of the i'th pixel in the desired image z and w_i is a K_i by 1 vector of weights. It is also possible to use the samples in one observation window to estimate multiple output pixels [22, 14, 15]. Consequently the observation window would move by multiple pixel positions at a time. This can have computational advantages including needing to compute or look-up fewer weight vectors. The minimum mean squared error weights are found using the well-known Wiener solution

$$w_i = R_i^{-1} p_i, \tag{2.7}$$

where $R_i = E\{g_i g_i^T\}$ and $p_i = E\{z_i g_i^T\}$. The weights are normalized so that they sum to 1 to eliminate potential artifacts from variable DC response of adaptive filter.

The required statistics are found empirically in [22] based on training images. A quantized HR grid and a full autocorrelation matrix and cross-correlation vector can be estimated from fully populated observation windows at the HR grid resolution. Then, when a partially populated observation windows are encountered, the full autocorrelation matrix and cross-correlation vector can be subsampled as needed. This allows us to compute the weights using (2.7). The method in [22] also goes one step farther, and uses vector quantization to partition on observation space and estimates a full autocorrelation matrix and cross-correlation vector for each partition [3, 33]. These statistics are tuned to specific structures such as edges, lines and flat regions. This is one way to deal with the nonstationarity of most image data for image restoration [3, 33]. The work in [14, 15] employs a parametric autocorrelation model for the desired underlying image and generates all of the necessary statistics from that. Both a global and spatially varying model are considered in [14]. The spatially varying model, like the vector quantization approach, seeks to treat the nonstationary nature of the image data. Both the vector quantization approach in [22] and the spatially varying approach in [14] are beneficial with moderate and high levels of noise. Under relatively high signal-to-noise ratio conditions, this added complexity generally does not improve performance [20]. Thus, for high SNR environments, a simple wide sense stationary (WSS) auto-correlation model may be the best choice.

The modeling of the autocorrelation matrix and cross-correlation vector is described in detail in [22, 14]. However, we repeat the key steps of the WSS model from [14] for convenience. First let $g_i = f_i + n_i$, where f_i is the noise-free version of the i'th observation vector g_i and n_i is a random noise vector. Assuming a zero-mean uncorrelated noise vector with independent and identically distributed elements of variance σ_n^2, the autocorrelation matrix for the observation vector is given by

$$R_i = E\{g_i g_i^T\} = E\{f_i f_i^T\} + \sigma_n^2 I. \tag{2.8}$$

The cross-correlation vector is given by

$$p_i = E\{z_i g_i^T\} = E\{z_i f_i^T\}. \tag{2.9}$$

Continuing to follow the analysis in [14], let us now assume a WSS autocorrelation function, $r_{dd}(x,y)$, for the desired image $d(x,y)$. The cross-correlation function between $d(x,y)$ and $f(x,y)$, as shown in Figure 2.2, can be expressed in terms of $r_{dd}(x,y)$ [37] as

$$r_{df}(x,y) = r_{dd}(x,y) * h(x,y). \tag{2.10}$$

The autocorrelation of $f(x,y)$ is given by

$$r_{ff}(x,y) = r_{dd}(x,y) * h(x,y) * h(-x,-y). \tag{2.11}$$

Evaluating (2.11) based on the distances between the samples in g_i yields $E\{f_i f_i^T\}$. Incorporating the noise term to this result yields R_i as expressed in (2.8). Similarly, evaluating (2.10) based on the distances between the samples in g_i and the desired sample position gives us p_i from (2.9). The desired image autocorrelations, $r_{dd}(x,y)$, can be obtained empirically from statistically representative training images or defined using a parametric model. Here we use the same circularly-symmetric parametric model as that used in [14]. This model is given by

$$r_{dd}(x,y) = \sigma_d^2 \rho^{\sqrt{x^2+y^2}}, \tag{2.12}$$

where σ_d^2 is the variance of the desired image and ρ controls the decay of the autocorrelation with distance.

It is interesting to observe how the AWF SR weights vary with the spatial distribution of samples within the partial observation window. Figure 2.6 shows weights for distinct spatial distributions of samples. For these results we have selected $L = 3$, $\rho = 0.75$, $\frac{\sigma_d^2}{\sigma_n^2} = 100$ and a PSF computed for Imaging System 2 in Table 2.1. The weights are shown with the same grayscale map where middle gray is 0. It can be seen that these minimum MSE weights change in nontrivial ways as the spatial distribution of samples changes. Note that Figure 2.6(a) shows the case where only the reference frame pixels are present in the observation window. This can occur when using only a single frame or when no motion in that area of the image is present. Here the AWF SR is effectively performing single frame interpolation and restoration. In Figure 2.6(d), the observation window is fully populated. In this case, with AWF weighting is equivalent to that of a standard FIR Wiener filter operating on the HR grid.

The number of possible spatial patterns in a given observation window on the quantized HR grid is 2^b, where $b = W_x W_y - \frac{W_x W_y}{L^2}$. For the 15×15 window shown in Figure 2.6, this amounts to 2^{200} patterns. For an unquantized HR grid, there are an infinite number of possible patterns. Thus, it is generally impractical to precompute all of the weights prior to processing video, even on a quantized HR grid. However, in [15], a partial observation window is proposed that includes the uniform reference frame samples within the $W_x \times W_y$ window plus the M closest samples to the output position. This is illustrated in Figure 2.7 where the positions outlined form the partial observation window. The M closest samples to the output are highlighted. Note that the reference pixels are guaranteed to be present, whereas the other samples may or may not be present, depending on the motion between frames. An upper bound on the number of filter weights to be precomputed using this partial observation window is given by $L^2 2^M$ [15]. Thus, with a suitable choice of M, it is possible to precompute and store all of the weights. This makes processing frames very fast as will be seen in Section 2.4.

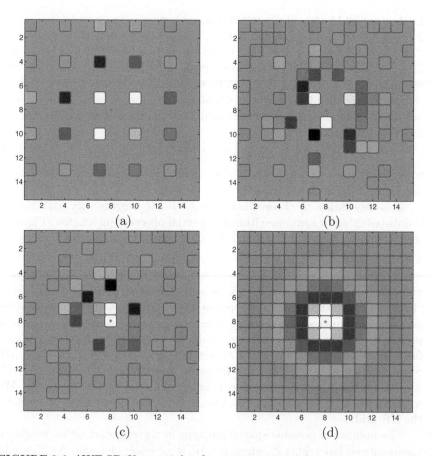

FIGURE 2.6: AWF SR filter weights for various spatial distribution of samples. Weights with only the reference frame samples (no motion) is shown in (a). Weights for a fully populated observation window are shown in (d). All weights are shown with the same grayscale map where middle gray is zero.

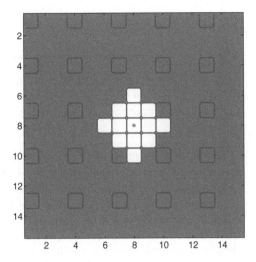

FIGURE 2.7: Partial observation window used by the Fast AWF SR algorithm in [15].

2.4   Experimental Results

In this section, a number of experimental results are presented where we compare the performance of the AWF SR methods to several other benchmark techniques. All of the SR methods applied are listed in Table 2.3. We use simulated data for quantitative analysis and real infrared imagery for subjective analysis in a true application.

2.4.1   SR Results for Simulated Data

The simulated LR frames are generated with different types of affine motion. We use 8 bit images with $L = 3$, the PSF for System 2 in Table 2.1, and a noise standard deviation of 2. The SR is done with $P = 9$ LR frames of size 180×134. For the affine motion, the translation parameters are Gaussian and have a mean of zero and standard deviation of 2 LR pixel spacings. The rotation angle is Gaussian and has a mean of zero and standard deviation of 5 degrees. The shear is horizontal only and the parameter is Gaussian with zero mean and a standard deviation of 0.05 and the zoom factor is Gaussian with a mean of 1 and standard deviation of 0.05. A three-level multiscale affine registration is employed with 5 iterations using bicubic interpolation at each level. The image results are shown in Figure 2.8 where the motion includes translation, rotation, shear and zoom. These image show a 250×250 region of interest (ROI) from the processed imagery. Figure 2.8(a) shows the true HR image. The first (and reference) frame zoomed using bicubic interpolation is shown

TABLE 2.3: SR Algorithm Table

Name	Description
Fast AWF	Adaptive Wiener filter method using quantized HR spatial grid and partial observation window with precomputed weights [15]. Simultaneous nonuniform interpolation and restoration.
Full AWF	Adaptive Wiener filter method in [14] adapted for affine motion and using a discrete (quantized) HR spatial grid. Full observation window is used and the optimum weights are computed for each window on the fly. Simultaneous nonuniform interpolation and restoration.
WNN	Weighted nearest neighbor method in [12, 2] adapted for affine motion. Nonuniform interpolation done using an inverse distance based weighting of the nearest 4 neighbors. Restoration is done with an FFT-based Wiener filter.
Delaunay	Based on the method in [19] adapted for affine motion. Nonuniform interpolation done using Delaunay triangulation. Restoration is done with an FFT-based Wiener filter.
RLS	Regularized least squares interative SR method in [17] adapted for affine motion. This method does not assume the PSF blurring and motion models commute.

in Figure 2.8(b). The partially populated HR grid is shown in Figure 2.8(c). The SR outputs for all of the SR methods in Table 2.3 make up the rest of Figure 2.8. The noise-to-signal ratios (NSRs) that provided the lowest MSE are used for each method. The AWF SR outputs use an NSR of 0.01. The WNN method uses an NSR of 0.04 and the Delaunay SR method uses an NSR of 0.02. The RLS method uses a regularization parameter of $\lambda = 0.01$ with 20 iterations [17].

Table 2.4 shows the MSE results for the SR methods along with the average run time (excluding registration). The processing was done on a Pentium 4 CPU with a clock speed of 2.8 GHz. Note that the registration of the 9 frames took 3.31 seconds. As noted in [15] this can be sped up using fewer levels and bilinear interpolation. When processing video, only one new frame needs to be registered to the previous frame to produce an output frame. This is because the affine parameters relating each frame to the reference frame can be determined by accumulating the incremental frame-to-frame registration parameters. It should be noted, however, that with this approach, registration errors may also accumulate. Notwithstanding this, we have observed that this can generally be a very effective video registration method. Note that single frame bicubic interpolation output has, by far, the highest MSE. All of the multiframe SR methods produce a much lower MSE. The RLS SR method with 20 iterations produces the lowest MSE in this experiment, but at the cost of a high run time. The Fast AWF has the lowest run time with MSEs comparable to WNN and RLS with 5 iterations.

TABLE 2.4: MSE for the various SR algorithms with affine motion ($L = 3$ and $P = 9$).

Algorithm	Translation	Rotation	Shear	Zoom	All	Time (s)
Bicubic	253.64	253.64	253.64	253.64	253.64	0.357
Fast AWF	141.20	157.25	185.58	158.46	152.23	0.811
Full AWF	127.36	143.65	181.50	144.51	139.26	57.238
WNN	141.93	162.82	211.37	163.95	164.85	1.784
Delaunay	121.36	137.44	191.06	141.20	144.29	21.105
RLS (20 Iterations)	107.40	118.20	167.74	120.40	114.55	275.337
RLS (5 Iterations)	141.32	151.64	182.73	150.88	149.50	67.204

2.4.2   SR Results for Infrared Video Data

The first infrared dataset used for evaluating the SR methods is one obtain with Imaging System 1 in Table 2.1. These results are shown in Figure 2.9. The camera is mounted on a stationary tripod and is manually panned and rotated acquiring a 30 Hz video sequence. The SR methods use $P = 20$ frames and form output images of size 256×256 with $L = 4$. Bicubic interpolation of the first (reference) frame is shown in Figure 2.9(a). The partially populated HR grid is shown in Figure 2.9(b). The output for the Fast AWF, Full AWF, WNN, and RLS methods are shown in Figure 2.9(c)–(f), respectively. The SNR for the AWF methods is 0.05. The NSR for the WNN method is 0.1. The RLS method uses 10 iterations and $\lambda = 0.01$. The tuning parameters have been chosen based on subjective image quality. Because of the high level of aliasing for this imaging system, the aliasing artifacts are rather obvious in Figure 2.9(a). Most prominently, the artifacts take on the form of jagged diagonal edges. Overall blurring of the image is also evident from the detector dominated PSF. It is clear that if single frame restoration is applied to this imagery, the aliasing artifacts would only be pronounced. The multiframe SR methods clearly reduce aliasing and sharpen the imagery. The RLS again appears to provide the best results. However, the fast SR methods like the Fast AWF and WNN provide a significantly enhanced image (compared with bicubic interpolation) in a small fraction of the run time of the RLS method.

Finally, results obtained with data from Imaging System 2 in Table 2.1 are shown in Figure 2.10. This camera is also mounted on a stationary tripod and is manually panned and rotated acquiring a 30 Hz video sequence. The

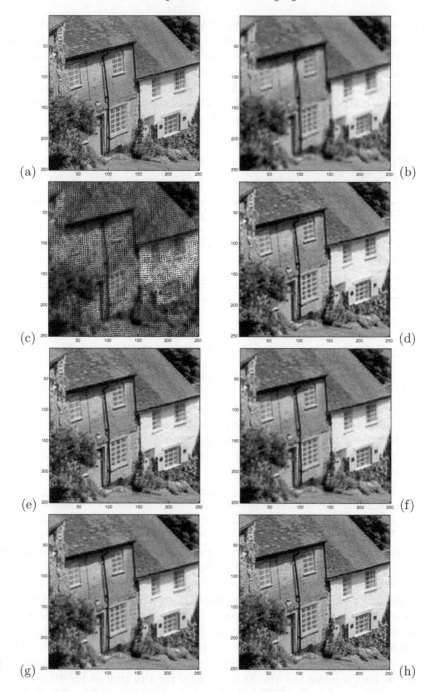

FIGURE 2.8: Output images for the simulated image sequence with $L = 3$ and $P = 9$. (a) Desired image, (b) single frame bicubic interpolation, (c) partially populated high-resolution grid after registration, (d) output of the fast AWF method with partial observation window and precomputed weights, (e) full AWF method with weights computed on the fly, (f) WNN with 4 nearest neighbors, (g) Delaunay triangulation based output, (h) RLS method.

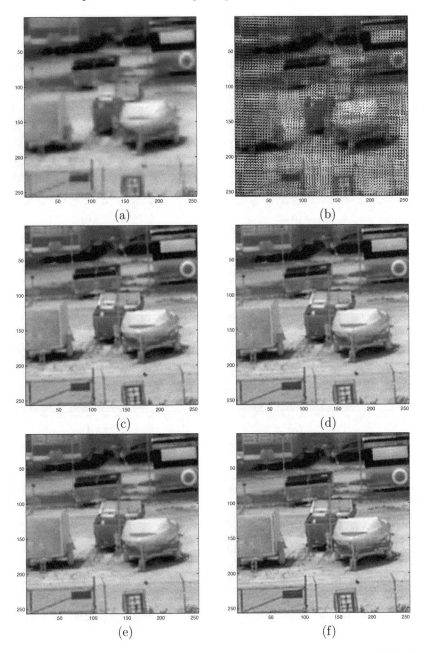

FIGURE 2.9: Outputs for image sequence from the tripod mounted Amber imager with $L = 4$ and $P = 20$. (a) Single frame bicubic interpolation, (b) partially populated high-resolution grid, (c) Fast AWF with partial observation window and precomputed weights, (d) Full AWF with weights computed on the fly, (e) WNN with 4 nearest neighbors, and (f) RLS method.

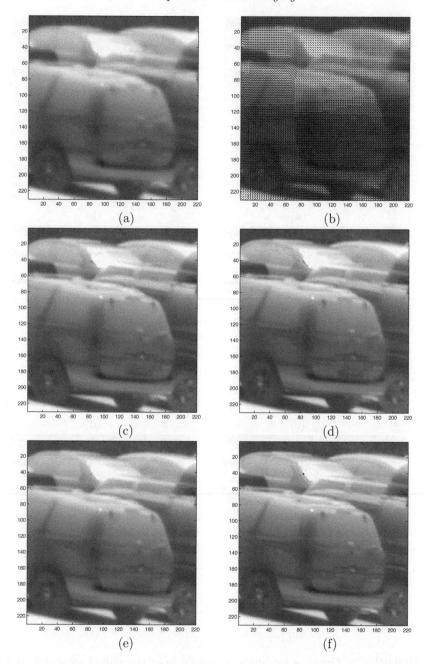

FIGURE 2.10: Outputs for image sequence from the tripod mounted L-3 imager with $L = 3$ and $P = 9$. (a) Single frame bicubic interpolation, (b) partially populated high-resolution grid, (c) Fast AWF with partial observation window and precomputed weights, (d) Full AWF with weights computed on the fly, (e) WNN with 4 nearest neighbors, and (f) RLS method.

SR methods use $P = 9$ frames and form output images of size 230×220 with $L = 3$. Bicubic interpolation of the first (reference) frame is shown in Figure 2.10(a). The partially populated HR grid is shown in Figure 2.10(b). The output for the Fast AWF, Full AWF, WNN and RLS methods are shown in Figure 2.10(c)–(f), respectively. The tuning parameters used on the previous dataset are also applied here. While the aliasing here is more subtle, the results appear to be consistent with those for the previous dataset.

2.5   Conclusions

The AWF SR methods are a type of nonuniform interpolation based SR algorithm. A distinctive feature of the AWF SR methods, however, is that the nonuniform interpolation and restoration are done simultaneously in a single weighted sum operation. The weights are determined based on FIR Wiener filter theory and they adapt to the specific spatial distribution of LR samples in the nonuniformly populated HR grid. Various methods have been explored in the literature for modeling the correlation statistics needed to determine the weights. Spatially varying statistics based on vector quantization [22] and local variance [14] have been used. The spatially varying statistics provide the most benefit in low signal-to-noise environments. In high signal-to-noise environments, like those considered here, a global statistical model is very effective and simpler to implement.

One of the main benefits of the AWF SR method is the potential for fast processing. Furthermore, the method is robust to the spatial distribution of LR pixels on the HR grid and it degrades gracefully towards the case where no motion is present (or only one frame is used). For translational motion, the number of distinct weight vectors needed for the Full AWF is small, and this allows for fast processing. The Fast AWF SR algorithm precomputes all of the weights for any motion model using a partial observation window [15].

The experimental results show that iterative SR methods, like the RLS, generally provide the best results. However these have a high computational complexity. The Fast AWF SR method is observed to have the shortest run time of any of the SR methods tested, with performance comparable to that of WNN and the RLS with 5 iterations. Delaunay SR was a notable performer in that the MSE was comparable to that of the Full AWF and it had a shorter run time. However, the run time for Delaunay SR was still much longer than that of the Fast AWF.

This chapter has also explored motion models for multiframe SR. It is noted that perhaps the most favorable conditions for SR are for a camera that is panning and tilting from a stationary location with a static scene. In that case, the 2-D optical flow is approximately quadratic, but can be effectively modeled as affine in small regions (or piecewise affine for large areas). With the stationary camera and static scene, no motion parallax, occulusion, or object pose variations are present and the optical flow is independent of the 3-D scene. Using the affine or piecewise affine model for this scenario, it is usually possible to form a highly overdetermined set of linear equations to solve for the motion parameters. With many equations and few motion parameters, it is often possible to get sufficiently accurate estimates, even in the presence of aliasing and noise.

2.6 Acknowledgments

The author would like to thank Dr. Doug Droege at L-3 Cincinnati Electronics for providing the L-3 infrared imagery used here. The Amber infrared images have been provided courtesy of the Air Force Research Laboratory at WPAFB in Dayton, OH.

Bibliography

[1] K. Aizawa, T. Komatsu, and T. Saito. Acquisition of very high resolution images using stereo cameras. In *Proc. SPIE Visual Communication and Image Processing*, volume 1, pages 318–328, Boston, MA, Nov 1991.

[2] M. S. Alam, J. G. Bognar, R. C. Hardie, and B. J. Yasuda. Infrared image registration using multiple translationally shifted aliased video frames. *IEEE Instrum. Meas. Mag.*, 49(5), Oct 2000.

[3] K. E. Barner, A. M. Sarhan, and R. C. Hardie. Partition-based weighted sum filters for image restoration. *IEEE Trans. Image Process.*, 8(5):740–745, May 1999.

[4] B. Bascle, A. Blake, and A. Zisserman. Motion deblurring and super resolution from an image sequence. *EECV*, 2:573–581, Apr 1996.

[5] James R. Bergen, P. An, Th J. Hanna, and Rajesh Hingorani. Hierarchical model-based motion estimation. In *Proceedings of the Second European Conference on Computer Vision*, pages 237–252. Springer-Verlag, 1992.

[6] S. Cain, R. C. Hardie, and E. E. Armstrong. Restoration of aliased video sequences via a maximum-likelihood approach. In *Infrared Information Symposium (IRIS) on Passive Sensors, Monterey, CA*, volume 1, pages 377–390, March 1996.

[7] R. Chan, T. Chan, M.Ng, W. Tang, and C. Wong. Preconditioned iterative methods for high-resolution image reconstruction from multisensors. *Adv. Signal Process. Algorithms, Architectures, Implementations*, 3461:348–357, 1998.

[8] P. Cheeseman, B. Kanefsky, R. Kraft, J. Stutz, and R. Hanson. Super-resolved surface reconstruction from multiple images. NASA Tech. Rep. FIA-94-12, Dec 1994.

[9] T. Connolly and R. Lane. Gradient methods for superresolution. In *Proc. Int. Conf. Image Processing*, volume 1, pages 917–920, 1997.

[10] M. Elad and Y. Hel-Or. A fast super-resolution reconstruction algorithm for pure translational motion and common space invariant blur. *IEEE Trans. Image Process.*, 10(8):1187–1193, Aug 2001.

[11] S. Farsiu, D. Robinson, M. Elad, and P. Milanfar. Fast and robust multiframe super-resolution. *IEEE Trans. Image Process.*, 13(10):1327–1344, Oct 2004.

[12] J. C. Gillette, T. M. Stadtmiller, and R. C. Hardie. Reduction of aliasing in staring infrared imagers utilizing subpixel techniques. *Opt. Eng.*, 34(11):3130–3137, Nov 1995.

[13] J.W. Goodman. *Introduction to Fourier Optics*. McGraw-Hill, 1968.

[14] R. C. Hardie. A fast super-resolution algorithm using an adaptive wiener filter. *IEEE Trans. Image Process.*, 16:2953–2964, Dec 2007.

[15] R. C. Hardie and K. J. Barnard. Fast adaptive wiener filter based super-resolution applied to affine motion. *Submitted to IEEE Transactions on Image Processing*, September 2009.

[16] R. C. Hardie, K. J. Barnard, and E. E Armstrong. Joint MAP registration and high-resolution image estimation using a sequence of undersampled images. *IEEE Trans. Image Process.*, 6(12):1621–1633, Dec 1997.

[17] R. C. Hardie, K. J. Barnard, J. G. Bognar, E. E Armstrong, and E. A. Watson. High-resolution image reconstruction from a sequence of rotated and translated frames and its application to an infrared imaging system. *Opt. Eng.*, 37(1):247–260, Jan 1998.

[18] M. Irani and S. Peleg. Improving resolution by image registration. *CHIP: Graph. Models Image Process.*, 53(3):231–239, May 1991.

[19] S. Lertrattanapanich and N. K. Bose. High resolution image formation from low resolution frames using delaunay triangulation. *IEEE Trans. Image Process.*, 11(12):1427–1441, Dec 2002.

[20] Y. Lin, R. C. Hardie, Q. Sheng, M. Shao, and K. E. Barner. Improved optimization of soft-partition-weighted sum filters and their application to image restoration. *Applied Optics*, 45(12):2697–2706, April 2006.

[21] S. Mann and R. Picard. Virtual bellows: construction of high quality stills from video. In *Proc. IEEE Int. Conf. Image Processing*, Austin, TX, Nov 1994.

[22] B. Narayanan, R. C. Hardie, K. E. Barner, and M. Shao. A computationally efficient super-resolution algorithm for video processing using partition filters. *IEEE Trans. Circuits Syst. Video Technol.*, 17(5):621–634, May 2007.

[23] N. Nguyen and P. Milanfar. A wavelet-based interpolation restoration method for superresolution. *Circuits, Syst., Signal Process.*, 19(4):321–338, Aug 2000.

[24] N. Nguyen, P. Milanfar, and G. Golub. A computationally efficient super-resolution image reconstruction algorithm. *IEEE Trans. Image Process.*, 10(4):573–583, Apr 2001.

[25] S. C. Park, M. K. Park, and M. G. Kang. Super-resolution image reconstruction: A technical overview. *IEEE Signal Process. Mag.*, pages 21–36, May 2003.

[26] A. Patti, M.Sezan, and A. Teklap. Superresolution video reconstruction with arbitrary sampling lattices and nonzero aperture time. *IEEE Trans. Image Process.*, 6(8):1064–1076, Aug 1997.

[27] S. Periaswamy and H. Farid. Elastic registration in the presence of intensity variations. *IEEE Transactions on Medical Imaging*, 22(7):865–874, 2003.

[28] T. Q. Pham, L. J. van Vliet, and K. Schutte. Robust fusion of irregularly sampled data using adaptive normalized convolution. *EURASIP Journal on Applied Signal Processing*, 2006(Article ID 83268):1–12, 2006.

[29] L. C. Pickup, D. P. Capel, S. J. Roberts, and A. Zisserman. Overcoming registration uncertainty in image super-resolution: Maximize or marginalize? *EURASIP Journal on Advances in Signal Processing*, 2007:Article ID 23565, 14 pages, 2007.

[30] K. Sauer and J. Allebach. Iterative reconstruction of band-limited images from non-uniformly spaced samples. *IEEE Trans. Circuits Syst.*, CAS-34:1497–1505, 1987.

[31] R. R. Schultz and R. L. Stevenson. Extraction of high-resolution frames from video sequences. *IEEE Trans. Image Process.*, 5:996–1011, June 1996.

[32] R. R. Shannon. Aberrations and their effect on images. In *Proceedings SPIE, Geometric Optics, Critical Review of Technology*, volume 531, pages 27–37.

[33] M. Shao, K. E. Barner, and R. C.Hardie. Partition-based interpolation for image demosaicking and super-resolution reconstruction. *Opt. Eng.*, 44:107003–1–107003–14, Oct 2005.

[34] H. Shekarforoush and R. Chellappa. Data-driven multi-channel super-resolution with application to video sequences. *J. Opt. Soc. Amer. A*, 16(3):481–492, Mar 1999.

[35] J. Shi, S. E. Reichenbach, and J. D. Howe. Small-kernel superresolution methods for microscanning imaging systems. *Applied Optics*, 45(6):1203–1214, Feb 2006.

[36] A. Tekalp, M. Ozkan, and M. Sezan. High resolution image reconstruction from lower-resolution image sequences and space-varying image restoration. In *Proc. ICASSP '92*, volume 3, pages 169–172, San Fransisco, CA, Mar 1992.

[37] C. W. Therrian. *Discrete Random Signals and Statistical Signal Processing*. Prentice Hall, 1992.

[38] E. Trucco and A. Verri. *Introductory Techniques for 3-D Computer Vision*. Prentice Hall, 1998.

[39] T. R. Tuinstra and R. C. Hardie. High resolution image reconstruction from digital video by exploitation on non-global motion. *Opt. Eng.*, 38(5), May 1999.

3

Locally Adaptive Kernel Regression for Space-Time Super-Resolution

Hiroyuki Takeda

University of California, Santa Cruz

Peyman Milanfar

University of California, Santa Cruz

CONTENTS

Abstract — In this chapter, we discuss a novel framework for adaptive enhancement and spatiotemporal upscaling of videos containing complex motions. Our approach is based on multidimensional kernel regression, where each pixel in the video sequence is approximated with a 3-D local (Taylor) series, capturing the essential local behavior of its spatiotemporal neighborhood. The coefficients of this series are estimated by solving a local weighted least-squares problem, where the weights are a function of the 3-D space-time orientation in the neighborhood. As this framework is fundamentally based upon the comparison of neighboring pixels in both space and time, it implicitly contains information about the local motion of the pixels across time, therefore rendering unnecessary an explicit computation of motions of modest size. When large motions are present, a basic, rough motion compensation step returns the sequence to a form suitable again for motion-estimation-free super-resolution. The proposed approach not only significantly widens the applicability of super-resolution methods to a broad variety of video sequences containing complex motions, but also yields improved overall performance. Using several examples, we illustrate that the developed algorithm has super-resolution capabilities that provide improved optical resolution in the output, while being able to work on general input video with essentially arbitrary motion.

3.1 Introduction

The emergence of high-definition displays in recent years (e.g. 720×1280 and 1080×1920 or higher spatial resolution, and up 240Hz in temporal resolution), along with the proliferation of increasingly cheaper digital imaging technology has resulted in the need for fundamentally new image processing algorithms. Specifically, in order to display relatively low quality content on such high resolution displays, the need for better space-time upscaling, denoising, and deblurring algorithms has become an urgent market priority, with correspondingly interesting challenges for the academic community. The existing literature on enhancement and upscaling (sometimes called super-resolution)[1] is vast and rapidly growing in both the single frame case [9, 17] and the multi-frame (video) case [6, 8, 10, 12, 16, 24, 35, 38, 42], and many new algorithms for this problem have been proposed recently. Yet, one of the most fundamental roadblocks has not been overcome. In particular, in order to be effective, essentially all the existing multi-frame super-resolution approaches must perform (sub-pixel) accurate motion estimation [6, 8, 10, 12, 16, 24, 35, 38, 42, 27]. As a result, most methods fail to perform well in the presence of complex motions that are quite common. Indeed, in most practical cases where complex motion and occlusions are present and not estimated with pinpoint accuracy, existing algorithms tend to fail catastrophically, often producing outputs that are of even worse visual quality than the low-resolution inputs.

In this work, we address the challenging problem of spatiotemporal video super-resolution in a fundamentally different way, which removes the need for explicit subpixel accuracy motion estimation. We present a methodology that is based on the notion of consistency between the estimated pixels, which is

---

[1]To clarify the use of words super-resolution and upscaling, we note that if the algorithm does not receive input frames that are aliased, it will still produce an output with a higher number of pixels and/or frames (i.e., "upscaled"), but which is not necessarily "super-resolved."

derived from the novel use of kernel regression [34], [31]. Classical kernel regression is a well studied, nonparametric point estimation procedure. In our earlier work [31], we generalized the use of these techniques to spatially adaptive (steering) kernel regression, which produces results that preserve and restore details with minimal assumptions on local signal and noise models [36]. Other related nonparametric techniques for multidimensional signal processing have emerged in recent years as well. In particular, the concept of normalized convolution [19], and the introduction of support vector machines [26] are notable examples. In the present work, the steering techniques in [31] are extended to 3-D where, as we will demonstrate, we can perform high fidelity space-time upscaling and super-resolution. Most importantly, this is accomplished without the explicit need for accurate motion estimation.

In a related work [28], we have generalized the nonlocal means (NLM) framework [2] to the problem of super-resolution. In that work, measuring the similarity of image *patches* across space and time resulted in "fuzzy" or probabilistic motions, as explained in the chapter by Protter and Elad. Such estimates also avoid the need for explicit motion estimation and give relatively larger weights to more similar patches used in the computation of the high resolution estimate. Another recent example of a related approach appears in [5] where Danielyan, et al. have presented an extension of the block-matching 3-D filter (BM3D) [4] for video super-resolution, in which explicit motion estimation is also avoided by classifying the image patches using a block matching technique. The objectives of the present work, our NLM-based approach [28], and Video-BM3D [5] just mentioned are the same: namely, to achieve super-resolution on general sequences, while avoiding explicit (subpixel-accurate) motion estimation. These approaches represent a new generation of super-resolution algorithms that are quite distinctly different from all existing super-resolution methods. More specifically, existing methods have required highly accurate subpixel motion estimation and have thus failed to achieve resolution enhancement on arbitrary sequences.

We propose a framework that encompasses both video denoising, spatiotemporal upscaling, and super-resolution in 3-D. This framework is based on the development of locally adaptive 3-D filters with coefficients depending on the pixels in a local neighborhood of interest in space-time in a novel way. These filter coefficients are computed using a particular measure of similarity and consistency between the neighboring pixels that uses the local geometric and radiometric structure of the neighborhood. To be more specific, the computation of the filter coefficients is carried out in the following distinct steps. First, the local (spatiotemporal) gradients in the window of interest are used to calculate a covariance matrix, sometimes referred to as the "local structure tensor" [18]. This covariance matrix, which captures a locally dominant orientation *at each pixel*, is then used to define a local metric for measuring the similarity between the pixels in the neighborhood. This local metric distance is then inserted into a (Gaussian) kernel which, with proper normalization, then defines the local weights to be applied in the neighborhood.

The above approach is based on the concept of *steering kernel regression* (SKR), earlier introduced in [31] for images. A specific extension of these concepts to 3-D signals for the express purpose of video denoising and resolution enhancement are the main subjects of this chpater. As we shall see, since the development in 3-D involves the computation of orientation in space-time [13], motion information is implicitly and reliably captured. Therefore, unlike conventional approaches to video processing, 3-D SKR does not require explicit estimation of (modestly sized but essentially arbitrarily complex) motions, as this information is implicitly captured within the locally "learned" metric. It is worth mentioning in passing here that the approach we take, while independently derived, is in the same spirit as the body of work known as *Metric Learning* in the machine learning community, e.g., [37].

Naturally, the performance of the proposed approach is closely correlated with the quality of estimated space-time orientations. In the presence of noise, aliasing, and other artifacts, the estimates of orientation may not be initially accurate enough, and as we explain in Section 3.2.3, we therefore propose an iterative mechanism for estimating the orientations, which relies on the estimate of the pixels from the previous iteration.

To be more specific, as shown in Figure 3.8, we can first process a video sequence with orientation estimates of modest quality. Next, using the output of this first step, we can re-estimate the orientations, and repeat this process several times. As this process continues, the orientation estimates are improved, as is the quality of the output video. The overall algorithm we just described will be referred to as the 3-D iterative steering kernel regression (3-D ISKR).

As we will see in the coming sections, the approach we introduce here is ideally suited for implicitly capturing relatively small motions using the orientation tensors. However, if the motions are somewhat large, the resulting (3-D) local similarity measure, due to its inherent local nature, will fail to find similar pixels in nearby frames. As a result, the 3-D kernels essentially collapse to become 2-D kernels centered around the pixel of interest within the same frame. Correspondingly, the net effect of the algorithm would be to do frame-by-frame 2-D upscaling. For such cases, as discussed in Section 3.2.4, some level of explicit motion estimation is unavoidable to reduce temporal aliasing and achieve resolution enhancement. However, as we will illustrate in this chapter, this motion estimation can be quite rough (accurate to within a whole pixel at best). This rough motion estimate can then be used to "neutralize" or "compensate" for the large motion, leaving behind a residual of small motions, which can be implicitly captured within the 3-D orientation kernel. In summary, our approach can accommodate a variety of complex motions in the input videos by a two-tiered approach: (i) large displacements are neutralized by rough motion compensation either globally or block-by-block as appropriate, and (ii) 3-D ISKR handles the fine-scale and detailed rest of the possibly complex motion present.

This chapter is organized as follows: in Section 3.2, first we briefly describe the fundamental concepts behind the SKR framework in 2-D and present the extension of the SKR framework to 3-D including discussions of how our method captures local complex motions and performs rough motion compensation, and explicitly describe its iterative implementation. In Section 3.3, we provide some experimental examples with both synthetic and real video sequences, and we conclude this chapter in Section 3.4.

3.2 Adaptive Kernel Regression

In this section, we first review the fundamental framework of *kernel regression* (KR) [36] and its extension, the *steering* kernel regression (SKR) [31], in 2-D. Then, we extend the steering approach to 3-D and discuss some important aspects of the 3-D extension.

3.2.1 Classic Kernel Regression in 2-D

The KR framework defines its data model as

$$y_i = z(\boldsymbol{x}_i) + \varepsilon_i, \quad \boldsymbol{x}_i \in \omega, \quad i = 1, \cdots, P, \tag{3.1}$$

where y_i is a noise-ridden sample measured at $\boldsymbol{x}_i = [x_{1i}, x_{2i}]^T$ (Note: x_{1i} and x_{2i} are spatial coordinates), $z(\cdot)$ is the (hitherto unspecified) *regression function* of interest, ε_i is an *i.i.d.* zero mean noise, and P is the total number of samples in an arbitrary "window" ω around a position $\boldsymbol{x}$ of interest as illustrated in Figure 3.1. As such, the KR framework provides a rich mechanism for computing pointwise estimates of the regression function with minimal assumptions about global signal or noise models.

While the particular form of $z(\cdot)$ may remain unspecified, we can develop a generic local expansion of the function about a sampling point $\boldsymbol{x}_i$. Specifically,

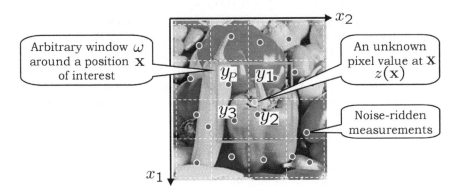

FIGURE 3.1: The data model for the kernel regression framework.

if the position of interest x is near the sample at x_i, we have the N-th order Taylor series

$$z(x_i) \approx z(x) + \{\nabla z(x)\}^T (x_i - x) + \frac{1}{2}(x_i - x)^T \{\mathcal{H}z(x)\}^T (x_i - x) + \cdots$$

$$\approx \beta_0 + \beta_1^T (x_i - x) + \beta_2^T \text{vech}\{(x_i - x)(x_i - x)^T\} + \cdots \quad (3.2)$$

where ∇ and $\mathcal{H}$ are the gradient (2×1) and Hessian (2×2) operators, respectively, and vech$(\cdot)$ is the half-vectorization operator that lexicographically orders the lower triangular portion of a symmetric matrix into a column-stacked vector. Furthermore, β_0 is $z(x)$, which is the signal (or pixel) value of interest, and the vectors β_1 and β_2 are

$$\beta_1 = \left[\frac{\partial z(x)}{\partial x_1}, \frac{\partial z(x)}{\partial x_2} \right]^T,$$

$$\beta_2 = \frac{1}{2} \left[\frac{\partial^2 z(x)}{\partial x_1^2}, \frac{\partial^2 z(x)}{\partial x_1 \partial x_2}, \frac{\partial^2 z(x)}{\partial x_2^2}, \right]^T. \quad (3.3)$$

Since this approach is based on *local* signal representations (i.e. Taylor series), a logical step to take is to estimate the parameters $\{\beta_n\}_{n=0}^N$ using all the neighboring samples $\{y_i\}_{i=1}^P$ while giving the nearby samples higher weights than samples farther away. A weighted least-square formulation of the fitting problem capturing this idea is

$$\min_{\{\beta_n\}_{n=0}^N} \sum_{i=1}^P \left[y_i - \beta_0 - \beta_1^T (x_i - x) - \beta_2^T \text{vech}\{(x_i - x)(x_i - x)^T\} - \cdots \right]^2$$

$$\cdot K_H(x_i - x)$$

$$(3.4)$$

with

$$K_H(x_i - x) = \frac{1}{\det(H)} K(H^{-1}(x_i - x)), \quad (3.5)$$

where N is the regression order, $K(\cdot)$ is the kernel function (a radially symmetric function such as a Gaussian), and H is the smoothing (2×2) matrix that dictates the "footprint" of the kernel function. The simplest choice of the smoothing matrix is $H = hI$, where h is called the *global smoothing parameter*. The contour of the kernel footprint is perhaps the most important factor in determining the quality of estimated signals. For example, it is desirable to use kernels with large footprints in the smooth local regions to reduce the noise effects, while relatively smaller footprints are suitable in the edge and textured regions to preserve the underlying signal discontinuity. Furthermore, it is desirable to have kernels that adapt themselves to the local structure of the measured signal, providing, for instance, strong filtering along an edge rather than across it. This last point is indeed the motivation behind the *steering* KR framework [31] that we will review Section 3.2.2.

Returning to the optimization problem (3.4), regardless of the regression order (N), and the dimensionality of the regression function, we can rewrite it as the weighted least squares problem:

$$\widehat{b} = \arg\min_{b} (y - Xb)^T K (y - Xb), \tag{3.6}$$

where

$$y = \begin{bmatrix} y_1, & y_2, & \cdots, & y_P \end{bmatrix}^T, \quad b = \begin{bmatrix} \beta_0, & \beta_1^T, & \cdots, & \beta_N^T \end{bmatrix}^T, \tag{3.7}$$

$$K = \mathrm{diag}\begin{bmatrix} K_H(x_1 - x), & K_H(x_2 - x), & \cdots, & K_H(x_P - x) \end{bmatrix} \tag{3.8}$$

and

$$X = \begin{bmatrix} 1, & (x_1 - x), & \mathrm{vech}\{(x_1 - x)(x_1 - x)^T\}, & \cdots \\ 1, & (x_2 - x), & \mathrm{vech}\{(x_2 - x)(x_2 - x)^T\}, & \cdots \\ \vdots & \vdots & \vdots & \vdots \\ 1, & (x_P - x), & \mathrm{vech}\{(x_P - x)(x_P - x)^T\}, & \cdots \end{bmatrix} \tag{3.9}$$

with "diag" defining a diagonal matrix. Using the notation above, the optimization (3.4) provides the weighted least square estimator:

$$\widehat{b} = (X^T K X)^{-1} X^T K \, y \tag{3.10}$$

and the estimate of the signal (i.e. pixel) value of interest β_0 is given by a weighted *linear* combination of the nearby samples:

$$\hat{z}(x) = \widehat{\beta}_0 = e_1^T \widehat{b} = \sum_{i=1}^{P} W_i(K, H, N, x_i - x) \, y_i \tag{3.11}$$

where e_1 is a column vector with the first element equal to one and the rest equal to zero, $\sum_i W_i = 1$, and we call W_i the *equivalent kernel* weight function for y_i (q.v. [31] or [36] for more detail). For example, for zero-th order regression (i.e. $N = 0$), the estimator (3.11) becomes

$$\hat{z}(x) = \widehat{\beta}_0 = \frac{\displaystyle\sum_{i=1}^{P} K_H(x_i - x) \, y_i}{\displaystyle\sum_{i=1}^{P} K_H(x_i - x)}, \tag{3.12}$$

which is the so-called *Nadaraya-Watson* estimator (NWE) [23], which is nothing but a space-varying convolution (if samples are irregularly spaced).

What we described above is the "classic" kernel regression framework, which as we just mentioned, yields a pointwise estimator that is always a local "linear," though possibly space-varying, combination of the neighboring samples. As such, it suffers from an inherent limitation. In the next sections, we describe the framework of *steering* KR in two and three dimensions, in which the kernel weights themselves are computed from the local window (or cube), and therefore we arrive at filters with more complex (nonlinear and space-varying) action on the data.

3.2.2   Steering Kernel Regression in 2-D

The steering kernel approach is based on the idea of robustly obtaining local signal structures by analyzing the radiometric (pixel value) differences locally, and feeding this structure information to the kernel function in order to affect its shape and size.

Consider the (2×2) smoothing matrix $\boldsymbol{H}$ in (3.5). As explained in Section 3.2.1, in the generic "classical" case, this matrix is a scalar multiple of the identity with the global parameter h. This results in kernel weights that have equal effect along the x_1- and x_2-directions. However, if we properly choose this matrix, the kernel function can capture local structures. More precisely, we define the smoothing matrix as a symmetric positive-definite matrix:

$$\boldsymbol{H}_i = h\boldsymbol{C}_i^{-\frac{1}{2}} \tag{3.13}$$

which we call the *steering matrix* and where, *for each given sample* y_i, the matrix $\boldsymbol{C}_i$ is estimated as the local covariance matrix of the neighborhood spatial gradient vectors. A naive estimate of this covariance matrix may be obtained as

$$\widehat{\boldsymbol{C}}_i^{\text{naive}} = \boldsymbol{J}_i^T \boldsymbol{J}_i, \tag{3.14}$$

with

$$\boldsymbol{J}_i = \begin{bmatrix} \vdots & \vdots \\ z_{x_1}(\boldsymbol{x}_j), & z_{x_2}(\boldsymbol{x}_j) \\ \vdots & \vdots \end{bmatrix}, \quad \boldsymbol{x}_j \in \xi_i, \quad j = 1, \cdots, Q, \tag{3.15}$$

where $z_{x_1}(\,\cdot\,)$ and $z_{x_2}(\,\cdot\,)$ are the first derivatives along x_1- and x_2-axes, ξ_i is the local analysis window around a sample position $\boldsymbol{x}_i$, and Q is the number of rows in $\boldsymbol{J}_i$. However, the naive estimate may in general be rank deficient or unstable. Therefore, instead of using the naive estimate, we obtain covariance matrices by using the (compact) singular value decomposition (SVD) of $\boldsymbol{J}_i$. A specific choice of $\boldsymbol{C}_i$ using the SVD for the 2-D case is introduced in [31], and we will show $\boldsymbol{C}_i$ for the 3-D case in Section 3.2.3.

With the above choice of the smoothing matrix and a Gaussian kernel, we now have the steering kernel function as

$$K_{\boldsymbol{H}_i}(\boldsymbol{x}_i - \boldsymbol{x}) = \frac{\sqrt{\det(\boldsymbol{C}_i)}}{2\pi h^2} \exp\left\{ -\frac{1}{2h^2} \left\| \boldsymbol{C}_i^{\frac{1}{2}}(\boldsymbol{x}_i - \boldsymbol{x}) \right\|_2^2 \right\}, \tag{3.16}$$

and the weighted least square estimator as

$$\widehat{\boldsymbol{b}} = \left(\boldsymbol{X}^T \boldsymbol{K}^{\text{s}} \boldsymbol{X} \right)^{-1} \boldsymbol{X}^T \boldsymbol{K}^{\text{s}} \boldsymbol{y} \tag{3.17}$$

where

$$\boldsymbol{K}^{\text{s}} = \text{diag}\left[\ K_{\boldsymbol{H}_1}(\boldsymbol{x}_1 - \boldsymbol{x}), \quad K_{\boldsymbol{H}_2}(\boldsymbol{x}_2 - \boldsymbol{x}), \quad \cdots, \quad K_{\boldsymbol{H}_P}(\boldsymbol{x}_P - \boldsymbol{x}) \ \right]. \tag{3.18}$$

Again, for example, for zeroth order (i.e., $N = 0$), the estimator (3.17) yields a pointwise estimator:

$$\hat{z}(\boldsymbol{x}) = \widehat{\beta}_0 = \frac{\displaystyle\sum_{i=1}^{P} K_{\boldsymbol{H}_i}(\boldsymbol{x}_i - \boldsymbol{x})\, y_i}{\displaystyle\sum_{i=1}^{P} K_{\boldsymbol{H}_i}(\boldsymbol{x}_i - \boldsymbol{x})}, \tag{3.19}$$

which is the data-adapted version of NWE. It is noteworthy that, as shown in the weight matrix $\boldsymbol{K}^s$ (3.18) involving the steering matrices $\{\boldsymbol{H}_i\}_{i=1}^{P}$ of all the neighboring samples $\{y_i\}_{i=1}^{P}$, the steering kernel function (3.16) effectively captures the local image structures. We will graphically show the steering kernels shortly in Figure 3.3.

Figure 3.2 illustrates a schematic representation of the estimates of local covariance matrices $\boldsymbol{C}_i$ in (3.13) at a local region with one dominant orientation. First, we compute gradients $z_{x_1}(\cdot)$ and $z_{x_2}(\cdot)$ of the given image, where the gradients are illustrated as vectors with red arrows. In this example, we set the size of the regression window ω to 5×5 and and the size of the window ξ for the calculation of the covariance estimate is set to 3×3. Therefore, the overall analysis window becomes 7×7. Next, sliding the window ξ_i for $i = 1, \cdots, 25$, we compute the covariance matrix $\boldsymbol{C}_i$ for each pixel y_i in the middle (5×5) portion. Once $\boldsymbol{C}_i$'s are available, we perform the steering kernel regression (3.17) with the weights given by the $\boldsymbol{C}_i$'s and estimate the pixel value $z(\boldsymbol{x})$ at the position of interest. Graphical representations of the steering kernel weights for noise-free (Pepper and Parrot) images are illustrated in

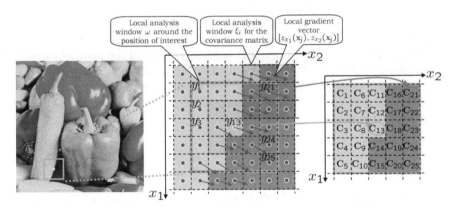

FIGURE 3.2: A schematic representation of the estimates of local covariance matrices at a local region with one dominant orientation: First, we estimate the gradients and compute the local covariance matrix $\boldsymbol{C}_i$ from the local gradient vectors for each pixel in the local analysis window ω_i around the position of interest (i.e., $\boldsymbol{x}_{13}$ in the figure).

Figure 3.3. Figures 3.3(c) and (d) show the steering weight K^s given by (3.16) without $\sqrt{\det C_i}/2\pi h^2$ at every 11 pixels in the horizontal and vertical directions. It should be noted that we compute the steering kernels $K_{H_i}(x_i - x)$ as a function of each x_i with the position of interest x held fixed. Thus, the kernel involves not only C_i at the position of interest but also its neighborhoods', and the steering kernel weights effectively take local image structures into account. Moreover, the steering weights spread wider in flat regions and spread along edges while staying small at the texture regions (for example, the region around Parrot's eye). Therefore, the steering kernel filtering smoothes pixels strongly along the local structures rather than across them. Figures 3.3(e) and (f) show the scalar values $\sqrt{\det C_i}/2\pi h^2$ of (3.16). The scalars become large at edges and textured regions and small at flat regions. We also note that even for the highly noisy case, we can obtain stable estimates of local structure [31].

3.2.3   Space-Time (3-D) Steering Kernel Regression

So far, we presented the KR framework in 2-D. In this section, we introduce the time axis and present *space-time* SKR to process video data. As mentioned in the introductory section, we explain how this extension can yield a remarkable advantage in that space-time SKR does not necessitate explicit (subpixel) motion estimation.

First, introducing the time axis, similar to the 2-D data model, we have the data model in 3-D as

$$y_i = z(x_i) + \varepsilon_i, \quad x_i \in \omega, \quad i = 1, \cdots, P, \tag{3.20}$$

where y_i is a noise-ridden sample at $x_i = [x_{1i}, x_{2i}, t_i]^T$, x_{1i} and x_{2i} are spatial coordinates, t_i $(= x_{3i})$ is the temporal coordinate, $z(\cdot)$ is the regression function of interest, ε_i is an *i.i.d.* zero-mean noise process, and P is the total number of nearby samples in a 3-D neighborhood ω of interest, which we will henceforth call ω a "cubicle". As in (3.2), we also locally approximate $z(\cdot)$ by a Taylor series in 3-D, where ∇ and $\mathcal{H}$ are now the gradient (3×1) and Hessian (3×3) operators, respectively. With a (3×3) steering matrix (H_i), the estimator takes the familiar form:

$$\hat{z}(x) = \widehat{\beta}_0 = \sum_{i=1}^{P} W_i(K, H_i, N, x_i - x)\, y_i. \tag{3.21}$$

It is worth noting that 3-D SKR is a pointwise estimator of the regression function $z(\cdot)$ and it is capable of estimating a pixel value at arbitrary space-time positions x. The derivation for the steering matrix is quite similar to the 2-D case. Indeed, we again define H_i as

$$H_i = h C_i^{-\frac{1}{2}}, \tag{3.22}$$

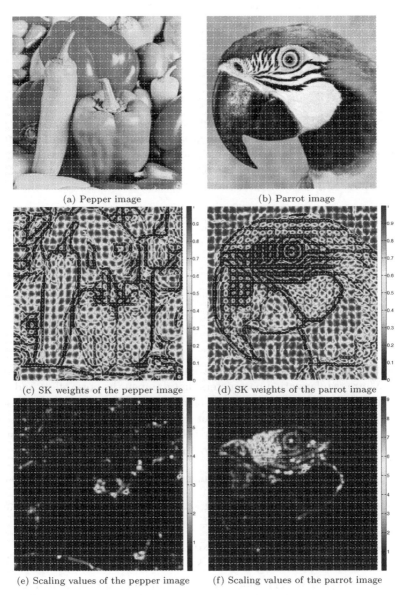

(a) Pepper image

(b) Parrot image

(c) SK weights of the pepper image

(d) SK weights of the parrot image

(e) Scaling values of the pepper image

(f) Scaling values of the parrot image

FIGURE 3.3: Graphical representations of steering kernel weights (3.18) for (a) Pepper and (b) Parrot images: The figures (c) and (d) illustrate the steering weight matrices K^s given by (3.16) without $\sqrt{\det C_i}/2\pi h^2$ at every 11 pixels in horizontal and vertical directions. For this illustration, we chose the analysis window sizes $\omega = 11 \times 11$ and $\xi = 5 \times 5$. The figures (e) and (f) shows the scalar values $\sqrt{\det C_i}/2\pi h^2$. The scalars becomes large at edge and textured regions and small at flat regions.

where the covariance matrix C_i can be naively estimated as $\widehat{C}_i^{\text{naive}} = J_i^T J_i$ with

$$J_i = \begin{bmatrix} \vdots & \vdots & \vdots \\ z_{x_1}(\boldsymbol{x}_j), & z_{x_2}(\boldsymbol{x}_j), & z_t(\boldsymbol{x}_j) \\ \vdots & \vdots & \vdots \end{bmatrix}, \quad \boldsymbol{x}_j \in \xi_i, \quad j = 1, \cdots, Q, \qquad (3.23)$$

where $z_{x_1}(\cdot)$, $z_{x_2}(\cdot)$, and $z_t(\cdot)$ are the first derivatives along x_1-, x_2-, and t-axes, ξ_i is a local analysis cubicle around a sample position at $\boldsymbol{x}_i$, and Q is the number of rows in J_i. Once again for the sake of robustness, as explained in Section 3.2.2, we compute a more stable estimate of C_i by invoking the SVD of J_i with regularization as:

$$\widehat{C}_i = \gamma_i \sum_{q=1}^{3} \varrho_q \boldsymbol{v}_q \boldsymbol{v}_q^T, \qquad (3.24)$$

with

$$\varrho_1 = \frac{s_1 + \lambda'}{s_2 s_3 + \lambda'}, \quad \varrho_2 = \frac{s_2 + \lambda'}{s_1 s_3 + \lambda'},$$

$$\varrho_3 = \frac{s_3 + \lambda'}{s_1 s_2 + \lambda'}, \quad \gamma_i = \left(\frac{s_1 s_2 s_3 + \lambda''}{Q} \right)^{\alpha}, \qquad (3.25)$$

where ϱ_q and γ_i are the *elongation* and *scaling* parameters, respectively, λ' and λ'' are regularization parameters that dampen the noise effect and restrict γ_i, the denominators of ϱ_q's from being zero (q.v., Appendix 3.5.1 for the derivations), and Q is the number of rows in J_i. We fix $\lambda' = 1$ and $\lambda'' = 0.1$ throughout this work. The singular values (s_1, s_2, and s_3) and the singular vectors ($\boldsymbol{v}_1$, $\boldsymbol{v}_2$, and $\boldsymbol{v}_3$) are given by the (compact) SVD of J_i:

$$J_i = U_i S_i V_i^T = U_i \, \text{diag} \, \{s_1, s_2, s_3\} \, [\boldsymbol{v}_1, \boldsymbol{v}_2, \boldsymbol{v}_3]^T. \qquad (3.26)$$

similar to the 2-D case, the steering kernel function in 3-D is defined as

$$K_{\boldsymbol{H}_i}(\boldsymbol{x}_i - \boldsymbol{x}) = \sqrt{\frac{\det(C_i)}{(2\pi h^2)^3}} \exp \left\{ -\frac{1}{2h^2} \left\| C_i^{\frac{1}{2}} (\boldsymbol{x}_i - \boldsymbol{x}) \right\|_2^2 \right\}, \qquad (3.27)$$

with $\boldsymbol{x} = [x_1, x_2, t]$. The main tuning parameters are the global smoothing parameter (h) in (3.27) and the structure sensitivity (α) in (3.25). The specific choices of these parameters are indicated in Section 3.3, and Appendix 3.5.2 gives more details about the parameters h and α.

Figure 3.4 shows visualizations of the 3-D weights given by the steering kernel function for two cases: (a) a horizontal edge moving vertically over time (creating a tilted plane in the local cubicle), and (b) a small circular dot also moving vertically over time (creating a thing tube in a local cubicle). Considering the case of denoising for the pixel located at the center of each data cube of Figures 3.4(a) and (b), we have the steering kernel weights illustrated in Figures 3.4(c)(d) and (e)(f). Figures 3.4(c)(d) and (e)(f) show the cross sections and the isosurface of the weights, respectively. As seen in these figures,

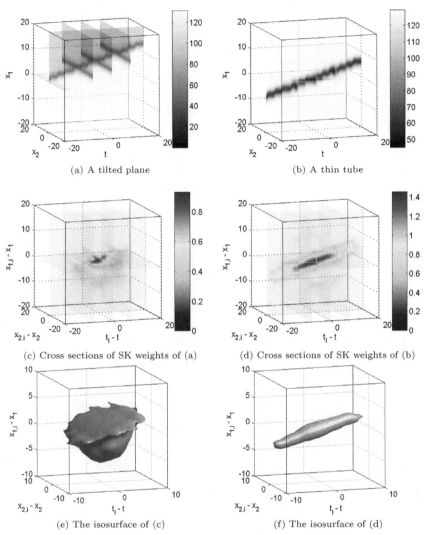

(a) A tilted plane

(b) A thin tube

(c) Cross sections of SK weights of (a)

(d) Cross sections of SK weights of (b)

(e) The isosurface of (c)

(f) The isosurface of (d)

FIGURE 3.4: Visualizations of steering kernels for (a) the case of one horizontal edge moving up (this creates a tilted plane in a local cubicle) and (b) the case of one small dot moving up (this creates a thin tube in a local cubicle). (a) and (b) show some cross sections of the 3-D data, and (b) and (c) show the cross sections of the weights given by the steering kernel function when we denoise the sample located at the center of the data cube, and (d) and (e) show the isosurface of the steering kernel weight for (a) and (b), respectively.

the weights faithfully reflect the local signal structure in space-time. Also, Figure 3.5 gives a graphical representation of the 3-D steering kernel weights for the Foreman sequence. In the figure, we show the cross sections (transverse, sagittal, and axial) of the video (3-D) data, and draw the cross sections of the steering kernel weights at every 15 pixels in every direction. For this example, we chose the analysis cubicle sizes $\omega = 15 \times 15 \times 15$ and $\xi_i = 5 \times 5 \times 5$. It is worth noting that the orientation structures that appear in the $x_1\text{-}t$ and $x_2\text{-}t$ cross sections are motion trajectories, and our steering kernel weights fit the local motion trajectories without explicit motion estimation.

As illustrated in Figures 3.4 and 3.5, the weights provided by the steering kernel function capture the local signal structures, which include both spatial and temporal edges. Here we give a brief description of how orientation information thus captured in 3-D contains the motion information implicitly. It is convenient in this respect to use the (gradient-based) optical flow framework [1, 11, 20] to describe the underlying idea. Defining the 3-D motion vector as $\widetilde{\boldsymbol{m}}_i = [m_1, m_2, 1]^T = [\boldsymbol{m}_i^T, 1]^T$ and invoking the brightness constancy equa-

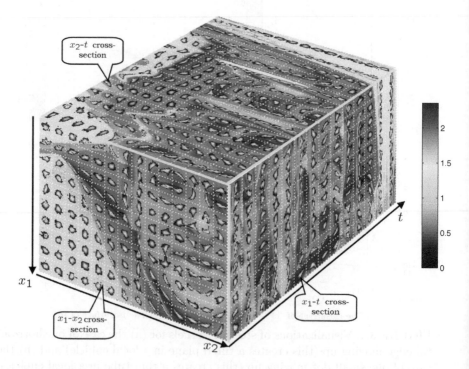

FIGURE 3.5: A graphical representation of 3-D steering kernel weights (3.27) for the Foreman sequence: The figure illustrate cross sections of the steering weight matrices $\boldsymbol{K}^s$ given by (3.27) at every 15 pixels in horizontal, vertical, and time. For the illustration, we chose the analysis cubicle sizes $\omega = 15 \times 15 \times 15$ and $\xi_i = 5 \times 5 \times 5$.

tion (BCE) [15] in a local cubicle centered at $\boldsymbol{x}_i$, we can use the matrix of gradients $\boldsymbol{J}_i$ in (3.23) to write the BCE as

$$\boldsymbol{J}_i \widetilde{\boldsymbol{m}}_i = \boldsymbol{J}_i \left[\begin{array}{c} \boldsymbol{m}_i \\ 1 \end{array} \right] = \underline{\boldsymbol{0}}. \tag{3.28}$$

Multiplying both sides of the BCE above by $\boldsymbol{J}_i^T$, we have

$$\boldsymbol{J}_i^T \boldsymbol{J}_i \widetilde{\boldsymbol{m}}_i = \widehat{\boldsymbol{C}}_i^{\text{naive}} \widetilde{\boldsymbol{m}}_i \approx \underline{\boldsymbol{0}}. \tag{3.29}$$

Now invoking the decomposition of $\widehat{\boldsymbol{C}}_i$ in (3.24), we can write

$$\sum_{q=1}^{3} \varrho_q \boldsymbol{v}_q \left(\boldsymbol{v}_q^T \widetilde{\boldsymbol{m}}_i \right) \approx \underline{\boldsymbol{0}}. \tag{3.30}$$

The above decomposition shows explicitly the relationship between the motion vector, and the principal orientation directions computed within the SKR framework. The most generic scenario in a small cubicle is one where the local texture or features move with approximate uniformity. In this generic case, we have $\varrho_1, \varrho_2 \gg \varrho_3$, and it can be shown that the singular vector $\boldsymbol{v}_3$ (which we do not directly use) corresponding to the smallest singular value ϱ_3 can be approximately interpreted as the total least squares estimate of the homogeneous optical flow vector $\frac{\widetilde{\boldsymbol{m}}_i}{\|\widetilde{\boldsymbol{m}}_i\|}$ [39, 3]. As such, the steering kernel footprint will therefore spread along this direction, and consequently assign significantly higher weights to pixels along this implicitly given motion direction. In this sense, compensation for small local motions is taken care of implicitly by the assignment of the kernel weights. It is worth noting that a significant strength of using the proposed implicit framework (as opposed to the direct use of estimated motion vectors for compensation) is the flexibility it provides in terms of smoothly and adaptively changing the elongation parameters defined by the singular values in (3.25). This flexibility allows the accommodation of even complex motions, so long as their magnitudes are not excessively large. When the magnitude of the motions is large (relative to the support of the steering kernels, specifically) a basic form of coarse but explicit motion compensation will become necessary.

There are two approaches that we can consider to compensate for large displacement. In our other work in [33], we presented the *motion-assisted steering kernel* (MASK) method, which explicitly feeds local motion vectors directly into 3-D kernels. More specifically, we construct 3-D kernels by shifting the 2-D (spatial) steering kernels by motion vectors. Moreover, in order to suppress artifacts in the estimated videos due to the errors in motion vectors, we compute the reliability of each local motion vector, and penalize the 2-D steering kernels accordingly. In the next section, we describe an alternative approach that does not require accurate motion vectors. In general, it is hard to estimate motions in the presence of occlusions and nonrigid transitions. As shown in Figure 3.5, the 3-D steering kernel effectively fits them. Therefore, all we need is to compensate large displacements by shifting the video frames with whole pixel accuracy, and the 3-D steering kernels implicitly take the leftover motions into account as local 3-D image structures.

3.2.4 Kernel Regression with Rough Motion Compensation

Before formulating the 3-D SKR with motion compensation, first, let us discuss how the steering kernel behaves in the presence of relatively large motions.[2] In Figures 3.6(a) and (b), we illustrate the contours of steering kernels the pixel of interest located at the center of the middle frame. For the small displacement case illustrated in Figure 3.6(a), the steering kernel ideally spreads across neighboring frames, taking advantage of information contained in the space-time neighborhood. Consequently, we can expect to see the effects of resolution enhancement and strong denoising. On the other hand, in the presence of large displacements as illustrated in Figure 3.6(b), similar pixels, though close in the time dimension, are found far away in space. As a result, the estimated kernels will tend not to spread across time. That is to say, the net result is that the 3-D SKR estimates in effect default to the 2-D case. However, if we can roughly estimate the relatively large motion of the block and compensate (or "neutralize") for it, as illustrated in Figure 3.6(c), and then compute the 3-D steering kernel, we find that it will again spread across neighboring frames and we regain the interpolation/denoising performance of 3-D SKR. The above approach can be useful even in the presence of aliasing

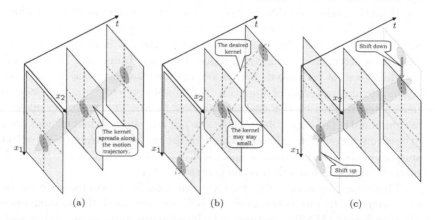

(a)   (b)   (c)

FIGURE 3.6: Steering kernel footprints for (a) a video with small displacements, (b) a video with large displacements, and (c) the video after neutralizing the large displacements.

---

[2]It is important to note here that by large motions we mean speeds (in units of pixels per frame), which are larger than the typical support of the local steering kernel window, or the moving object's width along the motion trajectory. In the latter case, even when the motion speed is slow, we are likely to see temporal aliasing locally.

when the motions are small but complex in nature. As illustrated in Figure 3.7(b), if we cancel out these displacements, and make the motion trajectory smooth, the estimated steering kernel will again spread across neighboring frames and result in good performance.

In any event, it is quite important to note that the above compensation is done for the sole purpose of computing the more effective steering kernel weights. More specifically, (i) this "neutralization" of large displacements is *not* an explicit motion compensation in the classical sense invoked in coding or video processing, (ii) it requires absolutely no interpolation, and therefore introduces no artifacts, and (iii) it requires accuracy no better than a whole pixel.

To be more explicit, 3-D SKR with motion compensation can be regarded as a two-tiered approach to handle a wide variety of transitions in video. Complicated transitions can be split into two different motion components: large whole-pixel motions (m_i^{large}) and small but complex motions (m_i):

$$m_i^{\text{true}} = m_i^{\text{large}} + m_i, \tag{3.31}$$

where m_i^{large} is easily estimated by, for instance, optical flow or block matching algorithms, but m_i is much more difficult to estimate precisely.

Suppose a motion vector $m_i^{\text{large}} = [m_{1i}^{\text{large}}, m_{2i}^{\text{large}}]^T$ is computed for each pixel in the video. We neutralize the motions of the given video data y_i by m_i^{large}, to produce a new sequence of data $y(\tilde{x}_i)$, as follows:

$$\tilde{x}_i = x_i + \left[\begin{array}{c} m_i^{\text{large}} \\ 0 \end{array} \right] (t_i - t), \tag{3.32}$$

where t is the time coordinate of interest. It is important to reiterate that since the motion estimates are rough (accurate to at best a single pixel) the formation of the sequence $y(\tilde{x}_i)$ does not require any interpolation, and therefore no artifacts are introduced. Rewriting the 3-D SKR problem for the new

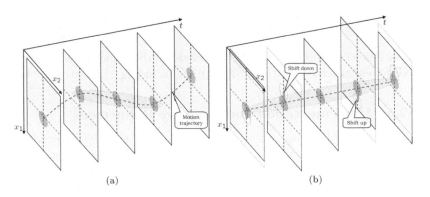

(a)   (b)

FIGURE 3.7: Steering kernel footprints for (a) a video with a complex motion trajectory, and (b) the video after neutralizing the relatively large displacements.

sequence $y(\tilde{\boldsymbol{x}}_i)$, we have:

$$\min_{\{\boldsymbol{\beta}_n\}_{n=0}^N} \sum_{i=1}^P \left[y(\tilde{\boldsymbol{x}}_i) - \beta_0 - \boldsymbol{\beta}_1^T (\tilde{\boldsymbol{x}}_i - \boldsymbol{x}) - \boldsymbol{\beta}_2^T \mathrm{vech}\{(\tilde{\boldsymbol{x}}_i - \boldsymbol{x})(\tilde{\boldsymbol{x}}_i - \boldsymbol{x})^T\} - \cdots \right]^2$$
$$\cdot K_{\widetilde{\boldsymbol{H}}_i}(\tilde{\boldsymbol{x}}_i - \boldsymbol{x})$$
(3.33)

where the steering matrix $\widetilde{\boldsymbol{H}}_i$ is computed from the motion-compensated sequence $y(\tilde{\boldsymbol{x}}_i)$. Similar to the 2-D estimator (3.11), the above minimization yields the following pixel estimator at the position of interest $(\boldsymbol{x})$ as

$$\hat{z}(\boldsymbol{x}) = \widehat{\beta}_0 = \boldsymbol{e}_1^T \left(\widetilde{\boldsymbol{X}}^T \widetilde{\boldsymbol{K}}^s \widetilde{\boldsymbol{X}} \right)^{-1} \widetilde{\boldsymbol{X}}^T \widetilde{\boldsymbol{K}}^s \, \tilde{\boldsymbol{y}}$$
$$= \sum_{i=1}^P W_i(K, \widetilde{\boldsymbol{H}}_i, N, \tilde{\boldsymbol{x}}_i - \boldsymbol{x}) \, y(\tilde{\boldsymbol{x}}_i),$$
(3.34)

where $\tilde{\boldsymbol{y}}$ is column-stacked vector of the given pixels $(y(\tilde{\boldsymbol{x}}_i))$, and $\widetilde{\boldsymbol{X}}$ and $\widetilde{\boldsymbol{K}}^s$ are the basis matrix and the steering kernel weight matrix constructed with the motion compensated coordinates $(\tilde{\boldsymbol{x}}_i)$; that is to say,

$$\tilde{\boldsymbol{y}} = \left[\ y(\tilde{\boldsymbol{x}}_1), \quad y(\tilde{\boldsymbol{x}}_2), \quad \cdots, \quad y(\tilde{\boldsymbol{x}}_P) \ \right]^T, \quad \boldsymbol{b} = \left[\ \beta_0, \quad \boldsymbol{\beta}_1^T, \quad \cdots, \quad \boldsymbol{\beta}_N^T \ \right]^T,$$

$$\widetilde{\boldsymbol{K}}^s = \mathrm{diag} \left[\ K_{\widetilde{\boldsymbol{H}}_1}(\tilde{\boldsymbol{x}}_1 - \boldsymbol{x}), \quad K_{\widetilde{\boldsymbol{H}}_2}(\tilde{\boldsymbol{x}}_2 - \boldsymbol{x}), \quad \cdots, \quad K_{\widetilde{\boldsymbol{H}}_P}(\tilde{\boldsymbol{x}}_P - \boldsymbol{x}) \ \right], \quad (3.35)$$

and

$$\widetilde{\boldsymbol{X}} = \begin{bmatrix} 1, & (\tilde{\boldsymbol{x}}_1 - \boldsymbol{x}), & \mathrm{vech}\{(\tilde{\boldsymbol{x}}_1 - \boldsymbol{x})(\tilde{\boldsymbol{x}}_1 - \boldsymbol{x})^T\}, & \cdots \\ 1, & (\tilde{\boldsymbol{x}}_2 - \boldsymbol{x}), & \mathrm{vech}\{(\tilde{\boldsymbol{x}}_2 - \boldsymbol{x})(\tilde{\boldsymbol{x}}_2 - \boldsymbol{x})^T\}, & \cdots \\ \vdots & \vdots & \vdots & \vdots \\ 1, & (\tilde{\boldsymbol{x}}_P - \boldsymbol{x}), & \mathrm{vech}\{(\tilde{\boldsymbol{x}}_P - \boldsymbol{x})(\tilde{\boldsymbol{x}}_P - \boldsymbol{x})^T\}, & \cdots \end{bmatrix}. \quad (3.36)$$

In the following section, we further elaborate on the implementation of the 3-D SKR for enhancement and super-resolution, including its iterative application.

3.2.5 Implementation and Iterative Refinement

As we explained earlier, since the performance of the SKR depends on the accuracy of the orientations, we refine it to derive an iterative algorithm we call *iterative SKR* (ISKR), which results in improved orientation estimates and therefore a better final denoising and upscaling result. The extension for upscaling is done by first interpolating or upscaling using some reasonably effective low-complexity method (say the "classic" KR method) to yield what we call a *pilot* initial estimate. The orientation information is then estimated from this initial estimate and the SKR method is then applied to the input video data y_i, which we embed in a higher resolution grid. To be more precise, the basic procedure, as shown in Figure 3.8 is as follow.

First, estimate the large motions (m_i^{large}) of the given input sequence ($\{y_i\}_{i=1}^P$). Then using m_i^{large}, we neutralize the large motions and generate a motion-compensated video sequence ($\{y(\tilde{x}_i)\}_{i=1}^P$). Next, we compute the gradients ($\widehat{\boldsymbol{\beta}}_1^{(0)} = [\hat{z}_{x_1}(\cdot), \hat{z}_{x_2}(\cdot), \hat{z}_t(\cdot)]^T$) at the sampling positions $\{\tilde{x}_i\}_{i=1}^P$ of the motion-compensated video. This process is indicated as the "pilot" estimate in the block diagram. After that, we create steering matrices ($\widetilde{\boldsymbol{H}}_i^{(0)}$) for all the samples $y(\tilde{x}_i)$ by (3.22) and (3.24). Once $\widetilde{\boldsymbol{H}}_i^{(0)}$ are available, we plug them into the kernel weight matrix (3.35) and estimate not only an unknown pixel value ($z(x)$) at a position of interest (x) by (3.34) but also its gradients ($\widehat{\boldsymbol{\beta}}_1^{(1)}$). This is the initialization process shown in Figure 3.8(a). Next, using $\widehat{\boldsymbol{\beta}}_1^{(1)}$, we re-create the steering matrices $\widetilde{\boldsymbol{H}}_i^{(1)}$. Since the estimated gradients $\widehat{\boldsymbol{\beta}}_1^{(1)}$ are also denoised and upscaled by SKR, the new steering matrices contain better orientation information. With $\widetilde{\boldsymbol{H}}_i^{(1)}$, we apply SKR to the embedded input video again. We repeat this process several times as shown in Figure 3.8(b). While we do not discuss the convergence properties of this approach here, it is worth mentioning that typically, no more than a few iterations are necessary to reach convergence.

Figure 3.9 illustrates a simple super-resolution example, where we created 9 of synthetic low-resolution frames from the image shown in Figure 3.9(a) by blurring with a 3×3 uniform PSF, shifting the blurred image by 0, 4,

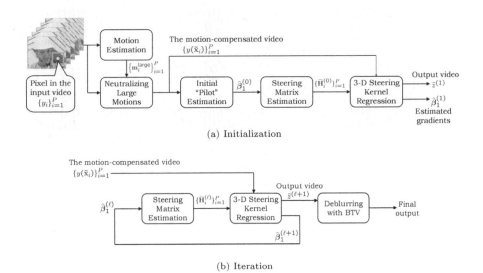

(a) Initialization

(b) Iteration

FIGURE 3.8: Block diagram representation of the 3-D iterative steering kernel regression with motion compensation: (a) the initialization process, and (b) the iteration process.

or 8 pixels[3] along the x_1- and x_2-axes, spatially downsampling with a factor
3 : 1, and adding white Gaussian noise with standard deviation 2. One of the
low-resolution frames is shown in Figure 3.9(b). Then we created a synthetic
input video by putting those low resolution images together in *random order*.
Thus, the motion trajectory of the input video is not smooth and the 3-D
steering kernel weights cannot spread effectively along time as illustrated in
Figure 3.7(a). The upscaled frames by Lanczos, robust super-resolution [8],
nonlocal mean based super-resolution [28], and 3-D ISKR with rough motion
estimation at time $t = 5$ are shown in Figures 3.9(c)–(f), respectively.

In the presence of severe temporal aliasing arising from large motions, the
task of accurate motion estimation becomes significantly harder. However,
rough motion estimation and compensation is still possible. Indeed, once this

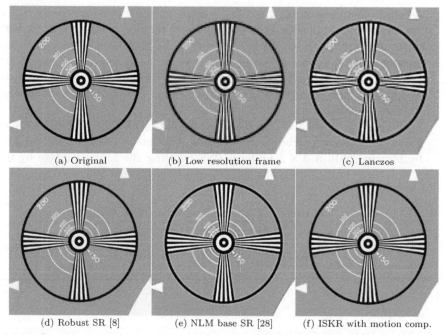

FIGURE 3.9: A simple super-resolution example: (a) the original image, (b)
one of 9 low resolution images generated by blurring with a 3×3 uniform PSF,
spatially downsampling with a factor of 3 : 1, and adding white Gaussian
noise with standard deviation 2, (c) an upscaled image by Lanczos (single
frame upscale), (d) an upscaled image by robust super-resolution (SR) [8],
(e) an upscaled image by nonlocal mean (NLM) based super-resolution [28],
and (f) an upscaled image by 3-D ISKR with rough motion compensation.
The corresponding PSNR values are (c) 19.67, (d) 30.21, (e) 27.94, and (f)
29.16[dB], respectively.

---

[3]Note: this amount of shift creates severe temporal aliasing.

compensation has taken place, the level of aliasing artifacts within the new data cubicle becomes mild, and as a result, the orientation estimation step is able to capture the true space-time orientation (and therefore implicitly the motion) quite well. This estimate then leads to the recovery of the missing pixel at the center of the cubicle, from the neighboring compensated pixels, resulting in true super-resolution reconstruction as shown in Figure 3.9.

It is worth noting that while in the proposed algorithm in Figure 3.8, we employ an SVD-based method for computing the 3-D orientations, other methods can also be employed such as that proposed by Farnebäck et al. using local tensors in [7]. Similarly, in our implementation, we used the optical flow framework [21] to compute the rough motion estimates. This step too can be replaced by other methods such as a block matching algorithm [40].

3.3   Examples

The utility and novelty of our algorithm lies in the fact that it is capable of both spatial and temporal (and therefore spatiotemporal) upscaling and super-resolution. Therefore, in this section we study the performance of our method in both spatial and spatiotemporal cases.

3.3.1   Spatial Upscaling Examples

In this section, we present some denoising/upscaling examples. The sequences in this section contain motions of relatively modest size due to the effect of severe spatial downsampling (we downsampled original videos spatially with the downsampling factor 3 : 1) and therefore motion compensation as we described earlier was not necessary. In Section 3.3.2, we illustrate additional examples of spatiotemporal video upscaling.

First, we degrade two videos (Miss America and Foreman sequences), using the first 30 frames of each sequence, blurring with a 3×3 uniform point spread function (PSF), spatially downsampling the videos by a factor of 3 : 1 in the horizontal and vertical directions, and then adding white Gaussian noise with standard deviation $\sigma = 2$. Two of the selected degraded frame at time $t = 14$ for Miss America and $t = 7$ for Foreman are shown in Figures 3.10(a) and 3.11(a), respectively. Then, we simultaneously upscale and denoise the degraded videos by Lanczos interpolation (frame-by-frame upscaling), the NL-means based approach of [28], and 3-D ISKR, which includes deblurring[4] the upscaled video frames using the BTV approach [8]. Hence, we used a radially symmetric Gaussian PSF that reflects an "average" PSF

---

[4]Note that the 3×3 uniform PSF is no longer suitable for the deblurring since the kernel regression gives its own blurring effects.

induced by the kernel function used in the reconstruction process. The final upscaled results are shown in Figures 3.10(b)–(d) and 3.11(b)–(d), respectively. The corresponding average PSNR values across all the frames for the

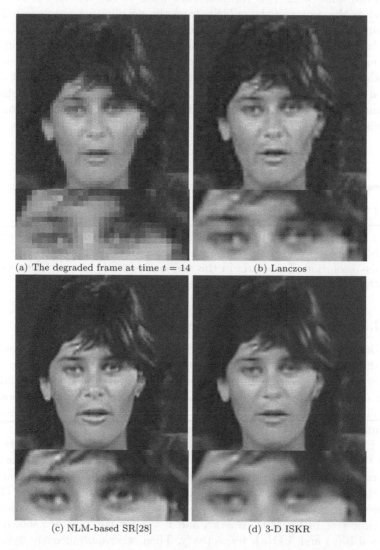

(a) The degraded frame at time $t = 14$ (b) Lanczos

(c) NLM-based SR[28]   (d) 3-D ISKR

FIGURE 3.10: A video upscale example using Miss America sequence: (a) the degraded frame at time $t = 14$, (b) the upscaled frame by Lanczos interpolation (PSNR = 34.25[dB]), (c) the upscaled frame by NLM-means based SR [28] (PSNR = 34.95[dB]), and (d) the upscaled frame by 3-D ISKR (PSNR = 35.65[dB]). Also, the PSNR values for all the frames are shown in Figure 3.12(a).

Miss America example are 34.05[dB] (Lanczos), 35.04[dB] (NL-means based SR [28]), and 35.60[dB] (3-D ISKR) and the average PSNR values for Foreman are 30.43[dB] (Lanczos), 31.87[dB] (NL-means based SR), and 32.60[dB] (3-D ISKR), respectively. The graphs in Figure 3.12 illustrate the PSNR values frame by frame. It is interesting to note that while the NL-means method appears to produce more crisp results in this case, the corresponding PSNR values for this method are surprisingly lower than that for the proposed 3-D ISKR method. We believe, as partly indicated in Figure 3.14, that this may

    (a) The degraded frame at time $t = 7$ (b) Lanczos

    (c) NLM-based SR[28]   (d) 3-D ISKR

FIGURE 3.11: A video upscaling example using Foreman sequence: (a) the degraded frame at time $t = 7$, (b) the upscaled frame by Lanczos interpolation (PSNR = 30.98[dB]), (c) the upscaled frame by NL-means based SR [28] (PSNR = 32.21[dB]), and (d) the upscaled frame by 3-D ISKR (PSNR = 33.58[dB]). Also the PSNR values for all the frames are shown in Figure 3.12(b)

be in part due to some leftover high frequency artifacts and possibly lesser denoising capability of the NL-means method.

As for the parameters of our algorithm, we applied SKR with the global smoothing parameter $h = 1.5$, the local structure sensitivity $\alpha = 0.1$ and a $5 \times 5 \times 5$ local cubicle and used an 11×11 Gaussian PSF with a standard deviation of 1.3 for the deblurring of Miss America and Foreman sequences. For the experiments shown in Figures 3.10 and 3.11, we iterated SKR 6 times.

The next example is a spatial upscaling example using a section of a real HDTV video sequence (300×300 pixels, 24 frames), shown in Figure 3.13(a), where no additional simulated degradation is added. As seen in the input frames, the video has real compression artifacts (i.e., blocking). In this example, we show the deblocking capability of the proposed method, and the upscaled results by Lanczos interpolation, NLM-based SR [28] and 3-D ISKR with a factor of $1 : 3$ (i.e., the output resolution is 900×900 pixels) are shown in Figures 3.13(b)–(d), respectively. The proposed method (applied to the luminance channel only) is able to remove the blocking artifacts effectively as well as to upscale the video.

3.3.2 Spatiotemporal Upscaling Examples

In this section, we present two spatiotemporal upscaling examples (also known as, frame interpolation and frame rate upconversion) by 3-D ISKR using the Carphone and Salesman sequences. Unlike the previous examples (Miss America and Foreman), in the next examples, the Carphone sequence has relatively large and more complex displacements between frames, and the Salesman sequence contains motion occlusions. In order to have better estimations of steering kernel weights, we estimate patchwise (4×4 block) translational mo-

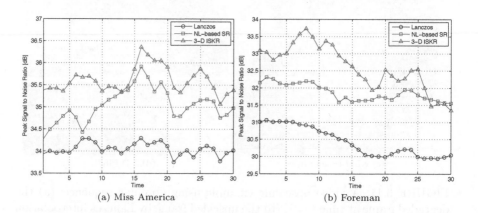

(a) Miss America   (b) Foreman

FIGURE 3.12: The PSNR values of each upscaled frame by Lanczos, NLM-based SR [28], and 3-D ISKR for (a) the results of Miss America shown in Figure 3.10 and (b) the results of Foreman shown in Figure 3.11.

tions by the optical flow technique [21], and apply 3-D ISKR to the roughly motion-compensated inputs.

Though we did not discuss temporal upscaling much explicitly in the text of this chapter, the proposed algorithm is capable of this functionality as well in a very straightforward way. Namely, the temporal upscaling is effected by

(a) The input frame at $t = 5$ (b) Lanczos

(c) NLM-based SR[28]   (d) 3-D ISKR

FIGURE 3.13: A spatial upscaling example of a real video: (a) Texas football sequence in luminance channel, and (b)–(d) the upscaled frames by Lanczos interpolation, NL-based SR [28] and 3-D ISKR, respectively. The input sequence has 24 frames in total and it is a real HD-TV content which carries compression artifacts, namely block artifacts. We upscale the video with the spatial upscaling factor of 1 : 3.

producing a pilot estimate and improving the estimate iteratively just as in the spatial upscaling case illustrated in the block diagrams in Figure 3.8. We note that this temporal upscaling capability, which essentially comes for free in our present framework, was not possible in the NL-means based algorithm [28].

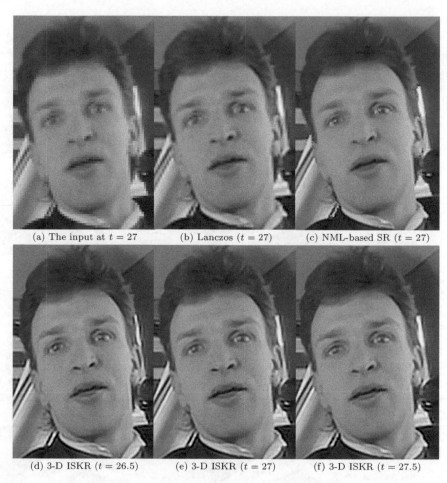

(a) The input at $t = 27$ (b) Lanczos ($t = 27$)   (c) NML-based SR ($t = 27$)

(d) 3-D ISKR ($t = 26.5$)   (e) 3-D ISKR ($t = 27$)   (f) 3-D ISKR ($t = 27.5$)

FIGURE 3.14: A Carphone example of video upscaling with spatial upscaling factor 1 : 2: (a) the input video frame at time $t = 27$ (144×176, 30 frames), (b)–(c) upscaled frames by Lanczos interpolation and NLM-based SR method [28], respectively, and (d)–(f) upscaled frames by 3-D ISKR at $t = 26.5$, 27, and 27.5, respectively.

The first example in Figure 3.14 is a real experiment[5] of space-time upscaling with a native Carphone sequence in QCIF format (144×176, 30 frames). Figure 3.14 shows (a) the input frame at time $t = 27$ and (b)–(c) the upscaled frames by Lanczos interpolation and NLM-based method [28], and (d)–(f) the upscaled frames by 3-D ISKR at $t = 26.5$, 27 and 27.5, respectively. The estimated frames in Figure 3.14(d)–(f) shows the application of 3-D ISKR, namely simultaneous space-time upscaling.

The final example shown in Figure 3.15 is also a real example of frame inter-

FIGURE 3.15: A Salesman example of frame interpolation: (a) original (input) video frames at time $t = 6$ to 7, (b) intermediate frames estimated by 3-D ISKR at $t = 6.5$ and $t = 7.5$.

[5]That is to say, the input to the algorithm was the native resolution video, which was subsequently upscaled in space and time directly. In other words, the input video is *not* simulated by downsampling a higher resolution sequence.

polation (temporal upscaling) using the Salesman sequence where, although there is no global (camera) motion, the both hands move toward different directions and occlusions can be seen around the tie as shown in the input frames (Figure 3.15(a)). In this example, we estimate intermediate frames at times $t = 6.5$ and $t = 7.5$, and the frames in Figure 3.15(b) are the results by 3-D ISKR. 3-D ISKR successfully generated the intermediate frames without producing any artifacts.

3.4   Conclusion

Traditionally, super-resolution reconstruction of image sequences has relied strongly on the availability of highly accurate motion estimates between the frames. As is well-known, subpixel motion estimation is quite difficult, particularly in situations where the motions are complex in nature. As such, this has limited the applicability of many existing upscaling algorithms to simple scenarios. In this chapter, we extended the 2-D steering KR method to an iterative 3-D framework, which works well for both (spatiotemporal) video upscaling and denoising applications. Significantly, we illustrated that the need for explicit subpixel motion estimation can be avoided by the two-tiered approach presented in Section 3.2.4, which yields excellent results in both spatial and temporal upscaling.

Performance analysis of super-resolution algorithm remains an interesting area of work, particularly with the new class of algorithms such as the proposed and NLM-based method [28], which can avoid subpixel motion estimation. Some results already exist that provide such bounds under certain simplifying conditions [29], but these results need to be expanded and studied further.

Reducing the computational complexity of 3-D ISKR is of great importance. Most of the computational load is due to (in order of severity): (i) the computations of steering (covariance) matrices (C_i) in (3.22), (ii) the generation of the equivalent kernel coefficients (W_i) in (3.34) from the steering kernel function with higher (i.e., $N = 2$), and (iii) iterations. For (i), to speed up the estimation of C_i, instead of application of SVD, which is computationally heavy, we can create a lookup table containing a discrete set of representative steering matrices (using, say, vector quantization), and choose an appropriate matrix from the table given local data. For (ii), computation of the second order ($N = 2$) filter coefficients (W_i) from the steering kernel weights (3.35) may be sped up by using an approximation using the lower order (e.g., zeroth order, $N = 0$) kernels. This idea was originally proposed by Haralick in [13] and may be directly applicable to our case as well. For (iii), we iterate the process of steering kernel regression in order to obtain better estimates of orientations. If the quantization mentioned above gives us fairly reasonable estimates of orientations, we may not need to iterate.

3.5 Appendix

3.5.1 Steering Kernel Parameters

Using the (compact) SVD (3.26) of the local gradient vector $\boldsymbol{J}_i$ (3.23), we can express the naive estimate of steering matrix as:

$$
\begin{aligned}
\widehat{C}_i^{\text{naive}} &= \boldsymbol{J}_i^T \boldsymbol{J}_i = \boldsymbol{V}_i \boldsymbol{S}_i^T \boldsymbol{S}_i \boldsymbol{V}_i^T \\
&= \boldsymbol{V}_i \operatorname{diag}\left\{s_1^2, s_2^2, s_3^2\right\} \boldsymbol{V}_i^T \\
&= s_1 s_2 s_3 \boldsymbol{V}_i \operatorname{diag}\left\{\frac{s_1}{s_2 s_3}, \frac{s_2}{s_1 s_3}, \frac{s_3}{s_1 s_2}\right\} \boldsymbol{V}_i^T \\
&= Q\gamma_i \left[\boldsymbol{v}_1, \boldsymbol{v}_2, \boldsymbol{v}_3\right] \operatorname{diag}\left\{\varrho_1, \varrho_2, \varrho_3\right\} \left[\boldsymbol{v}_1, \boldsymbol{v}_2, \boldsymbol{v}_3\right]^T \\
&= Q\gamma_i \sum_{q=1}^{3} \varrho_q \boldsymbol{v}_q \boldsymbol{v}_q^T,
\end{aligned}
\tag{3.37}
$$

where

$$
\varrho_1 = \frac{s_1}{s_2 s_3}, \quad \varrho_2 = \frac{s_2}{s_1 s_3}, \quad \varrho_3 = \frac{s_3}{s_1 s_2}, \quad \gamma_i = \frac{s_1 s_2 s_3}{Q}, \tag{3.38}
$$

and Q is the number of rows in $\boldsymbol{J}_i$. Since the singular values (s_1, s_2, s_3) may become zero, we regularized the elongation parameters (ϱ_q) and the scaling parameter (γ_i) as shown in (3.25) from being zero.

3.5.2 The Choice of the Regression Parameters

The parameters that have critical roles in steering kernel regression are the regression order (N), the global smoothing parameter (h) in (3.22) and (3.27), and the structure sensitivity (α) in (3.25). It is generally known that the parameters N and h control the balance between the variance and bias of the estimator [30]. The larger N and the smaller h, the higher the variance becomes and the loser the bias. In this work, we fix the regression order $N = 2$.

The structure sensitivity α (typically $0 \leq \alpha \leq 0.5$) controls how strongly the size of the kernel footprints is affected by the local structure. The product of the singular values (s_1, s_2, s_3) indicates the amount of the energy of the local signal structure: the larger the product, the stronger and the more complex the local structure is. A large α is preferable when the given signal carries severe noise. In this work, we focus on the case that the given video sequences have a moderate amount of noise and fix $\alpha = 0.1$.

Ideally, although one would like to automatically set these regression parameters using a method such as cross validation [14, 25], SURE (Stein's unbiased risk estimator) [22] or a no-reference parameter selection [41], this would add significant computational complexity to the already heavy load of the proposed method. So for the examples presented in the chapter, we make use of our extensive earlier experience to note that only certain ranges of values for the said parameters tend to give reasonable results. We fix the values of the parameters within these ranges to yield the best results, as discussed in Section 3.3.

3.5.3 Deblurring

Since we did not include the effect of sensor blur in the data model of the KR framework, deblurring is necessary as a post processing step to improve the outputs by 3-D ISKR further. Defining the estimated frame at time t as

$$\hat{\underline{z}}(t) = [\cdots, \hat{z}(\boldsymbol{x}_j), \cdots]^T, \tag{3.39}$$

where j is the index of the spatial pixel array and $\boldsymbol{u}(t)$ as the unknown image of interest, we deblur the frame $\underline{z}(t)$ by a regularization approach:

$$\hat{\underline{u}}(t) = \arg\min_{\underline{u}} \left\| \underline{u}(t) - \boldsymbol{G}\,\hat{\underline{z}}(t) \right\|_2^2 + \lambda C_{\mathrm{R}}(\hat{\underline{u}}(t)), \tag{3.40}$$

where $\boldsymbol{G}$ is the blur matrix, $\lambda(\geq 0)$ is the regularization parameter, and $C_{\mathrm{R}}(\,\cdot\,)$ is the regularization term. More specifically, we rely on our earlier work and employ the *bilateral total variation* (BTV) framework [8]:

$$C_{\mathrm{R}}(\underline{u}(t)) = \sum_{v_1=-\nu}^{\nu} \sum_{v_2=-\nu}^{\nu} \eta^{|v_1|+|v_2|} \left\| \underline{u}(t) - \boldsymbol{F}_{x_1}^{v_1} \boldsymbol{F}_{x_2}^{v_2} \underline{u}(t) \right\|_1 \tag{3.41}$$

where η is the smoothing parameter, ν is the window size, and $\boldsymbol{F}_{x_1}^{v_1}$ is the shift matrix that shifts $\underline{u}(t)$ v_1-pixels along x_1-axis.

In the present work, we use the above BTV regularization framework to deblur the upscaled sequences frame-by-frame, which is admittedly suboptimal. In our work [32], we have introduced a different regularization function called *adaptive kernel total variation* (AKTV). This framework can be extended to derive an algorithm that can simultaneously interpolate and deblur in one integrated step. This promising approach is part of our ongoing work and is outside the scope of this chapter.

Bibliography

[1] M. J. Black and P. Anandan. The robust estimation of multiple motions: Parametric and piecewise-smooth flow fields. *Computer Vision and Image Understanding*, 63(1):75–104, January 1996.

[2] A. Buades, B Coll, and J. M. Morel. A review of image denosing algorithms, with a new one. *Multiscale Modeling and Simulation, Society for*

Industrial and Applied Mathematics (SIAM) Interdisciplinary Journal, 4(2):490–530, 2005.

[3] S. Chaudhuri and S. Chatterjee. Performance analysis of total least squares methods in three-dimensional motion estimation. *IEEE Transactions on Robotics and Automation*, 7(5):707–714, October 1991.

[4] K. Dabov, A. Foi, V. Katkovnic, and K. Egiazarian. Image denoising by sparse 3D transform-domain collaborative filtering. *IEEE Transactions of Image Processing*, 16(8):2080–2095, August 2007.

[5] A. Danielyan, A. Foi, V. Katkovnik, and K. Egiazarian. Image and video super-resolution via spatially adaptive block-matching filtering. *Proceedings of International Workshop on Local and Non-Local Approximation in Image Processing (LNLA)*, August 2008. Lausanne, Switzerland.

[6] M. Elad and Y. Hel-Or. A fast super-resolution reconstruction algorithm for pure translational motion and common space-invariant blur. *IEEE Transactions on Image Processing*, 10(8):1187–1193, August 2001.

[7] Gunnar Farnebäck. *Polynomial Expansion for Orientation and Motion Estimation*. PhD thesis, Linköping University, Sweden, SE-581 83 Linköping, Sweden, 2002. Dissertation No 790, ISBN 91-7373-475-6.

[8] S. Farsiu, D. Robinson, M. Elad, and P. Milanfar. Fast and robust multi-frame super-resolution. *IEEE Transactions on Image Processing*, 13(10):1327–1344, October 2004.

[9] W. T. Freeman, T. R. Jones, and E. C. Pasztor. Example-based super resolution. *IEEE Computer Graphics*, 22(2):56–65, March/April 2002.

[10] H. Fu and J. Barlow. A regularized structured total least squares algorithm for high-resolution image reconstruction. *Linear Algebra and its Applications*, 391:75–98, November 2004.

[11] J. J. Gibson. *The Perception of the Visual World*. Houghton Mifflin, Boston, 1950.

[12] B. K. Gunturk, Y. Altunbasak, and R. M. Mersereau. Multiframe resolution enhancement methods for compressed video. *IEEE Signal Processing Letter*, 9:170–174, June 2002.

[13] R. M. Haralick. Edge and region analysis for digital image data. *Computer Graphic and Image Processing (CGIP)*, 12(1):60–73, January 1980.

[14] W. Hardle and P. Vieu. Kernel regression smoothing of time series. *Journal of Time Series Analysis*, 13:209–232, 1992.

[15] B. K. Horn. *Robot Vision*. MIT Press, Cambridge, 1986.

[16] M. Irani and S. Peleg. Super resolution from image sequence. *Proceedings of 10th International Conference on Pattern Recognition (ICPR)*, 2:115–120, 1990.

[17] S. Kanumuri, O. G. Culeryuz, and M. R. Civanlar. Fast super-resolution reconstructions of mobile video using warped transforms and adaptive thresholding. *Proceedings of SPIE*, 6696:66960T, 2007.

[18] H. Knutsson. Representing local structure using tensors. *Proceedings of the 6th Scandinavian Conference on Image Analysis*, pages 244–251, 1989.

[19] H. Knutsson and C. F. Westin. Normalized and differential convolution – methods for interpolation and filtering of incomplete and uncertain data. *Proceedings of IEEE Computer Society Conference on Computer Vision and Pattern Regocnition (CVPR)*, pages 515–523, June 1993.

[20] D. N. Lee and H. Kalmus. The optic flow field: the foundation of vision. *Philosophical Transactions of the Royal Society of London Series B-Biological Sciences*, 290(1038):169–179, 1980.

[21] B. Lucas and T. Kanade. An iterative image registration technique with an application to sterio vision. *Proceedings of DARPA Image Understanding Workshop*, pages 121–130, 1981.

[22] F. Luisier, T. Blu, and M. Unser. A new sure approach to image denoising: inter-scale orthonormal wavelet thresholding. *IEEE Transactions on Image Processing*, 16(3):593–606, March 2007.

[23] E. A. Nadaraya. On estimating regression. *Theory of Probability and its Applications*, pages 141–142, September 1964.

[24] M. K. Ng., J. Koo, and N. K. Bose. Constrained total least squares computations for high resolution image reconstruction with multisensors. *International Journal of Imaging Systems and Technology*, 12:35–42, 2002.

[25] N. Nguyen, P. Milanfar, and G. H. Golub. Efficient generalized cross-validation with applications to parametric image restoration and resolution enhancement. *IEEE Transactions on Image Processing*, 10(9):1299–1308, September 2001.

[26] K. S. Ni, S. Kumar, N. Vasconcelos, and T. Q. Nguyen. Single image superresolution based on support vector regression. *Proceedings of IEEE International Conference on Acoustics, Speech and Signal Processing (ICASSP)*, 2, May 2006.

[27] SC. Park, MK. Park, and MG. Kang. Super-resolution image reconstruction: A technical overview. *IEEE Signal Processing Magazine*, 20(3):21–36, May 2003.

[28] M. Protter, M. Elad, H. Takeda, and P. Milanfar. Generalizing the non-local-means to super-resolution reconstruction. *IEEE Transactions on Image Processing*, 16(2):36–51, January 2009.

[29] D. Robinson and P. Milanfar. Statistical performance analysis of super-resolution. *IEEE Transactions on Image Processing*, 15(6):1413–1428, June 2006.

[30] D. Ruppert and M. P. Wand. Multivariate locally weighted least squares regression. *The annals of statistics*, 22(3):1346–1370, September 1994.

[31] H. Takeda, S. Farsiu, and P. Milanfar. Kernel regression for image processing and reconstruction. *IEEE Transactions on Image Processing*, 16(2):349–366, February 2007.

[32] H. Takeda, S. Farsiu, and P. Milanfar. Deblurring using regularized locally-adaptive kernel regression. *IEEE Transactions on Image Processing*, 17(4):550–563, April 2008.

[33] H. Takeda, P. van Beek, and P. Milanfar. Spatiotemporal video upscaling using motion-assisted steering kernel (MASK) regression (Book chapter). *High-Quality Visual Experience: Creation, Processing and Interactivity of High-Resolution and High-Dimensional Video Signals*, Springer-Verlag, 2010.

[34] J. van de Weijer and R. van den Boomgaard. Least squares and robust estimation of local image structure. *Scale Space. International Conference*, 2695(4):237–254, 2003.

[35] P. Vandewalle, L. Sbaiz, M. Vetterli, and S. Susstrunk. Super-resolution from highly undersampled images. *Proceedings of International Conference on Image Processing (ICIP)*, pages 889–892, September 2005. Genova, Italy.

[36] M. P. Wand and M. C. Jones. *Kernel Smoothing*. Monographs on Statistics and Applied Probability. Chapman and Hall, London; New York, 1995.

[37] K. Q. Weinberger and G. Tesauro. Metric learning for kernel regression. *Proceedings of the Eleventh International Workshop on Artificial Intelligence and Statistics*, pages 608–615, 2007. (AISTATS-07), Puerto Rico.

[38] N. A. Woods, N. P. Galatsanos, and A. K. Katsaggelos. Stochastic methods for joint registration, restoration, and interpolation of multiple undersampled images. *IEEE Transactions on Image Processing*, 15(1):201–213, January 2006.

[39] J. Wright and R. Pless. Analysis of persistent motion patterns using the 3d structure tensor. *Proceedings of the IEEE Workshop on Motion and Video Computing*, 2005.

[40] S. Zhu and K. Ma. A new diamond search algorithm for fast block-matching motion estimation. *IEEE Transactions on Image Processing*, 9(2):287–290, February 2000.

[41] X. Zhu and P. Milanfar. Automatic parameter selection for denoising algorithms using a no-reference measure of image content. Accepted for publication in IEEE Transactions on Image Processing.

[42] A. Zomet, A. Rav-Acha, and S. Peleg. Robust super-resolution. *Proceedings of the International Conference on Computer Vision and Pattern Recognition (CVPR)*, 2001. Hawaii.

4

Super-Resolution with Probabilistic Motion Estimation

Matan Protter

The Technion – Israel Institute of Technology

Michael Elad

The Technion – Israel Institute of Technology

CONTENTS

Classic super-resolution has long relied on very exact motion estimation for the recovery of sub-pixel details. As a highly accurate motion field is hard to obtain for general scenes, classic super-resolution has been known to be limited to specific cases, where the motion is of a global nature. In this chapter, we present a recently developed family of algorithms that shatters this barrier. These novel algorithms relax the requirement of a one-to-one motion field, and replace it with a simple, probabilistic motion estimation. The probabilistic motion field is integrated into the classic (and heavily investigated) SR framework, and ultimately results in a very simple family of algorithms. The obtained paradigm gets an algorithmic structure that resembles that of the nonlocal means, and as such, leads to a localized and easily parallelizable procedure. Despite their simplicity, the obtained algorithms are nevertheless very powerful in handling the most general scenes, with the probabilistic motion estimation enabling the handling of challenging motion patterns. The resulting image sequences are of high quality, and contain few artifacts. These novel algorithms open the door to a new era in super-resolution that bypasses the limiting traditional reliance on explicit motion estimation for super-resolution.

4.1   Introduction

Super-Resolution Reconstruction (SRR) proposes a fusion of several low quality images $\{y_t\}_{t=1}^{T}$ into one higher quality result x, which has better optical resolution than the input images. A wide variety of SRR algorithms have been developed in the past two decades – see [13] for a list of representatives of this vast literature. A popular model used for relating the measurements to the super-resolved image, assumes that $\{y_t\}_{t=1}^{T}$ are generated from x through a sequence of operations that includes (i) geometrical warps F_t, (ii) a linear space-invariant blur H, (iii) a decimation step represented by D, and finally (iv) an additive zero-mean white and Gaussian noise n_t that represents both measurements noise and model mismatch[1] [7]. All of these operators are linear, each represented by a matrix multiplying the image they operate on. We assume hereafter that H and D are identical for all images in the sequence. Mathematically, the relationship between the high-quality image x and the measurements $\{y_t\}_{t=1}^{T}$ is given by

$$y_t = DHF_t x + n_t \quad \text{for} \quad t = 1, 2, \dots, T. \tag{4.1}$$

The recovery of x from $\{y_t\}_{t=1}^{T}$ is thus an inverse problem, combining denoising, deblurring, scaling-up operation, and fusion of the different images, all merged to one. We treat y_1 as our reference image, and aim to reconstruct x as its super-resolved version (this implies that $F_1 = I$).

SRR relies on the assumption that D, H, and F_t are known, or can be reliably estimated from the given data. In particular, such reconstruction relies on the ability to estimate the motion in the scene with a subpixel accuracy, so as to enable the merger of the different image sampling grids properly. Many SRR algorithms start with such an estimating of the motion in the sequence (e.g., [9, 15, 1, 7, 6]), or couple it with the recovery process, as a joint-estimation task [8, 19, 16].

Highly accurate general motion estimation, known as optical flow, is a severely under-determined problem. Various artifacts, and an output image that is even inferior to the given measurements, are often the result of using

---

[1]In [7], the model mismatches are represented as an iid Laplacian distribution, with L_1 penalization as to obtain robustness to outliers. In our work, we choose a Gaussian model, which simplifies the algorithmic development. Nevertheless, a robustness to outliers is obtained by the probabilistic approach, as will be discussed later, in Section 3.

an inaccurately estimated motion within one of the existing SRR algorithms. In order to estimate the motion with enough accuracy to lead to a successful reconstruction of a super-resolved image, some simplifying assumptions as to the structure of the motion field must be made, such as global warps or rigid bodies. This had led to the commonly agreed and unavoidable conclusion that general content movies are not likely to be handled well by classical SRR techniques.

Recently, several papers have tried to circumvent this problem by avoiding explicit motion estimation altogether [13, 17]. The method in [17] relies on extending the steerable kernel method to multiframe super-resolution. The method in [13] generalizes the very successful nonlocal means (NLM) [2] denoising method to performing super-resolution. The derivation of the SRR algorithm in [13], termed NLM-SR, is done by defining an energy functional that explains the NLM, and then modifying it to serve the SRR task. Both methods do not explicitly estimate the motion, and both are shown to be able to handle general content video sequences quite successfully.

In this chapter we approach the explicit-motion-estimation-free SRR from a different perspective. Our starting point is the classic SRR, as in [7]. We then replace the bijective motion between pixels in each pair of images with a probabilistic motion field. This simple and alternative derivation is shown to lead to the same line of algorithms that are proposed in [13]. Furthermore, the framework proposed here allows different extensions, such as a treatment of spatio-temporal re-sampling problems. We show this adaptation in general, and demonstrate its applicability on the de-interlacing problem.

The structure of the chapter is as follows. Section 4.2 describes a classic SRR formulation, as used in [9, 15, 1, 7, 6], on which we build our eventual algorithm. Section 4.3 presents the use of probabilistic motion within the framework of classic SRR, and develops the proposed algorithm. The adaptation to other re-sampling tasks is also described in this section. Section 4.4 provides results for SRR and de-interlacing, demonstrating the abilities of the proposed method. The key contributions of this work are outlined in Section 4.5, with several directions of possible future work also suggested. We note that a preliminary version of this chapter has appeared in [12].

4.2 Classic Super-Resolution: Background

Using the model in Equation (14.6), one can seek the most likely high resolution image, given the existing low-resolution images (and the known decimation, blur, and transformations). This image is called the Maximum-Likelihood

(ML) estimate of x, and is obtained by minimizing the penalty function

$$\epsilon_{ML}^2(x) = \frac{1}{2} \sum_{t=1}^{T} \|DHF_t x - y_t\|_2^2 \qquad (4.2)$$

with respect to x. Minimization of (4.2) leads to

$$\frac{\partial \epsilon_{ML}^2(x)}{\partial x} = \sum_{t=1}^{T} F_t^T H^T D^T (DHF_t x - y_t) = 0. \qquad (4.3)$$

Denoting $A = \sum_{t=1}^{T} F_t^T H^T D^T DHF_t$ and $b = \sum_{t=1}^{T} F_t^T H^T D^T y_t$, we face a linear system of equations $A\hat{x}_{ML} = b$.

In many cases the measurements are not sufficient for recovering x. In such cases, the constraints matrix A is singular or possibly ill-conditioned, and regularization is required. The Maximum A-posteriori Probability (MAP) estimation proposes a penalty of the form

$$\epsilon_{MAP}^2(x) = \epsilon_{ML}^2(x) + \lambda \cdot R(x), \qquad (4.4)$$

where the functional R is a regularization term that adds an algebraic stability to the inversion of A. Beyond the gained stability, R is also a way of incorporating prior knowledge about the sought x, such as spatial smoothness, sparsity of its wavelet representation, minimum entropy, etc.. In this work we force spatial smoothness, by choosing the Total Variation (TV) prior, that accumulates the gradients norms with ℓ^1 [14]. Thus, the MAP estimate in our case becomes the minimizer of

$$\epsilon_{MAP}^2(x) = \frac{1}{2} \sum_{t=1}^{T} \|DHF_t x - y_t\|_2^2 + \lambda \cdot TV(x), \qquad (4.5)$$

which is typically obtained by an iterative algorithm [9, 15, 8, 1, 7, 6, 19, 16]. This is the core technique we build upon.

The operators D, H, and F_t are assumed to be known in all of the above discussions. The decimation D is dependent on the resolution scale-factor we aim to achieve, and as such, it is easily fixed. In this work we shall assume that this resolution factor is an integer $s \geq 1$ in both axes. The blur H refers to the camera PSF in most cases, and therefore it is also accessible. Even if it is not, the blur is typically dependent on a small number of parameters, and those, in the worst case, can be manually set.

While D and H are relatively easy to obtain, this is not the case of F_t. The warp operators depend on the scene and require highly accurate motion estimation for their construction. Since such accuracy is hard to obtain in general, classical SRR algorithms often assume a simple motion pattern, such as pure translation or global affine warp. Such constraints stabilize the motion estimation, as they substantially reduce the number of parameters to be estimated, allowing greater accuracy in the estimation (if indeed the motion field obeys these assumptions). Attempts to embed the motion estimation (without assuming a specific structure) within the SRR process have been made, with little success [8, 19, 16]. As already mentioned, inaccurately estimated motion within SRR often leads to disturbing artifacts that cause the output to be inferior even when compared to a simple interpolated version of y_1. This fact motivated a quest for bypassing explicit motion estimation, as indeed practiced in [13, 17].

4.3 The Proposed Algorithm

4.3.1 The New Formulation

We now aim to integrate the notion of probabilistic motion estimation into the classic SRR formulation introduced in the previous section. Before we dive into the formulation, we note that when the motion is of a global nature, and therefore lends itself to an accurate estimation, motion-estimation-based techniques are likely to obtain better results than the proposed algorithm in many cases. In other cases, the usage of the proposed algorithm makes of intra-image redundancy may bring better results even compared to the motion-compensated algorithms. As such sequences comprise only a small subset of the sequences to be super-resolved, we don't continue this discussion further, and rather tackle general motion sequences by using the probabilistic motion estimation technique, which we now describe.

The starting point is the observation that the warp operator F_t considers a bijective (one-to-one) correspondence between pixels in the reference and the t-th image, and as such, it introduces sensitivity to errors. We replace this motion field with a probabilistic one that assigns each pixel in the reference image with *many* possible correspondences in all the images in the sequence (including itself), each with an assigned probability of being correct.

Can this become useful for super-resolution for handling general motion patterns? We now offer one possible way that illustrates that it can. We start by analyzing the operator F_t, which represents the motion field between the first image and image t, by indicating for each pixel in the first image its destination in image t. Equivalently, the motion field can be described by listing a single 2D translation vector for each pixel, independently of other pixels. Therefore, the entire motion field is represented as a collection of various displacement vectors, one for each pixel.

If the size of the maximal translation is at most D pixels, then the set of all the possible displacements are covered by a set of $M = (2D + 1)^2$ displacements. By defining $\{F_m\}_{m=1}^M$ to be this set of global translations,[2] we

---

[2]For simplicity, we shall use a set of integer displacements only.

can write the following equation

$$F_t x = \sum_{m=1}^{M} Q_{m,t} F_m x, \qquad (4.6)$$

which describes the action of warping the image x based on the operator F_t. The matrices $\{Q_{m,t}\}_1^M$ are diagonal weighting ones, containing ones along the main diagonal for pixels whose motion is the displacement F_m, namely $[dx(m), dy(m)]$, and zeros for the rest of the pixels. Using such a decomposition, even the most complicated of motion fields can be represented by a linear combination of global translations.

While we have replaced the single warping operator with a linear combination of global translation (representing the same general motion field), a one-to-one relationship between pixels in both images is still implied by this notation. The next natural step for introducing a probabilistic motion field is to relax the definition of $Q_{m,t}$, where varying confidences per pixel and per motion trajectory are reflected by continuous values. This leads to a newly defined super-resolution penalty that replaces the use of F_t by their decompositions as in (4.6).

While this seems like a worthy path to consider, we slightly divert from this approach, seeking yet a simpler algorithm. We modify the ML formulation posed in Equation (4.2) by proposing the following probabilistic ML (PML) penalty[3]

$$\epsilon_{PML}^2 (x) = \frac{1}{2} \sum_{m=1}^{M} \sum_{t=1}^{T} \|DHF_m x - y_t\|_{W_{m,t}}^2 . \qquad (4.7)$$

The same intuition, although applied differently, is used in proposing this penalty. Rather than accumulate the various global translations to form the effect of F_t as in Equation (4.6), we accumulate the least-squares errors that result from such global displacements,[4] and assign a weight matrix $W_{m,t}$ to each. Notice that the weights used in Equation (4.7) are different from those introduced in (4.6). Whereas $Q_{m,t}$ are defined for each pixel in the high resolution image, $W_{m,t}$ are also diagonal matrices, but defined over the low-resolution grid. We shall proceed with the assumption that $W_{m,t}$ are known, and revisit their computation in Section 4.3.5.

Even though this formulation contains only global translations, it should be noted that using the same rational that has led to Equation (4.6), it can represent any complex motion field. A known motion field can be re-created by properly assigning the values of $W_{m,t}$ to be 1s for those pixels whose motion is F_m and zeros for all others.

One particular interpretation of the above expression is a marginalization of the least-squared error term with respect to the motion probability density

---

[3]We use the notation $\|a\|_W^2 = a^T W a$.

[4]It is possible to use other sets of warps, such as ones that allow rotations as well.

function, in a way that resembles the concept proposed in [11]. However, the authors of [11] perform such a marginalization in order to avoid inaccuracies in the motion estimation, and their integration is only performed over the parameters of a global motion model. In our case, very similar to the video denoising scenario, we handle local motion, and the probabilistic viewpoint contributes both to a better handling of the estimated motion inaccuracies and also to the noise reduction.

As a final point in this section, we return to the matter of robustness. The usage of the above PML has another distinct advantage of robustifying the algorithm to outliers. Suppose one of the images in the low-resolution set is in fact an outlier, and does not belong in the sequence. Since this outlier image does not match the rest of the images, the pixels in it will be assigned zero weights. This can be understood qualitively from the weights reflecting the matching of the patch. In Section 4.3.5 the computation of the weights is discussed, demonstrating how outliers are indeed assigned zero (or negligible) weights. Effectively, since all pixels in an outlier image are ignored, and are not considered in the minimization – they are indeed treated as outliers. The same logic can be applied to local outliers, such as transmission errors, graphics, boundaries, and more.

4.3.2 Separating the Blur Treatment

Our task is the minimization of a functional that has two terms: $\epsilon_{PML}^2(x)$ and a regularization (e.g., TV). Rather than handling this problem directly, we decompose it, following the methods developed in [4, 7, 6]. Since both H and F_m are space-invariant operators, they can be assumed to have a block-circulant structure (assuming a cyclic boundary treatment), and as such, they commute. Thus, defining $z = Hx$, we separate the estimation into two stages, first concentrating on estimating the "blurry" high resolution image z by minimizing

$$\epsilon_{PML}^2(z) = \frac{1}{2} \sum_{m=1}^{M} \sum_{t=1}^{T} \|DF_m z - y_t\|_{W_{m,t}}^2 , \qquad (4.8)$$

which is the *fusion step*. The second step is applying a conventional *deblurring step*, that minimizes

$$\epsilon_{DB}^2(x) = \|Hx - z\|_2^2 + \lambda \cdot TV(x). \qquad (4.9)$$

This two-step process is sub-optimal to the joint treatment, but nevertheless leads to a simplified algorithm. As the second step is conventional and well-known, we focus hereafter on the fusion step. Note that the deblurring mechanism chosen here is relatively simple and could be replaced by more advanced techniques, thereby leading to better results.

4.3.3   The Algorithm: A Matrix-Vector Version

We now focus on the fusion step; the minimization of Equation (4.8). The derivative of this functional is given by

$$\frac{\partial \epsilon_{PML}^2(z)}{\partial z} = \sum_{m=1}^{M} \sum_{t=1}^{T} F_m^T D^T W_{m,t}(DF_m z - y_t), \qquad (4.10)$$

which leads to a linear system of equations. In order to simplify the obtained expressions, we introduce the following new notations:

$$\widetilde{W}_m = \sum_{t=1}^{T} W_{m,t} \quad \text{and} \quad \widetilde{y}_m = \sum_{t=1}^{T} W_{m,t} y_t. \qquad (4.11)$$

The matrix $\widetilde{W}_m$ is s sum of diagonal matrices, and therefore diagonal in itself. By rearranging and substituting $\widetilde{W}_m$ and $\widetilde{y}_m$, we obtain

$$\left[\sum_{m=1}^{M} F_m^T D^T \widetilde{W}_m D F_m \right] z = \sum_{m=1}^{M} F_m^T D^T \widetilde{y}_m. \qquad (4.12)$$

While this linear system of equations seems complicated, we show next that it can be rewritten for each pixel in z in a closed form, revealing a simple structure that leads to a stable solution.

4.3.4   The Algorithm: A Pixel-Wise Version

The Right-Hand-Side (RHS) in Equation (4.12) is an image of the same size as z. Furthermore, as we are about to show, the matrix multiplying z on the Left-Hand-Side (LHS) is a diagonal positive definite matrix. Thus, we can turn the above vector-matrix formulation into a pixel-wise one.

Since the RHS is an image of the same size as z, we start by looking at how a specific pixel at location $[i,j]$ in the RHS is constructed. A specific F_m shifts by $[dx(m), dy(m)]$. Therefore, the term $F_m^T v$ positions the $[i + dx(m), j + dy(m)]$-th element from the image v in the destination $[i,j]$ (since the transpose has the effect of an inverse displacement). The image $u = D^T \widetilde{y}_m$ is a scale-up version of the low-resolution image $\widetilde{y}_m$ by zero-filling. Combining the two implies that if the location $[i + dx(m), j + dy(m)]$ is not an integer multiple of s (the resolution ratio), this location has a zero entry. Otherwise, the entry is simply $\widetilde{y}_m[k,l]$, where $[k,l] = [i + dx(m), j + dy(m)]/s$. Accounting for all the displacements in the set and for all input images, we get that at location $[i,j]$

$$\text{RHS}[i,j] = \sum_{[k,l] \in N(i,j)} \widetilde{y}_m[k,l], \qquad (4.13)$$

where we have defined the neighborhood set

$$N(i,j) = \{[k,l] \mid \forall\, m \in [1,M],\ s \cdot k = i + dx(m),\ s \cdot l = j + dy(m)\} \qquad (4.14)$$

Plugging the definition of $\widetilde{y}_m$ from Equation (4.11) yields

$$\text{RHS}[i,j] = \sum_{[k,l] \in N(i,j)} \sum_{t=1}^{T} W_{m,t}[k,l] y_t[k,l]. \tag{4.15}$$

In this expression, $W_{m,t}[k,l]$ refers to the entry on the main diagonal in $W_{m,t}$ that multiplies the $[k,l]$ entry in y_t. This formula indicates that each pixel in the RHS is a weighted sum of pixels, in a neighborhood centered around its equivalent location in the low-resolution image.

We now turn to discuss the Left-Hand-Side (LHS) in (4.12). The operator $D^T \widetilde{W}_m D$ within this expression is a diagonal matrix that decimates an image by a factor of s in each axis, weights each pixel by the diagonal weight matrix $\widetilde{W}_m$, and then up-scales back the image using the same factor by zero-filling. When this operator is applied to an image v, a pixel in location $[i,j]$ is nulled if $[i,j]/s$ is a non-integer (since it is one of the pixels to be zero-filled by D^T), and is simply weighted otherwise, i.e., it becomes $\widetilde{W}_m[i,j] \cdot v[i,j]$.

When the full operator $F_m^T D^T \widetilde{W}_m D F_m$ is applied to the $[i,j]$-th pixel in z, it shifts it to the $[i + dx(m), j + dy(m)]$-th location, nulls it or weights it (based on whether $[i + dx(m), j + dy(m)]/s$ is an integer), and finally shifts the outcome back by $[-dx(m), -dy(m)]$ to its original place, $[i,j]$. Evidently, the operator $F_m^T D^T \widetilde{W}_m D F_m$ returns every pixel to its original location. Since every output pixel depends only on the value of the input pixel in the same location, this matrix is diagonal. Therefore, each pixel in the LHS is the pixel in z multiplied by a pixel-specific scalar, and can be computed by

$$\text{LHS}[i,j] = \sum_{[k,l] \in N(i,j)} \widetilde{W}_m[k,l] z[i,j] = \sum_{[k,l] \in N(i,j)} \sum_{t=1}^{T} W_{m,t}[k,l] z[i,j], \tag{4.16}$$

where we have substituted the definition of $\widetilde{W}_m$ in Equation (4.11). This expression is similar to Equation (4.15), summing only the weights and serving as a normalization term. Assuming that this sum is positive (i.e., at least one weight is non-zero), combining Equations (4.15) and (4.16) leads to a closed form expression for the $[i,j]$-th pixel in the estimated z,

$$\hat{z}[i,j] = \frac{\sum_{[k,l] \in N(i,j)} \sum_{t=1}^{T} W_{m,t}[k,l] y_t[k,l]}{\sum_{[k,l] \in N(i,j)} \sum_{t=1}^{T} W_{m,t}[k,l]}, \tag{4.17}$$

where m is related to $[k,l]$ through $[i,j] + [dx(m), dy(m)] = [k,l]$. The resemblance to the fusion algorithm in NLM-SR is evident (see Equation (30) in [13]). Just as explained there, the similarity of the final algorithm to the NLM stands out, but there is a subtle difference between the two, related to the domain of averaging. The proposed algorithm differs considerably from an interpolation followed by application of NLM. A visual comparison between the two in the experimental section will demonstrate the difference.

4.3.5   Computing the Weights

In the development of the closed-form formula for z, we assumed that the weighting matrices $W_{m,t}[i,j]$ are known. We now turn to explain how $W_{m,t}[i,j]$ are computed, in order to complete the description of the algorithm. Observing Equation (4.8), these weights are supposed to encompass the fit, per pixel, of the desired high resolution image z after being transformed by F_m and decimated by D, with the input image y_t. Thus, the weights could be related to the error $DF_m z - y_t$. Since the pixel value in itself is not enough to properly estimate the fit, we propose to use some spatial support for each pixel instead of computing the difference on a single pixel. Defining $R_{i,j}$ as an operator that extracts a patch of a fixed and predetermined size (say $q \times q$ pixels) from an image, the weights are computed by

$$
W_{m,t}[i,j] \;=\; \exp\left\{ -\frac{\|R_{i,j}\left(DF_m z - y_t\right)\|_2^2}{2\sigma^2} \right\}
$$

$$
\cdot \; f\left(\sqrt{(dx(m))^2 + (dy(m))^2 + (t-1)^2} \right). \tag{4.18}
$$

This formula is composed of the two independent parts. The first yields a value that is inversely proportional to the Euclidean distance between the transformed image $DF_m z$ and the input image y_t, computed over some support around each pixel. This term reflects the per-pixel fit of the displacement (after decimation). The second term reflects a decreasing confidence in large spatial and temporal displacements, and adds a decaying weight as a function of the displacement and time shift magnitudes versus the reference frame. The function f can be chosen as any monotonically non-increasing function (e.g., box function or Gaussian bell).

The computation of the weights relies on the knowledge of the unknown z. Instead, at the beginning, the weights are computed by using an estimated version of z, such as a scaled-up version of the reference frame y_1. This scale-up is done using a conventional image interpolation algorithm such as bilinear, bicubic, or the Lanczos method. As this is only a crude version of the desired outcome, the process can be iterated, using the newly estimated image $\hat{z}$ to obtain more accurate weights that contribute to an improved outcome. In our tests we employ two such iterations only.

The method in which the weights are computed is reminiscent of classic block-matching based SR algorithms (e.g., [3]). However, there is a key difference between these algorithms and the one proposed here. In both approaches, block-matching is used to crudely estimate the probability of each trajectory. However, in classic block-matching based SR, only the most likely of those trajectories is selected, while all other trajectories are ignored. In the proposed algorithm, all trajectories are considered together, in a probabilistic framework, reflecting the varying confidences of the trajectories. This difference is what enables the proposed algorithm to handle complex scenarios where highly accurate motion estimation is not currently possible.

As said earlier, outliers are to be assigned zero weights. Outliers are characterized by very different patches. Therefore, the block distance in Equation 4.18 is very large, which through the inverse exponent is translated to a negligible weight. Thus, the outliers are indeed assigned practically zero weight, and are effectively ignored.

4.3.6   Other Resampling Tasks

In this section we describe how the proposed framework can be adapted to other re-sampling tasks, such as de-interlacing, inpainting and more, and start by explaining this extension intuitively. Re-sampling tasks can be considered as computing pixel values for only some of the pixels in each image ("missing pixels"). For example, the de-interlacing task may be viewed as providing pixel values only for the even rows in the odd-numbered fields, as well as for the odd rows in the even-numbered fields. Formulating this idea, given each input image (or field) y_t, it can be linked to the original (unknown) image Y_t using a masking operator $M_t : y_t = M_t Y_t$. Simply put, M_t discards all unsampled pixels. It is a binary matrix, with as many rows as the number of pixels in y_t and as many columns as pixels in Y_t, with entries of ones indicating which pixels are to be kept. Note that y_t contains only sampled pixels. In the in-painting case, it contains only the unmasked pixels.

In line with the idea of the probabilistic motion estimation, Y_t can be constructed as a (pixel-wise) weighted average of different transformations of the target image x. The image x that we seek should be as similar as possible to each y_t, after undergoing each of the transformations and the relevant masking. This required similarity is weighted on a pixel-wise basis, according to the (local) probability of the specific transformation having taken place. Put into the maximum likelihood formulation, a penalty function very similar to Equation (4.7) arises, where the decimation operator is replaced by M_t,

$$\epsilon^2_{PML}(x) = \frac{1}{2} \sum_{m=1}^{M} \sum_{t=1}^{T} \| M_t H F_m x - y_t \|^2_{W_{m,t}}. \tag{4.19}$$

Minimizing this functional proceeds very similarly to the steps described before. The treatment of the blur is separated, and a pixel-wise formula for the values of z is given by Equation (4.17). The difference is in the order of summation, as the neighborhood $N(i, j)$ of a pixel is now time (and spatial) dependent. This is because the masking may be different for every image in the sequence.

The weights for this formula are computed very similarly to the SRR case, described in Equation (4.18). However, these tasks can benefit from computing the weights in high-resolution scale. Thus, if we consider that $W_{m,t}$ is for the coarse scale, we denote $W_{m,t} = M_t \tilde{W}_{m,t}$, with $\tilde{W}_{m,t}$ being the same size as Y_t. The formula for each entry of $\tilde{W}_{m,t}$ (when arranged as an image) is therefore the same as in Equation (4.18), but with $F_m z - Y_t$ replacing $D F_m z - y_t$.

In these weights, Y_t is an interpolated version of y_t (with the interpolation method depending on the specific task). Of course, these weights should be computed only for pixels that are kept after the masking $W_{m,t} = M_t \tilde{W}_{m,t}$.

4.4 Experimental Validation

4.4.1 Experimental Results

In this section we demonstrate the abilities of the proposed algorithm in super-resolving general content sequences. We start with one synthetic (text) sequence with global motion that comes to demonstrate the conceptual super-resolution capabilities of the proposed algorithms. Then we turn to several real-world sequences with a general motion pattern. The comparison we provide in most sequences is to a single image upsampling using the Lanczos algorithm [21, 18], that effectively approximates the Sinc interpolation. Finally, we demonstrate the adaptation of the algorithm to the de-interlacing problem.

The first test is a very simple synthetic test, that motion-estimated-based super-resolution algorithms are expected to resolve well, intended to show that the proposed algorithm indeed achieves super-resolution. A text image (in the input range $[0, 255]$) is used to generate a 9-image input sequence, by applying integer displacements prior to blurring (using a 3×3 uniform mask), decimation (by a factor of $1 : 3$ in each axis), and the addition of noise (with $std = 2$). The displacements are chosen so that the entire decimation space is covered (i.e., $dx = \{0, 1, 2\}$ and $dy = \{0, 1, 2\}$). The result for this test is shown in Figure 4.1, including a comparison to both Lanczos interpolation and the regularized shift-and-add algorithm [5, 7], which is a conventional motion-estimation-based super-algorithm resolution.

The block size used for computing the weights ($\hat{R}$) was set to 31×31, since the motion in the sequence is limited to displacements, and a larger block allows capturing the true displacement better (for real-world sequences, this size will be greatly reduced, as explained later). The value of σ that moderates the weights was set to 7.5 (due to the large differences between white and black values in the scene). Two iterations were ran on the entire sequence, the first iteration used for computing the weights for the second iteration.

The similarity between the quality of the classic SR result and the proposed algorithm is evident. This similarity stems from the large block size used in the proposed algorithm. This large block size, together with the exiting global translation, makes the proposed algorithm converge to classic motion-estimation-based format, as such large blocks basically identify the correct motion vector for each pixel. We note that such large blocks cannot be used

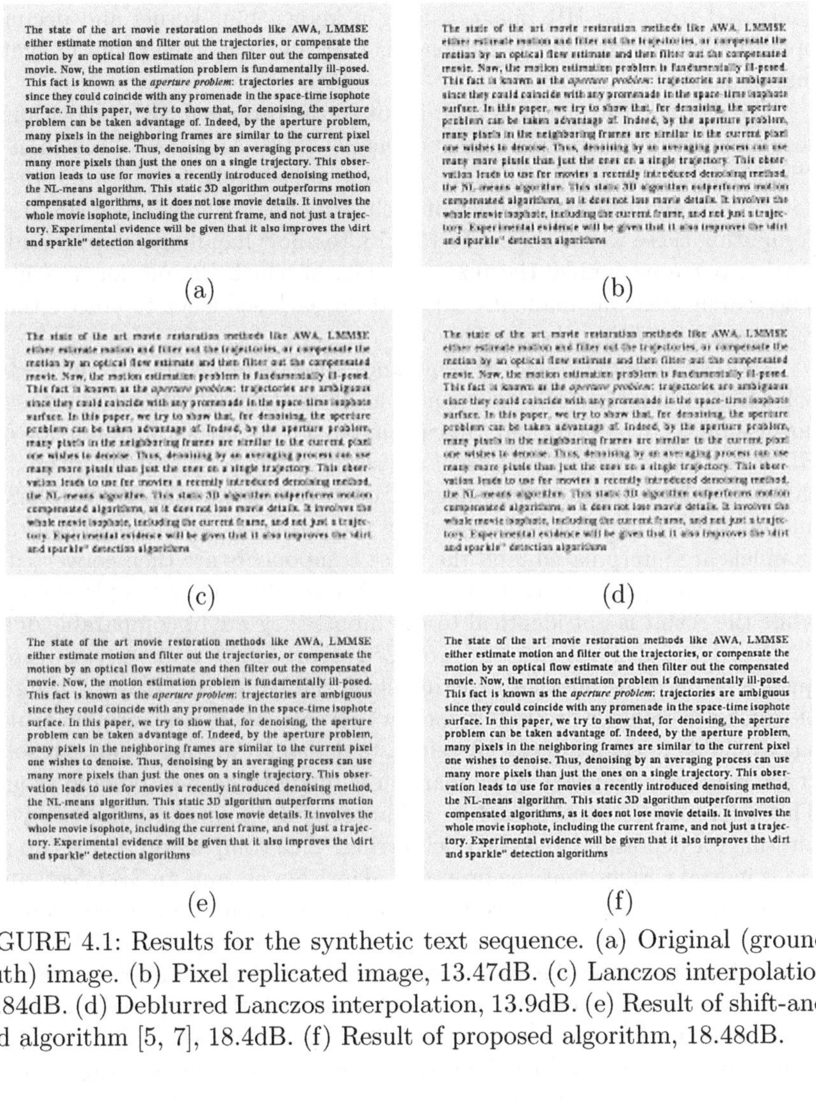

The state of the art movie restoration methods like AWA, LMMSE either estimate motion and filter out the trajectories, or compensate the motion by an optical flow estimate and then filter out the compensated movie. Now, the motion estimation problem is fundamentally ill-posed. This fact is known as the *aperture problem*: trajectories are ambiguous since they could coincide with any promenade in the space-time isophote surface. In this paper, we try to show that, for denoising, the aperture problem can be taken advantage of. Indeed, by the aperture problem, many pixels in the neighboring frames are similar to the current pixel one wishes to denoise. Thus, denoising by an averaging process can use many more pixels than just the ones on a single trajectory. This observation leads to use for movies a recently introduced denoising method, the NL-means algorithm. This static 3D algorithm outperforms motion compensated algorithms, as it does not lose movie details. It involves the whole movie isophote, including the current frame, and not just a trajectory. Experimental evidence will be given that it also improves the \dirt and sparkle" detection algorithms

(a)

(b)

(c)

(d)

The state of the art movie restoration methods like AWA, LMMSE either estimate motion and filter out the trajectories, or compensate the motion by an optical flow estimate and then filter out the compensated movie. Now, the motion estimation problem is fundamentally ill-posed. This fact is known as the *aperture problem*: trajectories are ambiguous since they could coincide with any promenade in the space-time isophote surface. In this paper, we try to show that, for denoising, the aperture problem can be taken advantage of. Indeed, by the aperture problem, many pixels in the neighboring frames are similar to the current pixel one wishes to denoise. Thus, denoising by an averaging process can use many more pixels than just the ones on a single trajectory. This observation leads to use for movies a recently introduced denoising method, the NL-means algorithm. This static 3D algorithm outperforms motion compensated algorithms, as it does not lose movie details. It involves the whole movie isophote, including the current frame, and not just a trajectory. Experimental evidence will be given that it also improves the \dirt and sparkle" detection algorithms

(e)

The state of the art movie restoration methods like AWA, LMMSE either estimate motion and filter out the trajectories, or compensate the motion by an optical flow estimate and then filter out the compensated movie. Now, the motion estimation problem is fundamentally ill-posed. This fact is known as the *aperture problem*: trajectories are ambiguous since they could coincide with any promenade in the space-time isophote surface. In this paper, we try to show that, for denoising, the aperture problem can be taken advantage of. Indeed, by the aperture problem, many pixels in the neighboring frames are similar to the current pixel one wishes to denoise. Thus, denoising by an averaging process can use many more pixels than just the ones on a single trajectory. This observation leads to use for movies a recently introduced denoising method, the NL-means algorithm. This static 3D algorithm outperforms motion compensated algorithms, as it does not lose movie details. It involves the whole movie isophote, including the current frame, and not just a trajectory. Experimental evidence will be given that it also improves the \dirt and sparkle" detection algorithms

(f)

FIGURE 4.1: Results for the synthetic text sequence. (a) Original (ground-truth) image. (b) Pixel replicated image, 13.47dB. (c) Lanczos interpolation, 13.84dB. (d) Deblurred Lanczos interpolation, 13.9dB. (e) Result of shift-and-add algorithm [5, 7], 18.4dB. (f) Result of proposed algorithm, 18.48dB.

in real-world sequences (shown next), as they do not allow enough adaptation to the various motion patterns, and therefore much smaller block sizes will be used.

We now turn to demonstrate the potential of the proposed SRR algorithm by presenting the results for image sequences with a general motion pattern. These sequences are also in the input range [0, 255]. Each Low Resolution (LR) frame is generated from one High Resolution (HR) frame. The HR frame is blurred using a 3×3 uniform mask, decimated by a factor of 1:3 (in each axis), and then contaminated by additive white zero-mean Gaussian noise with $STD = 2$. It is important to note the while the LR images are synthetically generated from the HR images (using a known blur kernel and decimation operator) the motion in the sequence is real, and is not the result of synthetic manipulations.

The degraded sequence is then input to the proposed SRR algorithm. The results for three such degraded sequences: "Miss-America," "Foreman" and "Suzie" appear in Figures 4.2, 4.3, and 4.4 respectively, for the 3rd, 8th, 13th, 18th, 23rd, and the 28th frames of each sequence.[5] The window size used for computing these weights is set to 13×13, to allow handling complex and local motion patterns (unlike the text example, in which the motion was global). The search area was manually adapted for each sequence to ensure that the real motion is within the search area.

Another example along the same lines appears in Figure 4.5. In this test, a color High-Definition (HD) sequence was blurred by a 2×2 uniform mask, and downsampled by a factor of 2 (in each axis). The figure shows a portion of one HD frame and the same portion of the result of the proposed algorithm. Since this is a color sequence, the images are converted into the YUV colorspace, and only the Y channel is processed by the proposed algorithm. The U and V channels are interpolated, and the three components are then converted back to RGB colorspace to create the final SR result. This example shows that while the result is not identical to the input, they are of comparable quality.

In order to demonstrate the proposed algorithm on a directly captured sequence, we provide another experiment on the sequence "Trevor," the results of which are displayed in Figure 4.6. In this case, there is no ground-truth image available to compare to. Therefore, to demonstrate that a super-resolution effect is indeed achieved, a comparison is made to an interpolated sequence. This interpolation is obtained by a Lanczos interpolation, followed by NLM filtering for denoising, and then deblurring. This comparison serves two goals: (1) It indeed verifies that the proposed algorithm obtains an SR effect; and (2) it demonstrates the difference between simply running NLM and deblurring after up-scaling, compared to running the proposed algorithm. This comparison is important, as the two schemes are confusingly similar (see Equation 4.17). Clearly, a far better image is obtained with the proposed algorithm.

In order to demonstrate the generalized algorithm, we apply it to an interlaced sequence. We used the Foreman sequence and composed each interlaced frame from a pair of original frames by taking the odd-numbered rows from one frame, and the even-numbered rows from the next, resulting in a sequence with half as many frames. This sequence was also contaminated by additive white zero-mean Gaussian noise with $STD = 2$. This generated sequence can be considered a true interlaced sequence, as no manipulation (e.g., simulated blurring) of the pixels has been made other than half the pixels being discarded.

The result of processing this sequence with the framework suggested in

---

[5]The sequences appearing in this section (input and output) and others from [13], along with the various parameters used to generate them, can be found at http://www.cs.technion.ac.il/~matanpr/NLM-SR.

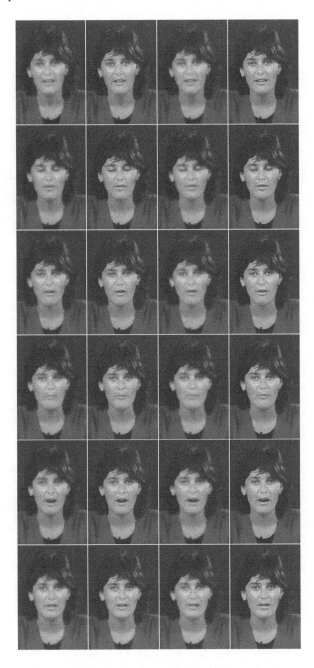

FIGURE 4.2: Results for the 3rd, 8th, 13th, 18th, 23rd, and the 28th frames from the "Miss America" sequence. From left to right: pixel-replicated low resolution image; original image (ground truth); Lanczos interpolation; result of the proposed algorithm.

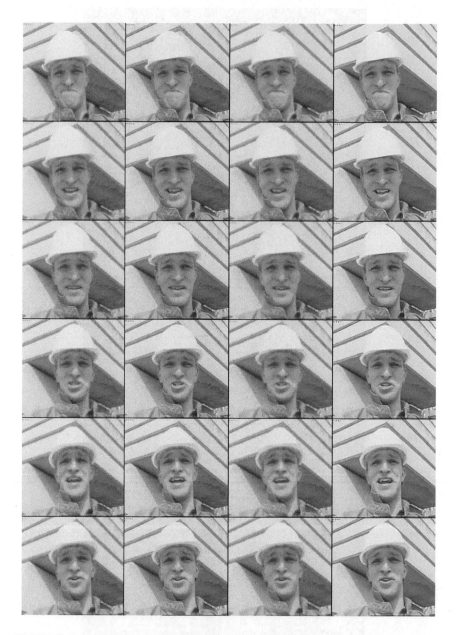

FIGURE 4.3: Results for the 3rd, 8th, 13th, 18th, 23rd, and the 28th frames from the "Foreman" sequence. From left to right: pixel-replicated low-resolution image; original image (ground truth); Lanczos interpolation; result of the proposed algorithm.

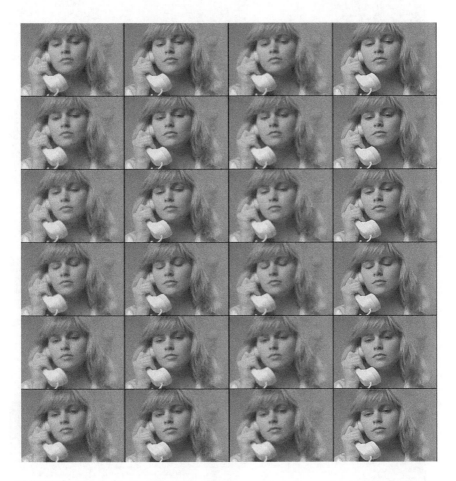

FIGURE 4.4: Results for the 3rd, 8th, 13th, 18th, 23rd, and the 28th frames from the "Suzie" sequence. From left to right: pixel-replicated low resolution image; original image (ground truth); Lanczos interpolation; result of the proposed algorithm.

FIGURE 4.5: A High-Definition sequence. Top: Portion of original HD image. Bottom: Same portion of SR result (from an input downscaled by 2 in each axis).

FIGURE 4.6: Original "Trevor" sequence. Top: Interpolated image. Middle: Interpolation, followed by NLM processing and deblurring. Bottom: Proposed algorithm. The right column offers a close-up of a portion of the images.

Section 4.3.6 appears in Figure 4.7. The initial interlaced sequence was split into fields, and each field was expanded by a factor of two in the vertical axis only. The missing rows were interpolated by averaging the rows immediately above and below each missing row. The masks M_t were designed to discard the even rows in the odd-numbered images, and the odd rows in the even-numbered images. 5 interlaced frames (10 fields) were used for processing, and the search area consisted of 10 pixels in every direction. We display the results for two iterations (where the first is used for computing the weights for the second), although the differences are much less dramatic than in the SRR case. As done above, we also show the results of directly filtering the re-scaled sequence with the NLM filter, to highlight the difference of the proposed approach. Note how the staircase effect (on the wall) is much decayed by the proposed algorithm. It should be noted that the purpose of this test is only to demonstrate the applicability of the proposed framework to other re-sampling tasks, without claiming that it out-performs other de-interlacing methods. Further work is required to compare the proposed technique to existing de-interlacing algorithms.

4.4.2 Computational Complexity

The complexity of the algorithm is essentially the same as that of the NLM algorithm, with the addition of a deblurring process, which is negligible compared to the fusion stage. The core of the algorithm, which also requires most of the computations, is computing the weights. In a nominal case in which a search area of 31×31 low-resolution pixels in the spatial domain, and 15 images in the temporal axis, we have $\approx 14,000$ pixels in this spatiotemporal window. For each pixel in the search area, the block difference is computed, with a block size of 13×13 (high-resolution) pixels. Thus, there is a total of almost $2,400,000$ operations per pixel. Performing this amount of calculations for every pixel makes the algorithm irrelevant for practical implementations, and therefore the computational load must be reduced. In this section we describe a few possible speed-up options for the proposed algorithm. Several of the methods to speed-up the NLM algorithm were suggested originally in [10], and were adopted in our simulations:

1. Computing the weights can be done using block differences in the low-resolution images, instead of on the interpolated images. This saves a factor of $\approx s^2$. This can be applied for the first iteration, resulting in only a small loss of quality.

2. Computing fast estimations for the similarity between blocks, such as the difference between the average gray level or the average direction of the gradient, can eliminate many nonprobable destinations from further processing. Such an approach was suggested in [10], and was found to be very effective for the original NLM algorithm.

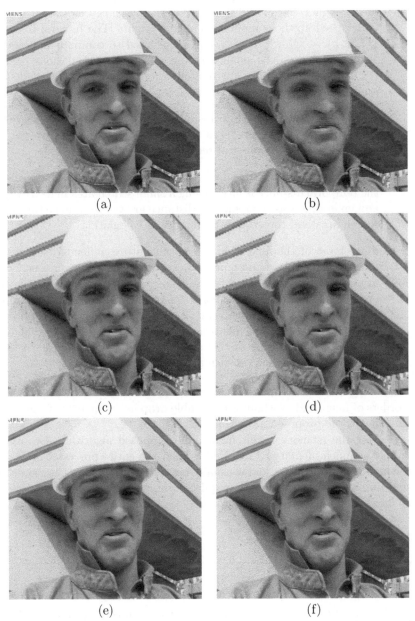

FIGURE 4.7: De-Interlacing Results. (a) Original (ground-truth) image. (b) Interlaced image. (c) Row Averaging, 29.87dB. (d) Row Averaging followed by NLM processing, 29.93dB. (e) Proposed algorithm first iteration, 30.69dB. (f) Proposed algorithm second iteration, 30.71dB.

3. If the patch used to compute the weights is rectangular and with uniform weights, the special structure of the patch can be used to dramatically speed up the computation of the weights. The Integral Image [20]. $(II(x, y) = \sum_{i=1}^{x} \sum_{j=1}^{y} I(i, j))$ can be used to compute the block differences using only a small constant number of calculations per pixel, regardless of block size. Fortunately, there is only a slight effect on the quality of the outputs of using such a patch structure.

4. A coarse-to-fine approach, transferring only high likelihood destinations from the coarse level, reduces the effective search area for each pixel, thus reducing the number of required calculations.

5. Since it is more likely that large spatial displacements will appear when the temporal distance is large, using a small search area in nearby frames and enlarging it as the temporal distance grows, can reduce the effective search area and thus the total number of calculations.

6. Since most of the algorithm is local in nature, it lends itself easily to parallelization. As 4 and 8 processor configurations are currently widely available, this can be used for speeding up the algorithm by about one order of magnitude. Furthermore, as parallel hardware such as Graphical Processing Units (GPUs) are very common and powerful, with programming tools making implementations on such hardware easier than before, the parallelistic nature of this algorithm might allow a great speed-up by an implementation on such processing units.

These suggested speed-up methods can reduce the complexity by at least 3 to 4 orders of magnitude without a noticeable drop in the quality of the outputs. This makes the proposed algorithm practical.

As for the memory requirements, the proposed algorithm uses approximately as much memory as required to hold the entire processed sequence in the high-resolution scale, and is usually not a limitation. However, some of the speed-up methods suggested do require more memory, so a trade-off between memory requirements and run-time may be needed.

4.5   Summary

In NLM-SR [13], an earlier work, an explicit-motion-estimation-free SRR algorithm was developed by extending the NLM filter to SRR reconstruction. This chapter approaches the same task from a different perspective, basing it on a probabilistic and crude motion estimation instead. Interestingly, this approach (under some assumptions) leads to the same algorithm as in NLM-SR. However, since the formulation described here relies on the classic super-resolution framework and on the imaging model, we believe it is more intuitive. We give several examples of the abilities of the proposed algorithm in super-resolving various sequences.

Another benefit of this formulation is that it allows for different extensions than those proposed in [13]. In this chapter, we have shown that this framework can in fact be adapted to any re-sampling task. We have given one example of de-interlacing, showing the validity of this adaptation. This example shows than even sequences with large, highly nonrigid motion patterns can be successfully de-interlaced by the proposed framework.

While the results of the proposed algorithm are encouraging, we believe further research is needed in order to extract the full potential of this family of algorithms. We note that in developing the algorithm, we have made several choices such as relying on integer displacements, only in order to simplify the development of the algorithm, and to arrive at an algorithm that is relatively simple to understand and implement. Avoiding such compromises will result in a more complex algorithm, but one that might also bring about better results.

Bibliography

[1] S. Baker and T. Kanade. Limits on super-resolution and how to break them. *IEEE Transactions on Pattern Analysis and Machine Intelligence*, 24(1):1167–1183, September 2002.

[2] A. Buades, B. Coll, and J.M. Morel. Denoising image sequences does not require motion estimation. In *Proc. IEEE Conference on Advanced Video and Signal Based Surveillance September (AVSS)*, pages 70–74, 2005.

[3] G.M. Callicó, S. López, O. Sosa, J.F. Lopez, and R. Sarmiento. Analysis of fast block matching motion estimation algorithms for video super-resolution systems. *IEEE Transactions. on Consumer Electronics*, 54(3):1430–1438, August 2008.

[4] M. Elad and Y. Hel-Or. A fast super-resolution reconstruction algorithm for pure translational motion and common space-invariant blur. *IEEE Transactions on Image Processing*, 10(8):1187–1193, August 2001.

[5] S. Farsiu, D. Robinson, M. Elad, and P. Milanfar. Robust shift and add approach to superresolution. In *SPIE Conference on Applications of Digital Signal and Image Processing*, pages 121–130, August 2003.

[6] S. Farsiu, D. Robinson, M. Elad, and P. Milanfar. Advances and challenges in superresolution. *International Journal of Imaging Systems and Technology*, 14(2):47–57, August 2004.

[7] S. Farsiu, D. Robinson, M. Elad, and P. Milanfar. Fast and robust multiframe superresolution. *IEEE Transactions on Image Processing*, 13(10):1327–1344, October 2004.

[8] R.C. Hardie, K.J. Barnard, and E.E. Armstrong. Joint map registration and high-resolution image estimation using a sequence of undersampled images. *IEEE Transactions on Image Processing*, 6(12):1621–1633, December 1997.

[9] M. Irani and S. Peleg. Improving resolution by image registration. *CVGIP: Graphical Models and Image Processing*, 53(3):231–239, May 1991.

[10] M. Mahamoudi and G. Sapiro. Fast image and video denoising via non-local means of similar neighbourhoods. *IEEE Signal Processing Letters*, 12(12):839–842, December 2005.

[11] L.C. Pickup, D.P. Capel, S.J. Roberts, and A. Zisserman. Overcoming registration uncertainty in image super-resolution: Maximize or marginalize? *EURASIP Journal on Advances In Signal Processing*, 2007:Article ID 23565, 14 pages, 2007.

[12] M. Protter and M. Elad. Super-resolution with probabilistic motion estimation. *IEEE Transactions on Image Processing*, 18(8):1899–1904, August 2009.

[13] M. Protter, M. Elad, H. Takeda, and P. Milanfar. Generalizing the non-local-means to super-resolution reconstruction. *IEEE Transactions on Image Processing*, 18(1):36–51, January 2009.

[14] L.I. Rudin, S.J. Osher, and E. Fatemi. Nonlinear total variation based noise removal algorithms. *Physica D*, 60:259–268, 1992.

[15] R.R. Schultz and R.L. Stevenson. Extraction of high-resolution frames from video sequences. *IEEE Transactions on Image Processing*, 5(6):996–1011, June 1996.

[16] H. Shen, L. Zhang, B. Huang, and P. Li. A map approach for joint motion estimation, segmentation, and super resolution. *IEEE Transactions on Image Processing*, 16(2):479–490, February 2007.

[17] H. Takeda, P. Milanfar, M. Protter, and M. Elad. Super-resolution without explicit subpixel motion estimation. *IEEE Transactions on Image Processing*, 18(9):1958–1975, September 2009.

[18] K. Turkowski. Filters for common resampling tasks. In *Graphical Gems*, pages 147–165. Academic Press Professional Inc., San Diego, CA, USA, 1990.

[19] P. van de Walle, S. Susstrunk, and M. Vetterli. A frequency domain approach to registration of aliased images with application to super-resolution. *EURASIP Journal On Applied Signal Processing*, 2006:1–14, March 2006.

[20] P.A. Viola and M.J. Jones. Rapid object detection using a boosted cascade of simple features. In *IEEE Computer Society Conference on Computer Vision and Pattern Recognition (CVPR)*, volume 1, pages I: 511–518, December 2001.

[21] G. Wolberg. *Digital Image Warping*. IEEE Computer Society Press, Los Alamitos, CA, 1990.

5

Spatially Adaptive Filtering as
Regularization in Inverse Imaging:
Compressive Sensing, Super-Resolution, and
Upsampling

Aram Danielyan

Tampere University of Technology

Alessandro Foi

Tampere University of Technology

Vladimir Katkovnik

Tampere University of Technology

Karen Egiazarian

Tampere University of Technology

CONTENTS

The recent developments in image and video denoising have brought a new generation of filtering algorithms achieving unprecedented restoration quality. This quality mainly follows from exploiting various features of natural images. The nonlocal self-similarity and sparsity of representations are key elements of the novel filtering algorithms, with the best performance achieved by adaptively aggregating multiple redundant and sparse estimates. In a very broad sense, the filters are now able, given a perturbed image, to identify its plausible representative in the space or manifold of possible solutions. Thus, they are powerful tools not only for noise removal, but also for providing accurate adaptive regularization to many ill-conditioned inverse imaging problems. In the case of image/video reconstruction from incomplete data, the general structure of the proposed approach is very simple and is based on iteratively refining the estimates alternating two procedures: image/video filtering (denoising) and projection on the observation-constrained subspace. In this chapter we give an overview of this versatile approach, with particular emphasis on three challenging and important imaging applications: inversion from sparse or limited-angle tomographic projections, image reconstruction from low-frequency or undersampled data, image and video super-resolution. This approach is especially appealing for the latter application, as the block-matching procedure, performed both in space and time, makes the explicit motion estimation unnecessary. The presented experimental results demonstrate an overall performance on the level of the state of the art.

5.1 Introduction

A priori assumptions on the image to be reconstructed are essential for any inverse imaging algorithm. In the standard variational approaches, these assumptions are usually given as penalty terms that serve as regularization in an energy criterion to be minimized.

A main limitation of these approaches is that the minimization usually involves the evaluation of the gradient of the penalty functional and its convexity. Therefore, if on the one hand the nonstationarity of natural images calls for locally adaptive nonconvex penalties, on the other hand, to obtain a feasible algorithm, the energy criterion and, thus, the penalty needs to be simple.

This limitation becomes evident particularly for image denoising, for which the recent years have witnessed the development of a number of spatially adaptive algorithms that dramatically outperform the established methods based on variational constraints (see, e.g., [20],[26]).

Our approach to inverse imaging essentially differs from these variational formulations and appeals to nonparametric regression techniques. We propose to replace the implicit regularization coming from the penalty by explicit filtering, exploiting spatially adaptive filters sensitive to image features and details. If these filters are properly designed, we have reasonable hopes to achieve better results by filtering than through the formal approach based on the formulation of imaging as a variational problem with imposed global constraints.

The presented framework is general and is applicable to a wide class of inverse problems for which the available data can be interpreted as a smaller portion of some transform spectrum of the signal of interest. This observation model has gained recently enormous popularity under the paradigm of compressive sensing [1].

We demonstrate application of our approach to the inversion from sparse or limited-angle tomographic projections, image upsampling, and image/video super-resolution. Depending on the particular way of sampling the spectral components and the assumptions about the complexity of the image, the recursive procedures can be improved by incorporating random search or staged reconstruction.

The chapter is organized as follows. In the next section we give formal definitions of the observation model and of the reconstruction algorithm for a rather general case. These are then reinterpreted in the context of compressive sensing in Section 5.3, where we also present inverse tomography and basic image upsampling experiments. Section 5.4 is devoted to image and video super-resolution. Concluding remarks are given in the last section.

5.2   Iterative Filtering as Regularization

Let us consider a general ill-posed inverse problem in the form

$$y = \mathcal{H}(x) \tag{5.1}$$

where x is the true image to be reconstructed and y is its observation through a linear operator $\mathcal{H} : X \to Y$, X and Y being Euclidean spaces with dimensions $\dim Y < \dim X$. We start from the naive pseudoinverse

$$\hat{x}^{(0)} = \underset{x : \mathcal{H}(x) = y}{\arg\min} \|x\|_2$$

of (5.1), which is usually very far from the solution one would like to obtain, particularly when $\dim Y$ is much smaller than $\dim X$. Nevertheless, due to the linearity of $\mathcal{H}$, any other solution of (5.1) should differ from $\hat{x}^{(0)}$ only by its component on the null space of $\mathcal{H}$, $\ker(\mathcal{H})$. To obtain an updated estimate $\hat{x}^{(1)}$, we first refine $\hat{x}^{(0)}$ with a filter Φ, project $\Phi\left(\hat{x}^{(0)}\right)$ on $\ker(\mathcal{H})$ and then add $\hat{x}^{(0)}$. The reason of the last two operations is that the refined estimate, while being closer to the desired solution may not satisfy (5.1) anymore. This procedure can be repeated iteratively leading to the following recursive scheme:

$$\begin{cases} \hat{x}^{(0)} = \underset{x:\mathcal{H}(x)=y}{\arg\min} \|x\|_2, \\ \hat{x}^{(k)} = \hat{x}^{(0)} + \mathcal{P}_{\ker(\mathcal{H})}\left(\Phi\left(\hat{x}^{(k-1)}\right)\right), \end{cases} \tag{5.2}$$

where $\mathcal{P}_{\ker(\mathcal{H})}$ is the projection operator on the null space $\ker(\mathcal{H})$ and the superscripts denote the corresponding iteration.

5.2.1 Spectral Decomposition of the Operator

There are several ways how the projection $\mathcal{P}_{\ker(\mathcal{H})}$ can be realized. A practical approach, which we follow throughout the chapter, relies on the following spectral decomposition of the operator $\mathcal{H}$ [8].

Let $\mathcal{T} : X \to \Theta$ be an orthonormal transform with basis elements $\{t_i\}_{i=1}^{\dim X}$ such that $\ker(\mathcal{H}) = \mathrm{span}\{t_i\}_{i\in\Omega^C}$, Ω being a subset of indices (subband) and Ω^C its complementary. Given such a transform $\mathcal{T}$, the projection of $x \in X$, $\mathcal{P}_{\ker(\mathcal{H})}(x)$, can be easily obtained as the zeroing out of the $\mathcal{T}$-spectrum of x on Ω followed by application of $\mathcal{T}^{-1}$. Moreover, we can choose $\mathcal{T}$ so that it performs an eigendecomposition, i.e., so that $\mathcal{H}$ can be rewritten as

$$\mathcal{H}(\cdot) = \mathcal{T}_0^{-1}(\mathcal{S}(\mathcal{T}(\cdot))), \tag{5.3}$$

where $\mathcal{S} : \Theta \to \Theta_\Omega$ is a diagonal operator, which scales each spectrum coefficient by its corresponding non-zero eigenvalue, and $\mathcal{T}_0 : Y \to \Theta_\Omega$ is an orthonormal transform. Here Θ_Ω is the space obtained by restricting the elements of Θ on Ω. Thus, the operator $\mathcal{S}$ is simply restricting the $\mathcal{T}$-spectra on Ω and scaling the retained coefficients, with $\ker(\mathcal{S}) = \mathcal{T}(\ker(\mathcal{H}))$. Unless the eigenvalues are all distinct, the basis elements $\{t_i\}_{i\in\Omega}$ of the transform $\mathcal{T}$ that satisfies the above requirements are not uniquely determined. Of course, as the $\{t_i\}_{i\in\Omega}$ are varied, also the transform $\mathcal{T}_0$ must vary accordingly, because $\mathcal{T}_0$ is essentially determined by them. In matrix form, the spectral decomposition (5.3) can be obtained through the singular value decomposition (SVD) of $\mathcal{H}$.

5.2.2 Nonlocal Transform Domain Filtering

In our implementations of the recursion (5.2), as the spatially adaptive filter Φ we utilize the Block-Matching 3D filtering (BM3D) denoising algorithm [10] and its extension to video V-BM3D [9]. Our choice of the BM3D algorithms is

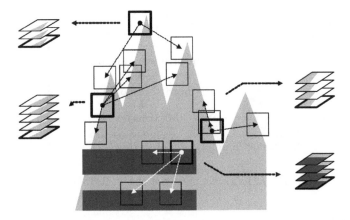

FIGURE 5.1: Illustration of grouping within an artificial image where for each reference block (with thick borders) there exist perfectly similar ones.

essentially determined by the fact that it is currently considered as one of the best denoising filters (see, e.g., [26],[20]) and that its remarkable performance is achieved at a competitive computational cost.

The BM3D algorithm exploits paradigms of nonlocal similarity [2],[20] in a blockwise fashion in order to obtain a highly sparse representation of the data. It means that the images to be processed are windowed/segmented into overlapping blocks and one looks for mutually similar blocks, which are collected into groups, so that the data in these groups can be processed jointly. In this way we arrive to a nonlocal estimator with varying adaptive support where the data used in the estimation can be located quite far from each other. This estimation can be treated as a sophisticated high-order generalization of nonlocal means (NLM) [2],[4],[22],[23],[24].

To clarify the idea of grouping, let us consider an illustrative example of blockwise nonlocal estimation of the image in Figure 5.1 from an observation (not shown) corrupted by additive zero-mean independent noise. In particular, let us focus on the already grouped blocks shown in the same figure. These blocks exhibit perfect mutual similarity, which makes the elementwise averaging (i.e., averaging between pixels at the same relative positions) an optimal estimator. In this way, we achieve an accuracy that cannot be obtained by processing the separate blocks independently.

However, perfectly identical blocks are unlikely in natural images. If nonidentical fragments are allowed within the same group, averaging is no longer optimal. Therefore, a filtering strategy more effective than averaging should be employed.

Here we give a brief description of the general V-BM3D algorithm, whose flowchart is presented in Figure 5.2. The BM3D algorithm can be then considered as equivalent to V-BM3D with the input sequence consisting of a single

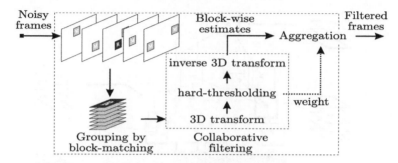

FIGURE 5.2: Flowchart of the the V-BM3D algorithm.

image. Detailed descriptions of the (V-)BM3D algorithms can be found in the corresponding references.

1. *Block-wise estimates.* Each image of an input sequence is processed in sliding-block manner. For each block the filter performs:

    (a) *Grouping by block-matching.* Searching within all images in the sequence, find blocks that are similar to the currently processed one, and stack them together in a 3-D array (group).

    (b) *Collaborative filtering.* Apply a 3-D transform to the formed group, attenuate the noise by hard-thresholding of the transform coefficients, invert the 3-D transform to produce estimates of all grouped blocks, and return the estimates of the blocks to their original place.

2. *Aggregation.* Compute the estimates of the output images by weighted averaging all of the obtained block-wise estimates that are overlapping.

What makes this algorithm very different from other nonlocal estimators is the use of a full-rank complete transform for modeling both the blocks and their mutual similarity or difference. Due to the similarity between the grouped blocks, the transform can typically achieve a highly sparse representation of the true signal so that the noise or small distortions can be well separated by shrinkage. In this way, the collaborative hard-thresholding reveals even the finest details shared by grouped fragments and at the same time it preserves the essential unique features of each individual fragment. Additionally, the adaptive aggregation of multiple redundant and sparse estimates allows us to significantly boost the performance and provide high level of robustness to the algorithm.

Let us note that, for the purposes of this work, we do not perform the additional collaborative Wiener filtering stage that is otherwise present in the original BM3D and V-BM3D denoising algorithms [10], [9].

In what follows, the (V-)BM3D filter will be denoted as $\Phi(z, \sigma)$, where

z is the degraded image/video to be filtered and $\sigma \geq 0$ is a parameter that originally corresponds to the standard-deviation of the noise in z and that here we mainly use to effectively control the filter strength: a larger σ implies a more aggressive filtering, thus enforcing a higher degree of sparsity, while a smaller σ provides better preservation of weak details at the expense of a milder noise suppression.

5.3 Compressed Sensing

During the last three years, compressed sensing (CS) has received growing attention, mainly motivated by the positive theoretical and experimental results shown in [5], [7], [15], [19], [31], [32], [33]. The basic settings of signal reconstruction under conditions of CS are as follows. An unknown signal of interest is observed (sensed) through a limited number of linear functionals. These observations can be considered as an incomplete portion of the spectrum of the signal with respect to a given linear transform $\mathcal{T}$. Thus, conventional linear reconstruction/synthesis (e.g., inverse transform) cannot in general reconstruct the signal. For example, when $\mathcal{T}$ is the Fourier transform, CS considers the case where the available spectrum is much smaller than what is required according to the Nyquist-Shannon sampling theory. It is generally assumed that the signal can be represented sparsely with respect to a different relevant basis (e.g., wavelets) or that, alternatively, it belongs to a specific class of functions (e.g., piecewise constant functions). Of particular importance is the so-called *incoherence* between the basis with respect to which the incomplete observations are given and the one with respect to which the signal is sparse [6]. In the publications cited above, it is shown that under such assumptions, stable reconstruction of the unknown signal is possible and that in some cases the reconstruction can be exact. These techniques typically rely on convex optimization with a penalty expressed by the ℓ_0 or ℓ_1 norm [34] which is exploited to enable the assumed sparsity [14]. It results in parametric modeling of the solution and in problems that are then solved by mathematical programming algorithms.

Based on the ideas discussed in the introduction, an alternative and novel approach to CS reconstruction can be developed by replacing the parametric modeling with a nonparametric one implemented by the use of spatially adaptive denoising filtering. The algorithm proposed in [17] extends the iterative scheme (5.2) by incorporating stochastic approximation to obtain stable recovery of the images. At every iteration, the current estimate is excited by injection of random noise in the unobserved portion of the spectrum. The denoising filter attenuates the noise and reveals new features and details out of the incomplete and degraded observations. Roughly speaking, we seek for the solution (reconstructed signal) by stochastic approximations whose search direction is driven by the denoising filter. It should be remarked that here we are concerned only in the operative way the solution is found, while we exploit essentially the same fundamental ideas of sparsity and basis incoherence undertaken by many other authors. As a matter of fact, the observation domains considered in what follows are much more incoherent with respect to the adaptive data-driven 3-D transform domain in which BM3D effectively forces the data to be sparse.

5.3.1 Observation Model and Notation

Let $\mathcal{T} : X \rightarrow \Theta$ be an orthonormal transform operating from the image domain X to the transform domain Θ. The unknown image $x \in X$ can be sensed through the linear operator $\mathcal{H} : X \rightarrow Y = \Theta_\Omega$, where Ω is a subset of the $\mathcal{T}$-spectral components we are able to sense and Θ_Ω is the corresponding space of the $\mathcal{T}$-spectra restricted on Ω. The space Θ_Ω can be identified with the subspace of Θ constituted by all spectra that are identically zero outside of Ω. To clarify these concepts and simplify the coming notation, we introduce two operators:

- the *restriction operator* $|_\Omega$ that, from a given $\mathcal{T}$- spectrum, extracts its smaller portion defined on Ω;

- the *zero-padding operator* $\mathcal{U}_\Omega$ that expands the part of $\mathcal{T}$- spectrum defined on Ω to the full $\mathcal{T}$-spectrum by introducing zeros in the complement of Ω, Ω^C.

Likewise, we can define $|_{\Omega^C}$ and $\mathcal{U}_{\Omega^C}$.

Using these operators, the sensing operator can be explicitly written as $\mathcal{H}(\cdot) = \mathcal{T}(\cdot)|_\Omega$. Referring to (5.3), we have $\mathcal{S} = |_\Omega$ and $\mathcal{T}_0$ is the identity.

Now, if $x \in X$ is the unknown image intensity and $\theta = \mathcal{T}\{x\} \in \Theta$ is its spectrum, the CS problem is to reconstruct θ (or equivalently x) from the measurements y,

$$y = \theta|_\Omega = \mathcal{T}(x)|_\Omega. \tag{5.4}$$

It means that given the known part of the spectrum defined on Ω we have to reconstruct the missing part defined on Ω^C.

5.3.2 Iterative Algorithm with Stochastic Approximation

Following (5.2), we obtain the initial estimate as $\hat{x}^{(0)} = \underset{x:\,\mathcal{H}(x)=y}{\arg\min} \|x\|_2 = \mathcal{T}^{-1}(\mathcal{U}_\Omega(y))$. When the observed data is highly undersampled, the initial estimate $\hat{x}^{(0)}$ may contain too little information to enable reconstruction by the simple scheme (5.2). An improved scheme can be obtained by incorporating stochasticity into (5.2) resulting in

$$\begin{cases} \hat{x}^{(0)} = \underset{x:\,\mathcal{H}(x)=y}{\arg\min} \|x\|_2 = \mathcal{T}^{-1}(\mathcal{U}_\Omega(y)), \\ \hat{x}^{(k)} = \hat{x}^{(0)} + \mathcal{P}_{\ker(\mathcal{H})}\left[(1-\gamma_k)\hat{x}^{(k-1)} + \gamma_k\left(\Phi\left(\hat{x}^{(k-1)},\sigma_k\right) + \eta_k\right)\right], k \geq 1. \end{cases} \tag{5.5}$$

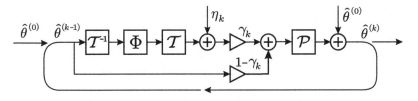

FIGURE 5.3: Flowchart of the recursive system (5.6).

Here, η_k is some pseudo-random white noise that is injected into the system in order to provide the stochastic excitation. Equivalently, we can rewrite (5.5) with respect to spectral variables as

$$
\begin{cases}
\hat{\theta}^{(0)} = \underset{\theta:\, \theta|_{\Omega}=y}{\arg\min} \|\theta\|_2 = \mathcal{U}_{\Omega}\left(y\right), \\
\hat{\theta}^{(k)} = \hat{\theta}^{(0)} + \mathcal{P}_{\ker(\,|_{\Omega})}\left[\left(1-\gamma_k\right)\hat{\theta}^{(k-1)} + \gamma_k\left(\mathcal{T}\left(\Phi\left(\mathcal{T}^{-1}\left(\hat{\theta}^{(k-1)}\right),\sigma_k\right)\right)+\eta_k\right)\right], k \geq 1,
\end{cases}
\tag{5.6}
$$

where $\mathcal{P}_{\ker(\,|_{\Omega})}\left(\,\cdot\,\right) = \mathcal{U}_{\Omega^C}\left(\left(\,\cdot\,\right)|_{\Omega^C}\right)$ is the corresponding projection operator in transform domain, which basically zeroes out all spectral coefficients defined on Ω.

The flowchart of the system (5.6) is shown in Figure 5.3. The system is initialized by setting $\hat{\theta}^{(0)} = \mathcal{U}_{\Omega}\left(y\right)$. Then, each iteration $(k \geq 1)$ comprises of the following steps:

- *Image-domain estimate filtering.* The spectrum estimate is inverted and the obtained image is filtered with the filter Φ. In this way, the coefficients from the unobserved part of the image spectrum are recreated from the given $\theta|_{\Omega} = y$. In our implementation of the algorithm the BM3D is used as the filter Φ.

- *Noise addition (excitation).* Pseudorandom noise η_k is introduced in the unobserved portion of the spectrum and works as a random generator of the missing spectral components. During the subsequent iterations, these components are attenuated or enhanced by the action of the filter Φ, depending to what extent they agree with the image features enabled by the observed spectrum $\theta|_{\Omega} = y$.

- *The updated* estimate $\hat{\theta}^{(k)}$ is obtained as the sum of the convex combination between the previous estimate $\hat{\theta}^{(k-1)}$ and the noise-excited prediction of the spectrum obtained after filtering, projected on the subspace defined by the unknown part of the spectra, and $\hat{\theta}^{(0)}$. The factor γ_k, in the convex combination, controls the rate of evolution (step size) of the algorithm.

5.3.2.1   Comments on the Algorithm

Stochastic approximation. The iterative algorithm (5.6) can be treated as the Robbins-Monro stochastic approximation procedure (see, e.g., [25]). If the step-size parameter γ_k satisfies the standard conditions

$$\gamma_k > 0, \qquad \sum \gamma_k = \infty, \qquad \sum \gamma_k^2 < \infty \qquad (5.7)$$

and some assumptions on the operator $\mathcal{T}\left(\Phi\left(\mathcal{T}^{-1}(\cdot)\right)\right)$ hold, as $k \to \infty$ the estimates $\hat{\theta}^{(k)}$ from the recursive system (5.6) converge in mean squared sense to a solution $\hat{\theta}$ of the equation

$$E\left\{\mathcal{P}_{\text{ker}(\,|_\Omega)}\left[-\hat{\theta}+\mathcal{T}\left(\Phi\left(\mathcal{T}^{-1}\left(\hat{\theta}\right),\sigma_k\right)\right)+\eta_k\right]\right\}=0,$$

i.e.

$$\mathcal{P}_{\text{ker}(\,|_\Omega)}\left[\hat{\theta}-\mathcal{T}\left(\Phi\left(\mathcal{T}^{-1}\left(\hat{\theta}\right)\right)\right)\right]=0$$

or equivalently

$$\hat{\theta}=\mathcal{U}_\Omega\left(y\right)+\mathcal{P}_{\text{ker}(\,|_\Omega)}\left[\mathcal{T}\left(\Phi\left(\mathcal{T}^{-1}\left(\hat{\theta}\right)\right)\right)\right]. \qquad (5.8)$$

If there is no smoothing in the filter Φ, the equation (5.8) becomes the identity. Thus, any $\hat{\theta}$ that satisfies observation (5.4) satisfies also to the equation (5.8), there is no image reconstruction and the algorithm does not work. Therefore, in order for the solution $\hat{\theta}$ to be nontrivial, the adaptive smoothing in (5.8) should be strong enough.

Excitation noise. The additive noise η_k used in the procedure (5.6) does not influence the equation (5.8). There are two arguments in favor of excitation of the algorithm by the random noise. First of all it improves the performance of the algorithm. It accelerates the transition process of the recursive procedure bringing it fast in the area of solution where the random walks steadies. The amplitude of these random walks decreases together with γ_k. It is well known (e.g., [21]) that the random search applied in optimization problems results in random walks well concentrated in areas of global extremum. Thus, the random search imposed by random excitation of the search trajectory can be useful for separation of local and global extrema. In a similar way, if the equation (5.8) has more than one solution, the randomness can help to find a "strong" solution with better quality of imaging or lower values of some hypothetical criterion function where the gradient (or quasi-gradient) can be defined as the vector corresponding to $\mathcal{P}_{\text{ker}(\,|_\Omega)}\left(\hat{\theta}-\mathcal{T}\left(\Phi\left(\mathcal{T}^{-1}\left(\hat{\theta}\right)\right)\right)\right)$. Further, by changing the variance of the additive noise η_k one can control the rate of evolution of the algorithm. Thus, in practice, the assumptions (5.7) can be relaxed and a fixed γ_k can be used provided that $\text{var}\{\eta_k\} \underset{k\to\infty}{\to} 0$.

Filter parameters. The parameter σ_k is used in place of the standard-deviation of the noise of the BM3D filter. This parameter controls the strength of collaborative hard-thresholding and thus affects the level of smoothing introduced by the filter. In order to prevent smearing of the small details the sequence $\{\sigma_k\}_{k=0,1,\ldots}$ should be decreasing with the progress of the iterations, and normally it is selected to be $\sigma_k^2 = \mathrm{var}\{\eta_k\}$. The last fact makes (5.6) to formally differ from the classical Robbins-Monro procedure, where the operator Φ is assumed to be fixed (thus without a second argument σ) with the overall aggressiveness of the recursion controlled instead by the step size parameter γ_k.

Stopping rule. The algorithm can be stopped when the estimates $\hat{\theta}^{(k)}$ approach numerical convergence or after a specified number of iterations.

Image estimates. An image estimate $\hat{x}^{(k)}$ can be obtained at k-th iteration as $T^{-1}\left(\hat{\theta}^{(k)}\right)$, although in practice $T^{-1}\left(\hat{\theta}^{(k)} - \mathcal{P}_{\mathrm{ker}\left(\cdot|_\Omega\right)}\left(\eta_{k-1}\right)\right)$ are better estimates because of the absence of excitation noise. All these estimates converge to $\hat{x} = T^{-1}\left(\hat{\theta}\right)$ as $k \to \infty$.

5.3.3 Experiments

The effectiveness of the proposed algorithm can be illustrated on two important inverse problems from computerized tomography: Radon inversion from sparse projections and limited-angle tomography. The former problem has been used as a benchmark for testing CS reconstruction algorithms (e.g., [5]). In particular, the results show that the presented algorithm allows us to achieve the exact reconstruction of synthetic phantom data even from a very limited number of projections. An example of image reconstruction from low-frequency data is also given. This particular example will serve as a bridge to the super-resolution problems considered in the next section.

The following experiments are carried out using a simplified form of the iterative scheme (5.6), where $\gamma_k \equiv 1$ and η_k is independent Gaussian noise with exponentially decreasing variance $\mathrm{var}\{\eta_k\} = \alpha^{-k-\beta}$. For the filter Φ, we use the block-matching and 3-D filtering algorithm (BM3D) [10], setting $\sigma_k^2 = \mathrm{var}\{\eta_k\}$. The separable 3-D Haar wavelet decomposition is adopted as the transform utilized internally by the BM3D algorithm.

FIGURE 5.4: Available portion Ω of the FFT spectrum for the three experiments: 22 radial lines, 11 radial lines, 90 degrees limited-angle with 61 radial lines.

We begin with illustrative inverse problems of compressed sensing for computerized tomography. As in [5], we simulate the Radon projections by "approximately" radial lines in the rectangular FFT domain. Note that the initial image estimate $\hat{x}^{(0)} = \underset{x:\mathcal{H}(x)=y}{\arg\min} \|x\|_2 = \mathcal{T}^{-1}\left(\underset{\theta:\theta|_\Omega=y}{\arg\min} \|\theta\|_2\right)$ coincides with the conventional back-projection estimate.

5.3.3.1 Radon Inversion from Sparse Projections

First, we reproduce exactly the same experimental setup from [5], where 22 radial lines are sampled from the FFT spectrum of the Shepp-Logan phantom (size 256×256 pixels), as shown in Figure 5.4 (left). Further, we reduce the number of available Radon projections from 22 to 11 (see Figure 5.4(center)). The initial back-projection estimates are shown in Figure 5.5. As the recursive algorithm progresses, the reconstruction error improves steadily until numerical convergence, as it can be seen from the plots in Figure 5.6. For both cases the reconstruction is *exact*, in the sense that the final reconstruction error (PSNR ~270dB) is comparable with the numerical precision of this particular implementation of the algorithm (double precision floating-point). We remark, however, that in practice such a high accuracy is never needed: already at a PSNR of about 60dB the image estimates can hardly be distinguished from the original.

5.3.3.2 Limited-Angle Tomography

In the two previous experiments, the available Radon projections were uniformly distributed with respect to the projection angle. A more difficult case arises when the angles under which the projections are taken are limited. Similarly to [27], we consider an overall aperture for the projections of 90 degrees. This restriction is essential, since all frequency information is completely missing along half of the orientations, which makes the reconstruction of, e.g., edges across these orientation extremely hard. We complicate the problem further, by taking only a smaller subset of 61 projections (a total of 256 properly-oriented projections would be required to cover a 90 degrees aperture). These sparse, limited-angle projections are illustrated in Figure 5.4 (right). Although the convergence is here much slower than in the previous two experiments, the algorithm eventually achieves exact reconstruction.

In the above three experiments, as soon as the estimate reaches a quality of about 70dB, the recursion enters a phase of improvement at a constant rate (linear in terms of PSNR since $\text{var}\{\eta_k\}$ decreases exponentially), which appears to be limited only by the used arithmetic precision.

5.3.3.3 Reconstruction from Low-Frequency Data

The proposed recursive procedure can be applied also to more conventional image-processing problems. As a prelude to the image/video super-resolution

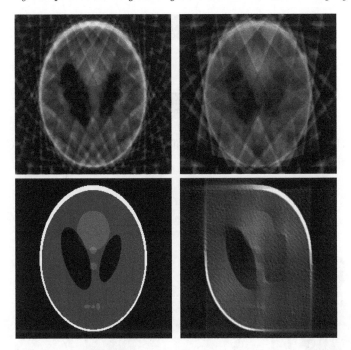

FIGURE 5.5: Clockwise from top-left: back-projection estimates for 22 radial lines, 11 radial lines, 61 radial lines with limited-angle (90 degrees), and original phantom (unknown and shown here only as a reference). For all three experiments, the estimates obtained after convergence of the algorithm coincide with the original image.

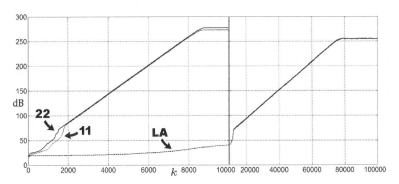

FIGURE 5.6: Progression of the PSNR (dB) of the reconstructed image estimate $\hat{x}^{(k)}$ with respect to the iteration count k for the three experiments: 22 and 11 sparse projections ("22" and "11") and limited-angle ("LA").

and upsampling algorithms, which are the subject of the next section, we

FIGURE 5.7: Available portion Ω of the FFT spectrum of the *Cameraman* image shown in Figure 5.8(left).

FIGURE 5.8: *Cameraman*: initial estimate $\hat{x}^{(0)}$ (zero-padding) (PSNR= 27.32dB), and reconstructed estimate $\hat{x}^{(62)}$ after 62 iterations (PSNR= 29.10dB).

present here a basic experiment that can be seen as a prototype of image up-sampling. In particular, we consider the reconstruction of a nonsynthetic test image, namely *Cameraman* (256×256 pixels), from the low-frequency portion of its Fourier spectrum, with the set Ω being a 128×128 square centered at the DC, as illustrated in Figure 5.7. In Figure 5.8 we show the initial estimate $\hat{x}^{(0)}$ (by zero-padding in FFT domain, thus minimum ℓ_2-norm) and the reconstructed image obtained after few iterations of the algorithm. Despite the reconstruction is not exact, the salient details of the image are properly restored and there are no significant artifacts (e.g., ringing) thanks to the adaptivity embedded in the BM3D filter.

5.4   Super-Resolution

Image *upsampling* or *zooming*, can be defined as the process of resampling a *single* low-resolution (LR) image on a high-resolution (HR) grid. Different resampling methods can be used to obtain zoomed images with specific desired properties, such as edge preservation, degree of smoothness, etc. In particular, we concluded the previous section with an example demonstrating how one can achieve image upsampling by means of iterative BM3D filtering. As a matter of fact, because of the Nyquist-Shannon theorem, the inverse Fourier transform of the central low-frequency portion of the spectrum, can be treated as the LR image obtained upon decimation of the low-passed HR image.

However, fine details missing or distorted in the low-resolution image cannot be reconstructed in the upsampled one. This is particularly evident when the downsampling ratio between the HR and LR images is high (in the above example this ratio was only 2). Roughly speaking, there is no sufficient information in the low-resolution image to do this. The situation changes when a number of LR images portraying slightly different views of the same scene are available. The reconstruction algorithm now can try to improve the spatial resolution by incorporating into the final HR result the additional new details revealed in each LR image. The process of combining a sequence of undersampled and degraded low-resolution images in order to produce a single high-resolution image is commonly referred to as *super-resolution (SR) image reconstruction*, or, simply, *image super-resolution*.

The standard formulation of the image SR problem assumes that the LR images are obtained through the observations given by the model

$$x_{\mathrm{low}r} = \mathcal{D}\left(\mathcal{B}\left(\mathcal{F}_r\left(x_{\mathrm{hi}}\right)\right)\right), \quad r = 1, \ldots, R, \tag{5.9}$$

where $x_{\mathrm{low}r}$ are LR images, $\mathcal{F}_r$, $\mathcal{B}$, $\mathcal{D}$ are the linear operators representing respectively warp, blur and decimaton, and x_{hi} is a high-resolution image of the scene subject to reconstruction. Blur and decimation operators are considered to be known.

Besides the difficulty involved in estimating the warping parameters, a principal drawback of the SR formulation (5.9) is that it assumes that all LR images $x_{\mathrm{low}r}$ can be mapped and upsampled to a unique HR image x_{hi}.

The observation model (5.9) then can be extended to the more general form

$$x_{\mathrm{low}r} = \mathcal{D}\left(\mathcal{B}\left(x_{\mathrm{hi}r}\right)\right), \quad r = 1, \ldots, R, \tag{5.10}$$

where the warp operators on a single HR image are replaced by a sequence of HR images. This allows consideration of arbitrary types of deformation between the frames, such as relative motion between different objects in the scene, and occlusions. The reconstruction of the HR sequence $\{x_{\mathrm{hi}r}\}_{r=1}^{R}$ from the LR sequence $\{x_{\mathrm{low}r}\}_{r=1}^{R}$ is termed *video super-resolution*.

The classical SR approach is loosely based on the following three steps:

(1) registration of the LR images to a HR coordinate grid, (2) warping of the LR images onto that grid by interpolation, and (3) fusion of the warped images into the final HR image. An additional deblurring step is sometimes considered to compensate the blur. Several algorithms based on such classical approach exist and detailed reviews can be found, e.g., in [18].

For successful reconstruction it is crucial to perform accurate registration between the features represented across different frames. Most of the existing SR methods rely either on a parametric global motion estimation, or on a computationally intensive optical flow calculation. However, an explicit registration of the LR frames is often not feasible: on the one hand, if the registration map has few degrees of freedom, it is too rigid to model the geometrical distortions caused by the lens system; on the other hand, when many degrees of freedom are available (e.g., a dense optical flow), reliable estimates of the registration parameters cannot be obtained. In either case, registration artifacts are likely to appear in the fusion, requiring heavy regularization (smoothing) for their concealment [18]. The situation becomes even more difficult when nonglobal motion is present in images, something that is typical of video SR. Modern SR methods depart from the classical approach and we specially mention the video SR reconstruction algorithms [16], [29], [28], [30] based on the nonlocal means (NLM) filtering paradigm [2]. In these algorithms, instead of trying to obtain an explicit registration as a one-to-one pixel mapping between frames, a one-to-many mapping is utilized, where multiple pixels can be assigned to a given one, with weights typically defined by the similarity of the patches/blocks surrounding the pixels. The HR image is estimated through a weighted average of these multiple pixels (or of their surrounding patches) with their corresponding weights. The increased redundancy of the NLM, which can exploit also multiple patches from a same frame, contributes significantly to the overall good performance of the methods [29].

Here, based on our previous works [12],[13] we present a unified algorithm for the upsampling and image/video SR based on iterative (V)-BM3D filtering. As discussed in Section 5.2.2, the (V)-BM3D algorithm shares with the NLM the idea of exploiting nonlocal similarity between blocks. However, thanks to its transform-domain modeling, the BM3D turns out to be a much more effective filter than the NLM, thus leading to outstanding SR results.

The remainder of this section is organized as follows. First, we show how the SR problem can be interpreted in terms of the spectral decomposition (5.3). Then we present an adaptation of the algorithm (5.5) to video SR reconstruction, with image upsampling treated as a particular case. Finally, we report experimental results demonstrating the effectiveness and superior performance of the proposed approach.

5.4.1 Spectral Decomposition for the Super-Resolution Problem

The observation model (5.10) for the video SR problem is a particular case of the general observation model (5.1) where X is the space of high-resolution images, Y that of the low-resolution frames, and

$$\mathcal{H}(\cdot) = \mathcal{D}(\mathcal{B}(\cdot)). \tag{5.11}$$

Let us show how the blur and decimation operator relate to the general spectral decomposition presented in Section 5.2.1.

The blur in (5.10) can be written as the integral

$$\mathcal{B}(x_{\mathrm{hir}})(s) = \int x_{\mathrm{hir}}(\xi)\, b(s, \xi - s)\, d\xi,$$

where b is a spatially varying point-spread function (PSF). Further, it can be represented as the inner products

$$\mathcal{B}(x_{\mathrm{hir}})(s) = \langle x_{\mathrm{hir}}, b_s \rangle, \tag{5.12}$$

where $b_s(\xi) = b(s, \xi - s)$. Now, the action of the decimation operator $\mathcal{D}$ in (5.10) can be seen as a retaining only a subset of inner products (5.12) corresponding to the given sampling points s.

Let us assume that the corresponding retained b_s constitute a basis $\{t_i\}_{i=1}^{\dim Y}$ for a linear subspace $\tilde{X} \subset X$ of dimension equal to the number of pixels of any of the LR frames x_{lowr}. For the sake of simplicity, we assume this basis to be orthonormal up to a scaling constant β. Hence, $\tilde{X}$ is the orthogonal complement of $\ker(\mathcal{H})$ in X.

The core of our modeling is to complete the basis $\{t_i\}_{i=1}^{\dim Y}$ with an orthonormal basis $\{t_i\}_{i=\dim Y+1}^{\dim X}$ of $\ker(\mathcal{H})$, thus obtaining a basis $\{t_i\}_{i=1}^{\dim X}$ for X. This constitutes an orthonormal transform $\mathcal{T} : X \to \Theta$ whose basis elements separate $\ker(\mathcal{H})$ from its orthogonal complement. By such a construction, the subset of basis indices $i = 1, \ldots, \dim Y$ correspond to the set Ω, $\Theta_\Omega = Y$. Hence, $\mathcal{T}_0$ is identity and $\mathcal{S}$ operates the division by β and restriction on Ω.

An evident example of this construction is when $\mathcal{B}$ is the blurring with a uniform kernel of size $n \times n$ and $\mathcal{D} = \downarrow_n$ is a decimation with rate n (along the rows as well as along the columns): the PSFs b_s, i.e., the basis elements $\{t_i\}_{i=1}^{\dim Y}$, are thus nonoverlapping (hence orthogonal) shifted copies of the same uniform kernel. This basis can be naturally seen as the subbasis of the $n \times n$ block-DCT composed by extracting the DC-basis elements of all blocks. In this sense, we can complete $\{t_i\}_{i=1}^{\dim Y}$ with the basis $\{t_i\}_{i=\dim Y+1}^{\dim X}$ composed by the AC-basis elements of all blocks.

However, we are not bound to use the transform $\mathcal{T}$ constructed from the PSFs in the above direct way. As observed at the end of Section 5.2.1, the uniqueness of the basis elements $\{t_i\}_{i \in \Omega}$ depends on the nonzero eigenvalues being distinct, while here all nonzero eigenvalues equal β^{-1}. In this case, any orthonormal transform that provides the same orthogonal separation between ker $(\mathcal{H})$ and its complementary can be used in the decomposition. Indeed, in the example in Section 5.3.3.3 we have used the FFT as the transform $\mathcal{T}$ while the blur PSFs were sinc functions. Of course, if the used $\mathcal{T}$ differs from that constructed directly from the PSFs, also the transform $\mathcal{T}_0$ must differ from the identity. Therefore, in all the following equations, $\mathcal{T}_0$ is always written explicitly, and (5.11) takes thus the form

$$\mathcal{H}\left(\,\cdot\,\right) = \mathcal{T}_0^{-1}\left(\beta^{-1}\,\mathcal{T}\left(\,\cdot\,\right)|_\Omega\right). \tag{5.13}$$

Before we proceed further, let us remark that the assumption on the orthonormality (up to scaling constant) of the basis $\{t_i\}_{i=1}^{\dim Y}$ for $\tilde{X}$ induced by the blur and decimation operators is mainly for simplicity of exposition. If $\{t_i\}_{i=1}^{\dim Y}$ were not orthonormal, the construction would have been similar, differing from (5.13) either by having a frame and its dual instead of the orthonormal transform $\mathcal{T}$, or, in accordance with the general spectral decomposition (5.3), by substituting the factor β^{-1} by a general diagonal operator (i.e., multiplying each transform coefficient by its own scaling factor).

5.4.2 Observation Model

Using the above representation of $\mathcal{H}$ (5.13), we reformulate the super-resolution observation model (5.10) as follows.

Let us be given a sequence of $R \geq 1$ *low-resolution images* $\{x_{\text{low}r}\}_{r=1}^R$ of size $n_0^{\text{h}} \times n_0^{\text{v}}$, with each $x_{\text{low}r}$ being obtained from the subband of the corresponding $\mathcal{T}$-spectra of original *higher-resolution* images $\{x_{\text{hir}}\}_{r=1}^R$ of size $n^{\text{h}} \times n^{\text{v}}$ as

$$y_r = x_{\text{low}r} = \mathcal{T}_0^{-1}\left(\beta^{-1}\,\mathcal{T}\left(x_{\text{hir}}\right)|_\Omega\right). \tag{5.14}$$

The problem is to reconstruct $\{x_{\text{hir}}\}_{r=1}^R$ from $\{x_{\text{low}r}\}_{r=1}^R$. Clearly, for a fixed r, any good estimate $\hat{x}_r$ of x_{hir} must have its Ω subband equal to $\beta\mathcal{T}_0\left(x_{\text{low}r}\right) = \mathcal{T}\left(x_{\text{hir}}\right)|_\Omega$. Under this restriction, the estimates constitute an affine subspace $\hat{X}_r$ of codimension $n_0^{\text{h}}n_0^{\text{v}}$ in the $n^{\text{h}}n^{\text{v}}$-dimensional linear space X: $\hat{X}_r = \{\hat{x}_r \in X : \mathcal{T}\left(\hat{x}_r\right)|_\Omega = \beta\mathcal{T}_0\left(x_{\text{low}r}\right)\}$.

For $R = 1$, the observation model (5.14) corresponds to the image upsampling problem. Whenever $R > 1$, we are instead in the image or video super-resolution setting.

5.4.3 Scaling Family of Transforms

The observation model (5.14) uses a pair of transforms $\mathcal{T}$ and $\mathcal{T}_0$ for relating the same image at two different resolutions. We can extend this to an arbitrary number of intermediate resolutions by means of the following scaling family of transforms.

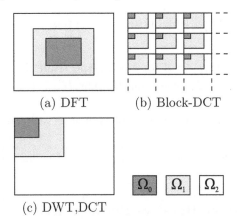

(a) DFT   (b) Block-DCT

(c) DWT,DCT

FIGURE 5.9: Nested support subsets: the subsets Ω_0 and Ω_1 are shown as the shaded subareas of the support $\Omega_M = \Omega_2$ of coefficients of the transform $\mathcal{T}_M$.

Let $\{\mathcal{T}_m\}_{m=0}^M$ be a family of orthonormal transforms of increasing sizes $n_m^{\mathrm{h}} \times n_m^{\mathrm{v}}$, $n_m^{\mathrm{h}} < n_{m+1}^{\mathrm{h}}$, $n_m^{\mathrm{v}} < n_{m+1}^{\mathrm{v}}$, such that for any pair m, m' with $m < m'$, up to a scaling factor $\beta_{m,m'}$, the whole $\mathcal{T}_m$-spectrum can be considered as a smaller portion of the $\mathcal{T}_{m'}$-spectrum. In particular, this means that the supports Ω_m of the $\mathcal{T}_m$-transform coefficients form a *nested sequence* of subsets (subbands) of Ω_M, i.e. $\Omega_0 \subset \cdots \subset \Omega_M$. The most notable examples of such $\{\mathcal{T}_m\}_{m=0}^M$ families are discrete cosine (DCT) or Fourier (DFT) transforms of different sizes, discrete wavelet transforms (DWT) associated to one common scaling function, as well as block DCT, DFT and DWT transforms. Figure 5.9 illustrates the nested sequence of supports for these families.

Depending on the specific transform family of choice, the sets Ω_m are commonly referred to as *lower-resolution, low-frequency,* or *coarser-scale subbands* of the $\mathcal{T}_M$-spectrum.

To use such scaling family for the observation model (5.14), we set $n_M^{\mathrm{h}} = n^{\mathrm{h}}$, $n_M^{\mathrm{v}} = n^{\mathrm{v}}$, $\mathcal{T}_M = \mathcal{T}$, $\Omega_0 = \Omega$, $\beta_{0,M} = \beta$, and for $m < m'$ we define the following three operators:

- the *restriction operator* $|_{\Omega_{m,m'}}$ that, from a given $\mathcal{T}_{m'}$-spectrum, extracts its smaller portion defined on Ω_m, which can be thus considered as the $\mathcal{T}_m$-spectrum of a smaller image;

- the *zero-padding operator* $\mathcal{U}_{m,m'}$ that expands a $\mathcal{T}_m$-spectrum defined on Ω_m to the $\mathcal{T}_{m'}$-spectrum defined on the superset $\Omega_{m'} \supset \Omega_m$ by introducing zeros in the complementary $\Omega_{m'} \setminus \Omega_m$;

- the *projection operator* $\mathcal{P}_{m,m'}$ that zeroes out all coefficients of $\mathcal{T}_{m'}$-spectrum defined on Ω_m. If $m = 0$ and $m' = M$, the operator $\mathcal{P}_{m,m'} =$

$\mathcal{P}_{0,M}$ coincides with the projection operator $\mathcal{P}_{\ker(\mathcal{H})}$ defined in Section 5.2.

Note that $\mathcal{U}_{m,m'}(A)|_{\Omega_m} = A$ for any $\mathcal{T}_m$-spectrum A, and $B = \mathcal{P}_{m,m'}(B) + \mathcal{U}_{m,m'}(B|_{\Omega_m})$ for any $\mathcal{T}_{m'}$-spectrum B. Thus, $\mathcal{U}_{m,m'}$ can be regarded as "dual" operator of $|_{\Omega_{m,m'}}$.

With this notation, the super-resolution observation model (5.14) becomes

$$y_r = x_{\text{lowr}} = \mathcal{T}_0^{-1}\left(\beta_{0,M}^{-1}\,\mathcal{T}_M\left(x_{\text{hir}}\right)|_{\Omega_{0,M}}\right). \qquad (5.15)$$

5.4.4   Multistage Iterative Reconstruction

In the iterative algorithm presented for the general CS reconstruction, at each iteration the noise excitation and the adaptive filtering are used to provide estimates for the whole spectrum. The algorithm presented below is different, in that it exploits the multiscale feature of the images and performs the scaling from $n_0^{\text{h}} \times n_0^{\text{v}}$ size to $n_M^{\text{h}} \times n_M^{\text{v}}$ gradually, using the transform family $\{\mathcal{T}_m\}_{m=0}^{M}$ across M *stages*, which are indicated using the subscript $m = 1, \ldots, M$.

The complete algorithm is defined by the iterative system

$$\begin{cases} \hat{x}_{r,0} = x_{\text{lowr}} \\ \hat{x}_{r,m} = \hat{x}_{r,m}^{(k_{\text{final}}(m))} \\ \hat{x}_{r,m}^{(0)} = \mathcal{T}_m^{-1}\left(\mathcal{U}_{m-1,m}\left(\beta_{m-1,m}\mathcal{T}_{m-1}\left(\hat{x}_{r,m-1}\right)\right)\right) \\ \hat{x}_{r,m}^{(k)} = \mathcal{T}_m^{-1}\left(\mathcal{U}_{0,m}\left(\beta_{0,m}\mathcal{T}_0\left(x_{\text{lowr}}\right)\right) + \mathcal{P}_{0,m}\left(\mathcal{T}_m\left(\Phi\left(r, \left\{\hat{x}_{r,m}^{(k-1)}\right\}_{r=1}^{R}, \sigma_{k,m}\right)\right)\right)\right). \end{cases}$$
$$(5.16)$$

At each stage, the images are being super-resolved from size $n_{m-1}^{\text{h}} \times n_{m-1}^{\text{v}}$ to $n_m^{\text{h}} \times n_m^{\text{v}}$. The sequence $\{x_{\text{lowr}}\}_{r=1}^{R}$ serves as input for the first stage, and the output of the current stage $\{\hat{x}_{r,m}\}_{r=1}^{R}$ becomes an input for the next one. At each stage, the initial estimate $\hat{x}_{r,m}^{(0)}$ is obtained from $\hat{x}_{r,m-1}$ by zero-padding its spectra following the third equation in (5.16). During the subsequent iterations, the estimates are obtained according to the last equation in (5.16), where the superscript $k = 0, 1, 2, \ldots$ corresponds to the iteration count inside each stage, $\hat{x}_{r,m}^{(k)}$ is a sequence of estimates for $\hat{x}_{r,m}$, Φ is a spatially adaptive filter and $\sigma_{k,m}$ is a parameter controlling the strength of this filter. In other words, at each iteration we jointly filter the images $\left\{\hat{x}_{r,m}^{(k-1)}\right\}_{r=1}^{R}$ obtained from the previous iteration, perform a transform $\mathcal{T}_m$ for each r, substitute the $n_0^{\text{h}} \times n_0^{\text{v}}$ coefficients defined on Ω_0 with $\beta_{0,m}\mathcal{T}_0\left(x_{\text{lowr}}\right)$, and take an inverse transform $\mathcal{T}_m^{-1}$ to obtain $\hat{x}_{r,m}^{(k)}$. The flowchart of the system (5.16) is presented in Figure 5.10. The iteration process stops at iteration $k_{\text{final}}(m)$ when the distance between $\left\{\hat{x}_{r,m}^{(k)}\right\}$ and $\left\{\hat{x}_{r,m}^{(k-1)}\right\}$ in some metric becomes less than a certain threshold δ_0, or if the maximum number of iterations $k_{\text{max}}(m)$ is reached.

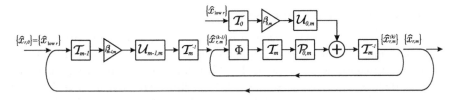

FIGURE 5.10: Multistage iterative reconstruction. The inner loop corresponds to an iteration process inside a stage; the outer loop corresponds to a transition to the next stage.

In order to prevent smearing of the small details, the sequence $\{\sigma_{k,m}\}_{k=0,1,\ldots}$ should be decreasing with the progress of the iterations.

There are a number of reasons in favor of the multistage reconstruction. First, since at every stage the complexity of each iteration depends on the size $n_m^h \times n_m^v$ of the image to reconstruct, we have that through multistage reconstruction we are able to decrease the computational cost of recovering the coarsest portions of the missing spectrum. Second, the recovery of the finest details becomes more stable, because when we arrive towards the final stages, most of the spectrum has already been reconstructed, with good approximation, in the earlier stages. It turns out that, in this way, the benefit of the stochastic excitation is also reduced, inasmuch as we do not include this element in our implementation of the SR reconstruction algorithm (1.12).

5.4.5   Experiments

We present results from three sets of experiments. First, using a synthetic image sequence, we assess how well our SR algorithm can deal with highly aliased data, provided that the set of LR images covers the whole HR sampling grid. Second, we consider video SR reconstruction. Finally, we demonstrate examples of image upsampling showing that successful reconstruction can be obtained even in the case when the exact blurring model is unknown.

As seen in the previous section, the algorithm formulations for upsampling and super-resolution coincide. In both cases, the algorithm performs a reconstruction for each image of the input sequence, and the output sequence always contains the same number of frames as the input one. Whether the algorithm performs upsampling or SR reconstruction, depends on the number of frames in which V-BM3D is allowed to search for similar blocks (the so-called "temporal search window"). When the search is restricted to the current frame only, the algorithm independently *upsamples* each frame of the input sequence.

5.4.5.1   Implementation Details

The filter's internal 3-D transform is implemented as a composition of a 2-D discrete sine transform (DST) applied to each grouped block and of a 1-D Haar wavelet transform applied along the third dimension of the group. The block size is decreasing with the stages, while within each stage $\sigma_{k,m}$ is decreasing linearly with respect to k. In terms both of smoothing and scale, this consistent with the "coarse-to-fine" approach analyzed in [3]. The temporal search window is set to be equal to the number of frames in the input sequence.

The block-matching in BM3D is implemented as a smallest l_2-difference search. It has been found that strong aliasing in the LR images can noticeably impair the block matching. To overcome this problem, during the first few iterations of the reconstruction algorithm, the block matching is computed over smoothed versions of the current image estimates. Furthermore, the BM3D-based α-rooting described in [11] can be embedded inside V-BM3D to enhance the high-frequency components during the last iterations. These simple adjustments effectively improve the overall numerical and visual quality of reconstruction, leading to sensibly better results than those reported in [13],[12].

In order to avoid the influence of border distortions and provide more fair comparisons, in all experiments the PSNR values are calculated over the central part of the images, omitting a border of 15-pixel width.

5.4.5.2   Super-Resolution

In our super-resolution experiments, we used the same four sequences as in [29] namely: *Text, Foreman, Suzie* and *Miss America*. Both LR and ground truth HR sequences are available at the Web site of the first author of [29] http://www.cs.technion.ac.il/~matanpr/NLM-SR/. All LR sequences were obtained using the observation model where the HR images are first blurred using a 3×3 uniform kernel, then decimated by factor 3 and contaminated with additive Gaussian noise with standard-deviation equal to 2. Since our observation model does not assume the presence of noise, we prefilter the noisy LR input sequence with the standard V-BM3D filter using default parameters [9].

A scaling family of transforms can be easily associated with the described observation model, noticing that the LR images can be treated (up to a scaling factor $\beta_{0,M} = 3$) to be composed of DC coefficients of some orthogonal 3×3 block transform. The transform family $\{\mathcal{T}_m\}_{m=0}^M$ has been chosen to consist of three 2-D block transforms with 1×1, 2×2, and 3×3 block sizes, which results in a progressive enlargement of 2 and 1.5 times, providing an overall enlargement of 3 times. As a particular family of transforms satisfying the above conditions, we choose the block DCT transforms.

Image super-resolution. For the image super-resolution experiment the *Text* sequence is used. The 9-image LR sequence has been obtained from a single ground-truth image (shown in Figure 5.11 (left)) by displacing it before

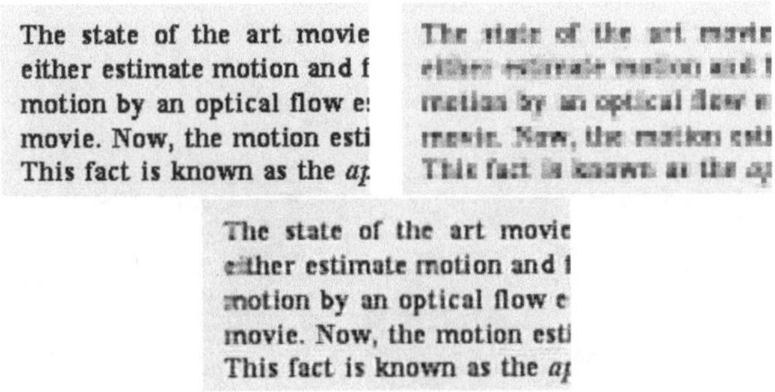

The state of the art movie
either estimate motion and f
motion by an optical flow e:
movie. Now, the motion esti
This fact is known as the a_f

The state of the art movie
either estimate motion and f
motion by an optical flow e:
movie. Now, the motion esti
This fact is known as the a_f

The state of the art movie
either estimate motion and f
motion by an optical flow e
movie. Now, the motion esti
This fact is known as the a_f

FIGURE 5.11: Super-resolution result for the *Text* image. From left to right: original high-resolution image (ground truth); pixel-replicated low-resolution image; image super-resolved by the proposed algorithm.

applying observation model described above. The displacements are chosen so that the entire HR grid is covered by the union of the LR grids (i.e., $d^h, d^v = 0, 1, 2,$). One of these nine LR images is shown in Figure 5.11 (center), while a super-resolved image obtained by the proposed algorithm is presented in Figure 5.11. We can see that despite the strong aliasing in the LR images, the algorithm succeeds in reconstructing a readable text.

Video super-resolution. We performed SR of the video sequences *Foreman*, *Suzie*, and *Miss America* mentioned above. For a comparison, we also present results obtained by the method [29].

The mean (over all 30 frames) PSNR values for the reconstructed sequences are summarized in Table 5.1. The numerical results obtained by our algorithm are superior to those of [29]. A visual comparison is provided in Figures 5.12 – 5.14. We can observe that while both methods provide roughly the same amount of reconstructed image details, in terms of artifacts, the results of the proposed method are much cleaner.

5.4.5.3 Image Upsampling

Let us present images upsampled of factors $q = 4$ or $q = 8$ from their original resolution. It should be emphasized that in this case we do not know which

	Nearest neighbor	[29]	Proposed
Foreman	29.0	32.9	**35.0**
Suzie	30.3	33.0	**34.2**
Miss America	32.0	34.7	**37.0**

TABLE 5.1
Mean (over all frames) PSNR (dB) values of the super-resolved video sequences (see Section 5.4.5.2).

FIGURE 5.12: Results for the 23rd frame from the *Foreman* sequence. From left to right and from top to bottom: pixel-replicated low-resolution image; original image (ground truth); super-resolved by the algorithm [29]; super-resolved by the proposed algorithm and their respective zoomed fragments.

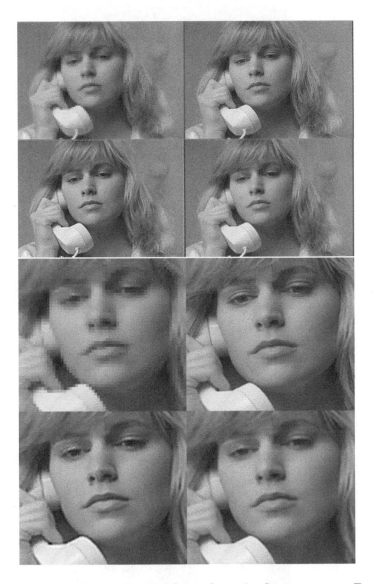

FIGURE 5.13: Results for the 23rd frame from the *Suzie* sequence. From left to right and from top to bottom: pixel-replicated low-resolution image; original image (ground truth); super-resolved by the algorithm [29]; super-resolved by the proposed algorithm and their respective zoomed fragments.

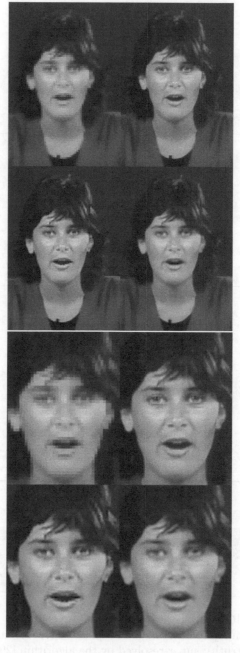

FIGURE 5.14: Results for the 23rd frame from the *Miss America* sequence. From left to right and from top to bottom: pixel-replicated low-resolution image; original image (ground truth); super-resolved by the algorithm [29]; super-resolved by the proposed algorithm and their respective zoomed fragments.

FIGURE 5.15: Fragments of the *Cameraman*, *Text*, and *Lighthouse* images.

FIGURE 5.16: Upsampling of the three fragments shown in Figure 5.15. From top to bottom: *Cameraman* (4 times), *Text* (4 times), and *Lighthouse* (8 times).

blurring and decimation operators have been used to obtain the given images. Instead, we assume that the blurring kernel is the low-pass analysis filter of a wavelet transform. Hence, we seek a high-resolution image whose wavelet approximation coefficients in the LL subband of the $\log_2(q)$-level decomposition coincide (up to a scaling factor $\beta_{0,M} = q$) to the pixel values of the given low-resolution image.

Figure 5.15 shows three fragments of the *Cameraman*, *Text*, and *Lighthouse* images at their original resolution. We upsample these fragments applying the $\log_2(q)$-stage algorithm with the Symlet-8 wavelet transform. The obtained high-resolution images are shown in Figure 5.16. It is interesting to notice that the results are quite reasonable, despite our fictitious assumptions about the blurring and decimation operators. The visual quality is particularly good, due to the sharp edges and because of the virtual absence of ringing artifacts typical of transform-domain upsampling.

5.5 Conclusions

In this chapter we discussed the application of spatially adaptive filters as regularization constraint in inverse imaging problems. Using BM3D as the leading example of such filters, we demonstrated that even simple iterative schemes, when coupled with a good filter, can be turned into powerful and competitive reconstruction methods.

Overall, in the context of compressing sensing, our method introduces a new and alternative view on the reconstruction strategy from undersampled data. In super-resolution applications, the algorithm proposed in Section 5.4 stands in line with the best super-resolution algorithms, possessing registration-free reconstruction, and showing a state-of-the-art performance.

From a general perspective, the presented material expands the breadth of filtering in the modern image processing.

Bibliography

[1] R.G. Baraniuk. Compressive sensing [lecture notes]. *Signal Processing Magazine, IEEE*, 24(4):118–121, July 2007.

[2] A. Buades, B. Coll, and J. M. Morel. A review of image denoising algorithms, with a new one. *Multiscale Modeling and Simulation*, 4(2):490–530, 2005.

[3] A. Buades, B. Coll, J.-M. Morel, and C. Sbert. Self-similarity driven color demosaicking. *Image Processing, IEEE Transactions on*, 18(6):1192–1202, June 2009.

[4] A. Buades, B. Coll, and J.M. Morel. Nonlocal image and movie denoising. *Int. J. Computer Vision*, 76(2):123–139, February 2008.

[5] E. J. Candes and J. Romberg. Practical signal recovery from random projections. In *Wavelet Applications in Signal and Image Processing XI, Proc. SPIE Conf.*, volume 5914, 2005.

[6] E. J. Candes and J. Romberg. Sparsity and incoherence in compressive sampling. *Inverse Problems*, 23:969–985, 2007.

[7] E.J. Candes, J. Romberg, and T. Tao. Robust uncertainty principles: exact signal reconstruction from highly incomplete frequency information. *Information Theory, IEEE Transactions on*, 52(2):489–509, Feb. 2006.

[8] J.B. Conway. *A course in functional analysis.* Springer, 1990.

[9] K. Dabov, A. Foi, and K. Egiazarian. Video denoising by sparse 3D transform-domain collaborative filtering. In *Proc. European Signal Process. Conf.*, Poznan, Poland, Sep. 2007.

[10] K. Dabov, A. Foi, V. Katkovnik, and K. Egiazarian. Image denoising by sparse 3D transform-domain collaborative filtering. *IEEE Trans. Image Process.*, 16(8):2080–2095, Aug. 2007.

[11] K. Dabov, A. Foi, V. Katkovnik, and K. Egiazarian. Joint image sharpening and denoising by 3d transform-domain collaborative filtering. In *Proc. 2007 Int. TICSP Workshop Spectral Meth. Multirate Signal Process.*, Moscow, Russia, Sep. 2007.

[12] A. Danielyan, A. Foi, V. Katkovnik, and K. Egiazarian. Image and video super-resolution via spatially adaptive block-matching filtering. In *Proc. Int. Workshop Local Non-Local Approx. Image Process.*, Lausanne, Switzerland, Aug. 2008.

[13] A. Danielyan, A. Foi, V. Katkovnik, and K. Egiazarian. Image upsampling via spatially adaptive block-matching filtering. In *Proc. European Signal Process. Conf.*, Lausanne, Switzerland, Aug. 2008.

[14] D. L. Donoho and M. Elad. Maximal sparsity representation via l_1 minimization. *Proc. Nat. Aca. Sci.*, 100:2197–2202, 2003.

[15] D.L. Donoho. Compressed sensing. *Information Theory, IEEE Transactions on*, 52(4):1289–1306, Apr. 2006.

[16] M. Ebrahimi and E. R. Vrscay. Multi-frame super-resolution with no explicit motion estimation. In *Proc. Computer Vision and Pattern Recognition*, pages 455–459, Las Vegas, Nevada, USA, July 2008.

[17] K. Egiazarian, A. Foi, and V. Katkovnik. Compressed sensing image reconstruction via recursive spatially adaptive filtering. In *Proc. IEEE Int. Conf. on Image Process., ICIP 2007*, pages 549–552, Sept. 2007.

[18] S. Farsiu, D. Robinson, M. Elad, and P. Milanfar. Advances and challenges in super-resolution. *International Journal of Imaging Systems and Technology*, 14:47–57, 2004.

[19] R. Gribonval and M. Nielsen. Sparse representation in unions of bases. *Information Theory, IEEE Transactions on*, 49(12):3320–3325, 2003.

[20] V. Katkovnik, A. Foi, K. Egiazarian, and J. Astola. From local kernel to nonlocal multiple-model image denoising. *International Journal of Computer Vision*, 86(1):1–32, Jan. 2010.

[21] V. Katkovnik and O. Yu. Kulchitskii. Parametric statistical estimates for multiextremum optimization problems. *Engineering Cybernetics*, (6), 1975.

[22] C. Kervrann and J. Boulanger. Local adaptivity to variable smoothness for exemplar-based image denoising and representation. Technical Report 5624, INRIA, July 2005.

[23] C. Kervrann and J. Boulanger. Unsupervised patch-based image regularization and representation. In *Proc. Eur. Conf. Computer Vision (ECCV'06)*, volume 4, pages 555–567, 2006.

[24] C. Kervrann and J. Boulanger. Local adaptivity to variable smoothness for exemplar-based image regularization and representation. *Int. J. Comput. Vision*, 79(1):45–69, 2008.

[25] H. Kushner and G. G. Yin. *Stochastic approximation and recursive algorithms and applications, 2nd ed.* Springer, 2003.

[26] S. Lansel. DenoiseLab. http://www.stanford.edu/~slansel/DenoiseLab.

[27] T. Olson. Stabilized inversion for limited angle tomography. *Engineering in Medicine and Biology Magazine, IEEE*, 14(5):612–620, Sep/Oct 1995.

[28] M. Protter and M. Elad. Super resolution with probabilistic motion estimation. *Image Processing, IEEE Transactions on*, 18(8):1899–1904, Aug. 2009.

[29] M. Protter, M. Elad, H. Takeda, and P. Milanfar. Generalizing the nonlocal-means to super-resolution reconstruction. *Image Processing, IEEE Transactions on*, 18(1):36–51, Jan. 2009.

[30] H. Takeda, P. Milanfar, M. Protter, and M. Elad. Super-resolution without explicit subpixel motion estimation. *Image Processing, IEEE Transactions on*, 18(9):1958–1975, Sept. 2009.

[31] J.A. Tropp. Just relax: convex programming methods for identifying sparse signals in noise. *Information Theory, IEEE Transactions on,* 52(3):1030–1051, March 2006.

[32] J.A. Tropp and A.C. Gilbert. Signal recovery from random measurements via orthogonal matching pursuit. *Information Theory, IEEE Transactions on,* 53(12):4655–4666, Dec. 2007.

[33] Y. Tsaig and D. L. Donoho. Extensions of compressed sensing. *Signal Process.,* 86(3):549–571, March 2006.

[34] M.J. Wainwright. Sharp thresholds for high-dimensional and noisy sparsity recovery using ell_1 -constrained quadratic programming (lasso). *Information Theory, IEEE Transactions on,* 55(5):2183–2202, May 2009.

6

Registration for Super-Resolution: Theory, Algorithms, and Applications in Image and Mobile Video Enhancement

Patrick Vandewalle

Philips Research

Luciano Sbaiz

Google

Martin Vetterli

Ecole Polytechnique Fédérale de Lausanne

CONTENTS

The relation between sensor resolution and the optics of a digital camera is determined by the Nyquist sampling theorem: the sampling frequency should be larger than twice the maximum frequency of the image content coming out of the optical system. If a lower resolution is used, the output is aliased. Aliasing in digital images is often considered as a nuisance and (both optical and digital) filters are designed to avoid aliasing in digital cameras. However, aliasing also contains extra high-frequency information with additional details about the scene. Super-resolution algorithms extract the information present in the aliasing to reconstruct a higher-resolution image.

Super-resolution algorithms typically combine multiple aliased images with small relative motion, and create a single high-resolution image. The input can be a set of pictures taken with a digital camera from approximately the same point of view. An application could be to use a low-resolution camera (with a good optical system), capture a set of images while holding the camera manually in approximately the same position, and use the small movements of the camera to reconstruct a high-resolution image. This would allow to take multiple images with a cheap camera, and combine them to a higher resolution image as if it had been taken with a more expensive camera. Other applications can be found for example in situations where a camera sensor cannot be easily replaced, such as in satellites. It is (almost) impossible to install a new camera sensor, while a modification of the software enables a series of images of approximately the same subject to be taken.

The set of images used in a super-resolution algorithm can also be (part of) a video sequence, where the motion between subsequent frames is typically small. An application can be found in upscaling of low resolution videos, such as those acquired with handheld devices. These devices are typically able to acquire videos with low-resolution (such as CIF, i.e., 288×352 pixels, or QCIF, i.e., 144×176 pixels) and the videos are coded at relatively low bit rates (such as 128 Kb/s). Despite the low quality of these videos, we will show later that super-resolution algorithms can be applied, under certain hypotheses, to increase the resolution and obtain additionally a significant reduction of coding artifacts. The procedure to apply a super-resolution algorithm to a video sequence is represented in Figure 6.1. The input frames are combined in groups of N consecutive frames. One of the frames, for example I_0, is estimated at higher resolution to produce the output frame I_0'. The result is added to the output sequence and the procedure is repeated taking the next input frame as a reference. A simple way to manage the input frames is to use a circular buffer containing the N most recent input frames. A more flexible approach, not considered here, is to use a buffer of variable size. This would allow processing sequences with varying speed and scene changes.

Super-resolution has been a very active research topic over the past few decades. In 1984, Tsai and Huang introduced a first super-resolution algorithm to reconstruct a high-resolution image from multiple shifted low-resolution images using a frequency domain approach [18]. A good overview of existing super-resolution algorithms is given by Borman and Stevenson [2] and Park et

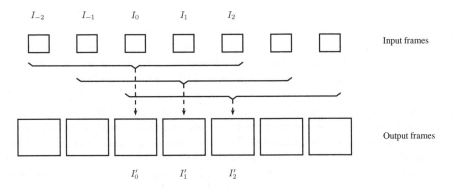

FIGURE 6.1: Super-resolution can be applied to video sequences by combining the frames in (overlapping) groups of N consecutive frames.

al. [14], or in special issues on the topic in *IEEE Signal Processing Magazine* (edited by Kang and Chaudhuri [10]), *EURASIP Journal of Applied Signal Processing* (edited by Ng et al. [4]) and *The Computer Journal* [3].

Most super-resolution algorithms consist of two main parts: image registration, where the images are precisely aligned, and image reconstruction, where the aligned images are combined to estimate a higher resolution image. In this chapter, we will concentrate on the first part, as precise, subpixel image registration is needed in order to be able to correctly reconstruct any high resolution information. For the reconstruction, we will use existing approaches.

In the next section, we first discuss our camera model, and how super-resolution can be applied to images captured with such cameras, followed by a definition of what we understand by the term "resolution." We will then present super-resolution as a multichannel sampling problem with unknown offsets. Using this description, an analysis can be made about the nature of the problem and conditions under which a solution can be found. Next, we describe a few solution methods using subspace approaches. We describe two solution methods for registration of totally aliased signals, followed by two more efficient methods that take advantage of aliasing-free parts of the input images to perform subpixel registration. The work presented in Sections 6.1-6.5.1 was already presented earlier [19, 20, 21] and is reproducible. The code and data to reproduce those results can be downloaded from the cited websites.

6.1 Camera Model

The pinhole model is the simplest model for a camera. An image of an infinitesimally small point light source taken with a pinhole camera contains a single, infinitesimally narrow peak at the corresponding pixel location. We can model its frequency response as a Dirac function.

In a real camera several non idealities contribute to a significant deviation from the pinhole model. The linear distortion introduced by the optics is represented by the point spread function (PSF). This is the impulse response of the imaging system, i.e. the image obtained when a point light source of infinitesimal size is placed in front of the system. Even when the system is perfectly focused, the image is not a point of infinitesimal size, but rather a disk of nonnegligible diameter. This measure describes the quality of the optical system. For example, lenses that are not ideal or are not precisely placed, result in an increase of the size of the point spread function. However, even in the ideal case, the point spread function has a non-negligible size. For an ideal lens with circular aperture, the point spread function is also called the Airy disk [8]. Its size is determined by the diffraction of the system, which is proportional to the wavelength of the light source and the aperture value (or f-number). Note that higher f-numbers correspond to a smaller aperture area, or less incident light. A large f-number corresponds to a large Airy disk and a strong low-pass effect (and at the same time a large depth of field). Conversely, a small f-number corresponds to a smaller Airy disk and sharper images.

Similarly to the point spread function, an additional low-pass effect is introduced by the sensor. In fact, it is not possible to measure light intensity on a sampling point of infinitesimal size. Instead, a sensor integrates the amount of photons hitting the pixel surface. Such an integration (along space and time coordinates) corresponds to a low-pass effect that is proportional to the size of the integration surface. In the continuing quest for higher-resolution, pixel sizes are reduced, and therefore the low-pass filtering effect is decreased. However, to increase light sensitivity, sensor manufacturers increase the fill factor, i.e. the active part of the pixels. Unfortunately, increasing the fill factor reduces the bandwidth of the system and limits the advantage of applying super-resolution algorithms.

In Figure 6.2 and Figure 6.3, we compare the frequency response $H(\omega)$ for some imaging systems. The frequency response is the Fourier transform of the imaging system's response to an infinitesimally small point light source. It includes the effect of both the optical system and the sensor. Typically, only the magnitude of the frequency response is given. The frequency scale is normalized with respect to the sampling frequency. Assuming circular uniformity of the system, we average horizontal and vertical responses, resulting in a one-dimensional function. The frequency responses in Figure 6.2 are computed analytically, taking only the diffraction of the (ideal) optics and the spatial integration of the sensor into account (assuming a fill factor of 100%). For the sensor, we consider the case of a 10.1 Mpixel $4/3''$ sensor[1] (with pixel pitch 4.6 μm), typical of a high quality camera, and a 3.2 Mpixel, $1/2.5''$ sensor with

---

[1]Note that such size designations in fractional inches do not represent actual sensor sizes. This notation dates back to the 50s and TV camera tubes, where the size gives the

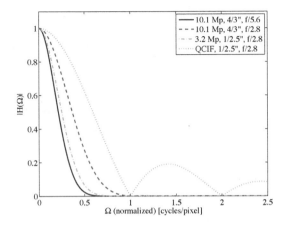

FIGURE 6.2: Comparison of frequency responses (PSF and sensor integration) for different imaging systems. Simulated frequency responses are shown for camera systems with a 10.1 Mpixel 4/3″ sensor and $f/5.6$ and $f/2.8$ aperture of the optics, a 3.2 Mpixel 1/2.5″ sensor with lens aperture $f/2.8$, and a QCIF video obtained by combining blocks of pixels from the 3.2 Mpixel 1/2.5″ sensor.

a pixel pitch of 2.8 μm, which can be found in a handheld device, such as a mobile phone. In the case of the high-resolution sensor, we consider two (circular) lens aperture values, namely $f/5.6$ and $f/2.8$, and for the low-resolution sensor we use a lens aperture of $f/2.8$. Aliasing is necessary to any algorithm for super-resolution. Therefore, the frequency response has to be nonnegligible for normalized frequency values larger than 0.5 cycles/pixel. This occurs for the case of the high-resolution sensor and larger aperture optics, while for the other cases the aperture area is small with respect to the sensor resolution, resulting in a large Airy disk, such that the amount of aliasing is not significant. Moreover, we see that for the 10.1 Mpixel camera with aperture $f/2.8$, the response vanishes for normalized frequencies larger than 1 cycle/pixel.

An interesting case is the one where the low-resolution sensor is used to acquire a QCIF video. In this case, the sensor resolution (3.2 Mpixel) is larger than the output resolution (144×176 pixels). Blocks of sensor pixels are combined in order to give an effect equivalent to a reduction of resolution (such that the pixel pitch would become 32.7 μm). This operation is equivalent to filtering the image with an averaging filter and then downsampling. Note that in order to avoid aliasing, an (additional) low-pass filter should be applied prior to downsampling. However, handheld devices normally do not include such a filter, and directly subsample the image. It results in a higher level of aliasing

outer diameter of the long glass envelope of the tube. The sensor diagonal is typically approximately 2/3 of this distance.

on the final images that compose the video. The equivalent response function is shown in Figure 6.2. We remark that a significant amount of aliasing can be present in the range of frequencies between 0.5 and 1 (and even up to 2) cycles/pixel.

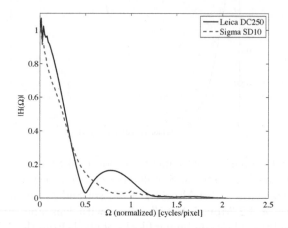

FIGURE 6.3: Measured frequency responses for a Leica DC250 camera with a Nikon 85 mm optical system and a Sigma SD10 camera with a Sigma $18-50$ mm lens.

Finally, these modeled frequency responses were compared to the measured frequency responses of some practical systems (see Figure 6.3). Frequency responses were measured using a test chart with a slanted edge (see Figure 6.4) and the method described in the ISO standard [11]. First, we took a Leica DC250 grayscale digital camera that is often used in microscopy. It has a 8.6×6.9 mm (or equivalently, $2/3''$) sensor with pixel pitch 6.7 μm, producing images of 1280×1024 pixels. We combined this camera with a Nikon 85mm lens using a C mount to Nikon adapter (no f-number available for this experiment). A considerable amount of aliasing is obtained in this setup, up to normalized frequencies of 1.2 cycles/pixel. As a second test camera, we used a Sigma SD10 digital camera with a Foveon X3 sensor (20.7×13.8 mm, or equivalently about $4/3''$ with a pixel pitch of 9.12 μm). This sensor measures the red, green, and blue channel at each pixel position, taking away the need for a demosaicing color interpolation. We used a Sigma $18-50$ mm lens at a focal length of 35 mm with this camera with aperture $f/10$. The images captured are 2268×1512 pixels. The Sigma camera has a lower (relative) cutoff frequency, but still shows a nonnegligible frequency response up to 0.8 cycles/pixel.

6.2   What Is Resolution?

Before we dive into super-resolution algorithms to "increase" the resolution, let us define what we understand by the term "resolution." There is definitely more to resolution than a simple pixel count, which is often simplistically used to indicate camera 'resolution'. Applying a low-pass filter to an image does much more to its resolution than merely increasing its number of pixels by repeating each pixel. Resolution relates to the ability to distinguish details in an image, in other words, to the resolving power.

In optics, the term "optical resolution" is used as a measure of the ability of a camera system, or a component of a camera system, to depict picture detail [8]. Assuming a diffraction-limited lens, two point light sources are said to be just resolved if the center of one Airy disk coincides with the first minimum of the other Airy disk. This is called Rayleigh's criterium. Actually, this criterium slightly underestimates the resolution, and a better condition is given by Sparrow. It says that two point light sources can be resolved until their two Airy disks overlap such that the second derivative at the center of one of the Airy disks is zero: the dip between the two Airy disks has disappeared. Similar criteria can also be applied to nonideal optical systems. The role of the Airy disk is then taken over by the point spread function.

In imaging, we talk about image resolution as a measure of the amount of detail that is visible in an image. The International Organization for Standardization (ISO) has developed a precise method to measure the resolution of a digital camera system [11]. The visual resolution can be measured as the highest frequency pattern of black and white lines where the individual black and white lines can still be visually distinguished in the image. It is expressed in line widths per picture height (LW/PH). The standard also describes a method to compute the spatial frequency response of a digital camera. It describes the variation between the maximum and minimum values that is visible as a function of the spatial frequency (the number of black and white lines per millimeter). It can be measured using an image of a slanted black and white edge, and is expressed in relative spatial frequencies (relative to the sampling frequency), line widths per picture height, or cycles per millimeter on the image sensor. Figure 6.4 shows the resolution chart used in the ISO standard. Examples of measured spatial frequency responses using the horizontally and vertically slanted edges at the center of the test chart are shown in Figure 6.3.

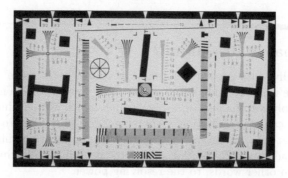

FIGURE 6.4: Resolution chart used to measure image resolution according to the ISO standard [11].

6.3 Super-Resolution as a Multichannel Sampling Problem

Let us now analyze super-resolution reconstruction mathematically, and formulate it as a multichannel sampling problem with unknown offsets. In order to keep the equations and analysis as simple as possible, we will present most of the material for 1D signals, and only use 2D notations where needed. The extension to 2D signals is straightforward. For simplicity, we will also assume a pinhole camera model, except when explicitly specified.

Consider a (continuous-time) input signal $f(x)$ in an L-dimensional Hilbert space. We can write $f(x)$ as a linear combination of the Hilbert space basis functions:

$$f(x) = \sum_{l=0}^{L-1} \alpha_l \varphi_l(x), \tag{6.1}$$

with α_l the expansion coefficient corresponding to the l-th basis function $\varphi_l(x)$. In many cases, the Hilbert space will be the space of truncated Fourier series, but it can also be applied to other spaces such as splines, wavelets, etc. Such a different basis can be useful if we know that the signal can be well approximated using that particular basis. In general, if no particular information about the signal characteristics is available, a Fourier series will be used as a basis.

We now sample this signal $f(x)$ with N sample sets y_n $(0 \leq n < N)$ at a rate K, where each set is taken with an arbitrary offset t_n:

$$y_n(k) = f\left(\frac{k+t_n}{K}\right) = \sum_{l=0}^{L-1} \alpha_l \varphi_l\left(\frac{k+t_n}{K}\right). \tag{6.2}$$

This results in N sets of K uniformly spaced samples with offsets $t = (t_0, t_1, \dots t_{N-1})$. We can combine the samples for each set in a sample vector $\boldsymbol{y}_n$, the expansion coefficients α_l into a coefficient vector $\boldsymbol{\alpha}$, and the sampled basis functions $\varphi_l(x)$ into a matrix $\boldsymbol{\Phi}_{t_n}$, where $\boldsymbol{\Phi}_{t_n}(k, l) = \varphi_l\left(\frac{k+t_n}{K}\right)$. Equation (6.2) can then be rewritten as

$$\boldsymbol{y}_n = \boldsymbol{\Phi}_{t_n}\boldsymbol{\alpha}. \tag{6.3}$$

Stacking the different sample vectors and basis function matrices vertically, we obtain

$$\boldsymbol{y} = \boldsymbol{\Phi}_t\boldsymbol{\alpha}. \tag{6.4}$$

In super-resolution imaging we typically want to reconstruct the original signal $f(x)$ (or equivalently, its coefficients $\boldsymbol{\alpha}$) from the images y_n ($0 \leq n < N$). There are NK equations (6.2) in the L unknown signal coefficients and the $N-1$ offsets (without loss of generality, we can set $t_0 = 0$). This is exactly the same configuration as in multichannel sampling with unknown offsets. A reconstruction method for multichannel sampling with known offsets was presented by Papoulis [13]. We will mainly concentrate here on an accurate estimation of the offset values t, which is an essential first step in accurate super-resolution.

As can be seen from (6.2), for general basis functions these equations are linear in the signal coefficients, but nonlinear in the offset values. It can be shown that if $NK > L + N - 1$, the solution to this set of equations is unique (except in some degenerate cases) [19]. If less samples are available (either by taking less sample sets or by using sets of lower resolution), the problem is ill-posed, and an additional regularization is typically needed. In this chapter, we will mainly consider cases for which $NK > L + N - 1$.

6.3.1 Fourier Series

Let us now analyze the above setup for the specific case of a truncated Fourier series:

$$f(x) = \sum_{l=-M}^{M} \alpha_l\varphi_l(x), \tag{6.5}$$

where the index l is now numbered from $-M$ to M because of the usual numbering for Fourier series ($L = 2M + 1$). The samples from (6.2) now become

$$y_n(k) = f\left(\frac{k + t_n}{K}\right) = \sum_{l=-M}^{M} \alpha_l e^{j2\pi\frac{l(k+t_n)}{K}} = \sum_{l=-M}^{M} \alpha_l W^{lk}z_n^l, \tag{6.6}$$

with $W = e^{j2\pi/K}$ and $z_n = e^{j2\pi t_n/K}$. As before, we can rewrite a sample set in matrix notation as

$$\boldsymbol{y}_n = \boldsymbol{F}^*\boldsymbol{D}_{t_n}\boldsymbol{\alpha}, \tag{6.7}$$

with

$$
\boldsymbol{F} = \begin{pmatrix}
1 & W^M & \cdots & W^{(K-1)M} \\
\vdots & \vdots & & \vdots \\
1 & W & \cdots & W^{K-1} \\
1 & 1 & \cdots & 1 \\
1 & W^{-1} & \cdots & W^{-(K-1)} \\
\vdots & \vdots & & \vdots \\
1 & W^{-M} & \cdots & W^{-(K-1)M}
\end{pmatrix}
\tag{6.8}
$$

$$
\boldsymbol{D}_{t_n} = \begin{pmatrix}
z_n^{-M} & & & & & 0 \\
& \ddots & & & & \\
& & z_n^{-1} & & & \\
& & & 1 & & \\
& & & & z_n & \\
& & & & & \ddots & \\
0 & & & & & & z_n^M
\end{pmatrix}.
\tag{6.9}
$$

$$\tag{6.10}$$

Note that $\boldsymbol{F}$ is an $L \times K$ forward DFT matrix, and the notation $\boldsymbol{F}^*$ is used to indicate the Hermitian transpose of $\boldsymbol{F}$, indicating the inverse DFT matrix. Due to the undersampling ($K < L$), some of the rows in $\boldsymbol{F}$ are repeated. The matrix $\boldsymbol{D}_{t_n}$ is an $L \times L$ diagonal matrix with its diagonal elements depending on the offset t_n. Just like for the general case, we can combine the different sample sets into one vector, resulting in

$$
\boldsymbol{y} = \begin{pmatrix}
\boldsymbol{y}_0 \\
\boldsymbol{y}_1 \\
\vdots \\
\boldsymbol{y}_{N-1}
\end{pmatrix}
= \begin{pmatrix}
\boldsymbol{F}^* \\
\boldsymbol{F}^* \boldsymbol{D}_{t_1} \\
\vdots \\
\boldsymbol{F}^* \boldsymbol{D}_{t_{N-1}}
\end{pmatrix} \boldsymbol{\alpha}.
\tag{6.11}
$$

The Fourier transform $\boldsymbol{y}_n^F$ of a sample set $\boldsymbol{y}_n$ can be computed as

$$
\boldsymbol{y}_n^F = \frac{1}{K} \boldsymbol{F}_K \boldsymbol{y}_n = \frac{1}{K} \boldsymbol{F}_K \boldsymbol{F}^* \boldsymbol{D}_{t_n} \boldsymbol{\alpha},
\tag{6.12}
$$

with $\boldsymbol{F}_K$ a square $K \times K$ (non-aliased) DFT matrix. As we know from sampling theory, $\boldsymbol{y}_n^F$ is an aliased and phase shifted version of the original Fourier coefficient vector $\boldsymbol{\alpha}$. This can be seen if we take for example $L = 3K$:

$$
\begin{aligned}
\boldsymbol{y}_n^F &= \frac{1}{K} \boldsymbol{F}_K \boldsymbol{F}^* \boldsymbol{D}_{t_n} \boldsymbol{\alpha} = \frac{1}{K} \boldsymbol{F}_K \begin{pmatrix} \boldsymbol{F}_K^* & \boldsymbol{F}_K^* & \boldsymbol{F}_K^* \end{pmatrix} \boldsymbol{D}_{t_n} \boldsymbol{\alpha} \\
&= \begin{pmatrix} \boldsymbol{I} & \boldsymbol{I} & \boldsymbol{I} \end{pmatrix} \boldsymbol{D}_{t_n} \boldsymbol{\alpha} = \sum_{i=-1}^{1} z_n^{iK} \boldsymbol{D}_{t_n}' \boldsymbol{\alpha}_i,
\end{aligned}
\tag{6.13}
$$

where $\boldsymbol{D'_{t_n}}$ is the $K \times K$ central part of the $L \times L$ matrix $\boldsymbol{D_{t_n}}$, and $\boldsymbol{\alpha}_i$ is the i-th block of K coefficients from $\boldsymbol{\alpha}$. In general, when L is not a multiple of K, we can still do the same decomposition by adding zeros to the vector $\boldsymbol{\alpha}$ up to the next multiple of K.

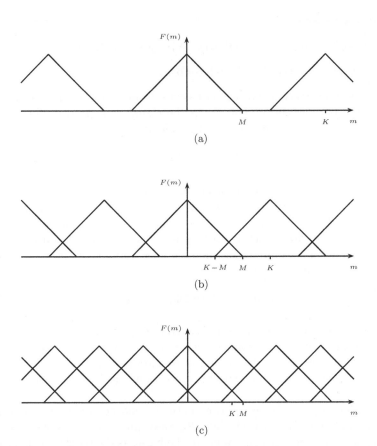

FIGURE 6.5: Three sampling situations can be distinguished: (a) Nyquist sampling $(K > 2M)$, (b) Partially aliased signals $(M < K < 2M)$, and (c) Totally aliased signals $(K < M)$.

Depending on the sampling frequency K, we can consider three different cases (see Figure 6.5). If $K > L$, the signal/image is sampled according to the Nyquist sampling theorem, and no aliasing is present. From a super-resolution point of view, this case is not interesting, as there is no aliasing from which to extract additional high frequency information. If $M < K < 2M$, part of the frequency spectrum is aliased, leaving also part of the spectrum free of aliasing. Solutions for such a case will be discussed in Section 6.5. In the next section, we will analyze the third case, $K < M$, where the entire frequency spectrum is aliased. In such a situation, the registration parameters can only be determined accurately by jointly estimating them from the full set of images (as opposed to the common pairwise registration in other cases).

6.4   Registration of Totally Aliased Signals

6.4.1   Variable Projection Method

As discussed in Section 6.3, the equations from (6.4) are nonlinear in the registration parameters, and linear in the signal expansion coefficients. We can write (6.4) as an l_2 minimization problem:

$$\min_{\alpha,t} \|y - \Phi_t \alpha\|_2^2, \qquad (6.14)$$

which is exactly the template problem in nonlinear least squares [7]. It can be solved using a variable projection method. For the correct values t, the sample vector y is a linear combination of the sampled basis functions, represented by the columns of Φ_t. In other words, y is in the subspace spanned by Φ_t, so we can estimate t by minimizing the difference between y and its projection onto the estimated subspace:

$$\min_{t} \|y - \Phi_t (\Phi_t^* \Phi_t)^{-1} \Phi_t^* y\|_2^2. \qquad (6.15)$$

This method for solving a nonlinear least squares problem is called the variable projection method [7]: the sample vector is actually projected on a variable subspace that depends on the minimization parameters. The link between the super-resolution problem and variable projections was first made by Robinson et al. [15].

Note that the variable projection method could be applied to any type of basis: not only the Fourier basis, but also for example splines or wavelets. The method does not add any constraints to the basis, as the basis only determines the space onto which the signal is projected.

For the Fourier basis, this space can be split in a number of orthogonal subspaces corresponding to the different Fourier vectors. Aliased frequencies appear in the same subspace. The minimization can therefore be applied independently on each of those subspaces (where each subspace will have periodically repeating minima), and these can then be combined to obtain the joint minimum. Moreover, the subspaces containing as many aliased components as their dimensionality can be skipped, as any set of offsets will work here. The minimization over independent subspaces is illustrated in Figure 6.6.

This registration method can be generalized to any type of motion model, as it only requires the basis functions $\varphi_l(t)$ to be sampled according to the model. Of course, more complex motion models will increase the dimensionality of the optimization problem.

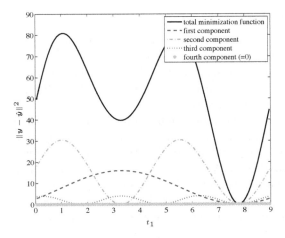

FIGURE 6.6: Example of the decomposition of the objective function into its different components belonging to orthogonal subspaces.

6.4.2 Frequency Analysis Method

Let us now consider again an image in Fourier space. The equation from (6.13) can be generalized for arbitrary L to

$$y_n^F = D'_{t_n} \sum_{i=\lceil -(S-1)/2 \rceil}^{\lceil (S-1)/2 \rceil} z_n^{iK} \alpha_i, \tag{6.16}$$

where $S = \lceil L/K \rceil$, or in other words, L was increased to the next multiple of K, and α was split up accordingly into S parts α_i of length K. If we multiply both sides of the above equation by $D'^{-1}_{t_n}$, we obtain modified sample vectors

$$D'^{-1}_{t_n} y_n^F = \sum_{i=\lceil -(S-1)/2 \rceil}^{\lceil (S-1)/2 \rceil} z_n^{iK} \alpha_i. \tag{6.17}$$

From this equation, it is clear that each modified sample vector $D'^{-1}_{t_n} y_n^F$ is part of the same S-dimensional subspace spanned by the spectrum vectors α_i. If we therefore take $N > S$ sample vectors, the matrix

$$\left(\begin{array}{cccc} y_0^F & D'^{-1}_{t_1} y_1^F & \cdots & D'^{-1}_{t_{N-1}} y_{N-1}^F \end{array} \right) \tag{6.18}$$

should be rank-deficient. These modified sample vectors depend on the offset values, and therefore we can estimate the registration parameters by searching the parameters that minimize the rank of the matrix in (6.18), or equivalently, minimize its $S + 1$-th singular value σ_{S+1}:

$$\min_{t} \sigma_{S+1} \left(\begin{array}{cccc} y_0^F & D'^{-1}_{t_1} y_1^F & \cdots & D'^{-1}_{t_{N-1}} y_{N-1}^F \end{array} \right) \tag{6.19}$$

6.4.3   Results

The above algorithms were tested in numerical simulations. From an original image, we created a set of 5 low-resolution images at half the resolution with relative (periodic) shifts (such that the entire spectrum is aliased). Two such input images can be seen in Figures 6.7a and 6.7c. Both of the above techniques give correct estimation of the motion parameters, and allow perfect reconstruction of the original high-resolution images 6.7b and 6.7d.

The minimization required in the two described algorithms has a high computational complexity. In both cases, an $N-1$-dimensional function has to be minimized (or $2(N-1)$ for images with horizontal and vertical shifts), which has a large number of local minima. Examples of such minimization functions are shown in Figure 6.8. For some approaches to search algorithms for finding the optimum and a more detailed complexity analysis, we refer the reader to our earlier work [19].

6.5   Registration of Partially Aliased Signals

The algorithms presented in the previous section are generally applicable to super-resolution from aliased images. However, as discussed above, they also have a high computational complexity. In this section we present two methods with lower complexity in case only part of the frequency spectrum is aliased.

6.5.1   Super-Resolution Using Frequency Domain Registration

If we assume $M < K < 2M$ (or equivalently $L/2 < K < L$), the signal is aliased, but not over the entire spectrum (see Figure 6.5). In such a case, we can use the aliasing-free part of the spectrum to estimate the registration parameters. Using these registration parameters, the aliased part can then be disambiguated and we can reconstruct a higher-resolution signal.

6.5.1.1   Image Registration

As $N > L/2$, part of the spectrum is free of aliasing. In (6.13) we can see that for certain frequencies,

$$\boldsymbol{y}_n^F(l) = z_n^l \boldsymbol{\alpha}(l), \tag{6.20}$$

or in other words, those frequency coefficients are phase shifted versions of the same coefficients for other images:

$$\frac{\boldsymbol{y}_n^F(l)}{\boldsymbol{y}_{n'}^F(l)} = \frac{z_n^l \boldsymbol{\alpha}(l)}{z_{n'}^l \boldsymbol{\alpha}(l)} = \frac{z_n^l}{z_{n'}^l}. \tag{6.21}$$

We can therefore estimate the registration parameters from all frequencies $l < K - L/2$. This can be done robustly by fitting a plane through the phase differences for each of those frequencies. Note that phase wrapping needs to be taken into account in such an approach. Such a registration method is equivalent to applying a low-pass filter to the images (removing all aliased frequencies) prior to registration.

The above approach can be used to estimate horizontal and vertical motion in a plane parallel to the image plane. We will now extend the frequency domain registration to rotations in the same image plane. In order to do this, we need to use two-dimensional notations:

$$f_1(\boldsymbol{x}) = f_0(\boldsymbol{R}(\boldsymbol{x} + \boldsymbol{x}_1)), \tag{6.22}$$

$$\text{with } \boldsymbol{x} = \begin{pmatrix} x_h \\ x_v \end{pmatrix}, \boldsymbol{x}_1 = \begin{pmatrix} x_{1,h} \\ x_{1,v} \end{pmatrix}, \boldsymbol{R} = \begin{pmatrix} \cos\theta_1 & -\sin\theta_1 \\ \sin\theta_1 & \cos\theta_1 \end{pmatrix}.$$

This can be expressed in Fourier domain as

$$\begin{aligned}
f_1^F(\boldsymbol{u}) &= \iint_{\boldsymbol{x}} f_1(\boldsymbol{x}) e^{-j2\pi \boldsymbol{u}^T \boldsymbol{x}} d\boldsymbol{x} \\
&= \iint_{\boldsymbol{x}} f_0(\boldsymbol{R}(\boldsymbol{x} + \boldsymbol{x}_1)) e^{-j2\pi \boldsymbol{u}^T \boldsymbol{x}} d\boldsymbol{x} \\
&= e^{j2\pi \boldsymbol{u}^T \boldsymbol{x}_1} \iint_{\boldsymbol{x}'} f_0(\boldsymbol{R}\boldsymbol{x}') e^{-j2\pi \boldsymbol{u}^T \boldsymbol{x}'} d\boldsymbol{x}',
\end{aligned} \tag{6.23}$$

with $f_1^F(\boldsymbol{u})$ the two-dimensional Fourier transform of $f_1(\boldsymbol{x})$ and the coordinate transformation $\boldsymbol{x}' = \boldsymbol{x} + \boldsymbol{x}_1$.

The rotation can be estimated independently before the shift estimation, as the amplitude of the Fourier transforms does not depend on the shift values (for the aliasing-free part of the spectrum):

$$\begin{aligned}
|f_1^F(\boldsymbol{u})| &= \left| e^{j2\pi \boldsymbol{u}^T \boldsymbol{x}_1} \iint_{\boldsymbol{x}'} f_0(\boldsymbol{R}\boldsymbol{x}') e^{-j2\pi \boldsymbol{u}^T \boldsymbol{x}'} d\boldsymbol{x}' \right| \\
&= \left| \iint_{\boldsymbol{x}'} f_0(\boldsymbol{R}\boldsymbol{x}') e^{-j2\pi \boldsymbol{u}^T \boldsymbol{x}'} d\boldsymbol{x}' \right| \\
&= \left| \iint_{\boldsymbol{x}''} f_0(\boldsymbol{x}'') e^{-j2\pi \boldsymbol{u}^T (\boldsymbol{R}^T \boldsymbol{x}'')} d\boldsymbol{x}'' \right| \\
&= \left| \iint_{\boldsymbol{x}''} f_0(\boldsymbol{x}'') e^{-j2\pi (\boldsymbol{R}\boldsymbol{u})^T \boldsymbol{x}''} d\boldsymbol{x}'' \right| \\
&= |f_0^F(\boldsymbol{R}\boldsymbol{u})|,
\end{aligned} \tag{6.24}$$

using the transformation $\boldsymbol{x}'' = \boldsymbol{R}\boldsymbol{x}'$. We can see that $|f_1^F(\boldsymbol{u})|$ is a rotated version of $|f_0^F(\boldsymbol{u})|$ over the same angle θ_1 as the spatial domain rotation (see also Figure 6.9). $|f_0^F(\boldsymbol{u})|$ and $|f_1^F(\boldsymbol{u})|$ do not depend on the shift values $\boldsymbol{x}_1$, because the spatial domain shifts only affect the phase values of the Fourier transforms. Therefore, we can first estimate the rotation angle θ_1 from the amplitudes of the Fourier transforms $|f_0^F(\boldsymbol{u})|$ and $|f_1^F(\boldsymbol{u})|$. After compensation for the rotation, the shift $\boldsymbol{x}_1$ can be computed from the phase difference between $f_0^F(\boldsymbol{u})$ and $f_1^F(\boldsymbol{u})$.

One option to estimate the rotation angle between two images is to compute the spectral differences of the aliasing-free frequencies for the reference image $f_0^F(\boldsymbol{u})$ with various rotations $f_1^F(\boldsymbol{Ru})$ of the image to register. However, this is a computationally very intensive method, as we need to compute rotations of the image spectra over a large number of rotation angles to find the optimal value. Instead, we will project the (aliasing-free) frequency content of the image onto a circular line and estimate the rotation angle by estimating the shift between two such one-dimensional functions. This is equivalent to transforming the image into polar coordinates and projecting this transform onto the axis associated to the angular coordinate.

6.5.1.2 Image Reconstruction

After image registration, we reconstruct a high-resolution image from the set of images using a nonuniform interpolation method implemented in Matlab [12]. Assuming the PSF is very narrow, and can be approximated by a Dirac, we compute the precise locations of all pixel coordinates on the high-resolution grid. Next, we perform a Delaunay triangulation using the Quickhull algorithm [1]. The high-resolution pixel values are then non-uniformly interpolated using bicubic interpolation. Such a reconstruction method provides good precision, with very low computational complexity. For more advanced reconstruction methods, we refer the reader to other chapters in this book.

6.5.1.3 Results

Some results using the above algorithm are presented in Figures 6.10 and 6.11. Figure 6.10 shows a high-resolution image reconstructed from 4 grayscale input images obtained using the Leica digital camera measured in Section 6.1. From the detail images, it is clear that more details can be observed in the reconstructed image than in any of the input images.

In a second experiment, we reconstructed a high-resolution image from 4 color images taken with the Sigma camera from Section 6.1. Results for a patch of the image can be seen in Figure 6.11. Again, we can see that the aliasing has been accurately removed in the horizontal grids of the building. At the same time, a small mismatch can be seen on the branches of the tree in front. This is due to errors in the motion model: the planar motion parameters found for the building do not apply for the branches of the tree that move in the wind.

FIGURE 6.7: Simulation results of the algorithms for totally aliased images (noiseless). (a) One of the five 16×16 images used as input. (b) Reconstructed 31×31 image using the algorithm from Section 6.4.1. (c) One of the five 32×32 images used as input. (d) Reconstructed 63×63 image using the algorithm from Section 6.4.2.

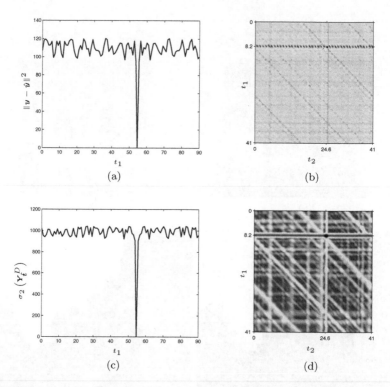

FIGURE 6.8: Examples of the objective functions in (6.15) and (6.19). (a) Two sets of 91 samples, with 81 unknown coefficients used in (6.15). The exact offset is $t_1 = 54.6$. Next to the global minimum, they also contain many local minima. (b) Three sets of 41 samples, with 81 unknown coefficients used in (6.15). The exact offsets are $t_1 = 8.2$ and $t_2 = 24.6$. Small values are represented by dark pixels. (c) Two sets of 91 samples, with 81 unknown coefficients used in (6.19). The exact offset is $t_1 = 54.6$. (d) Three sets of 41 samples, with 81 unknown coefficients used in (6.19). The exact offsets are $t_1 = 8.2$ and $t_2 = 24.6$. Small values are represented by dark pixels.

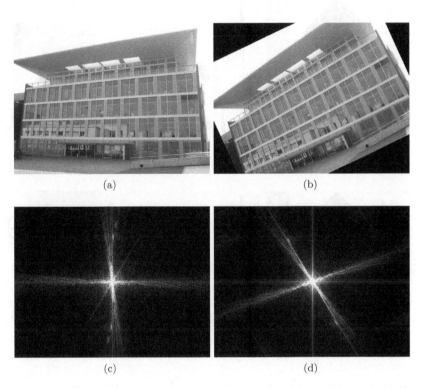

(a) (b)

(c) (d)

FIGURE 6.9: The amplitude of the Fourier transform of an image is rotated over the same angle ($\theta_1 = 25°$) as the spatial domain image. (a) Original image. (b) Rotated image. (c) Fourier transform amplitude of the original image. (d) Fourier transform amplitude of the rotated image.

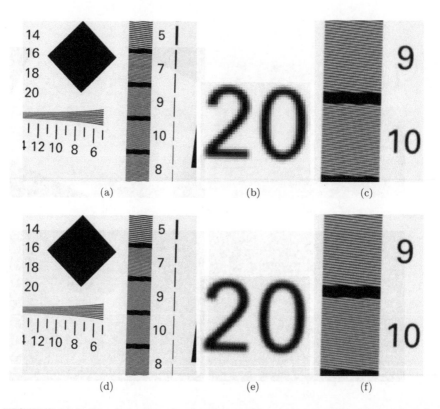

FIGURE 6.10: Results using the frequency domain registration algorithm for partially aliased images on a set of still images taken with a Leica DC250 camera. (a) Part of one of the input images, with a detail showing the aliasing in (b) and (c). (d) high-resolution image reconstructed from part of four Leica DC250 input images. Details are shown in (e) and (f) to display the differences better.

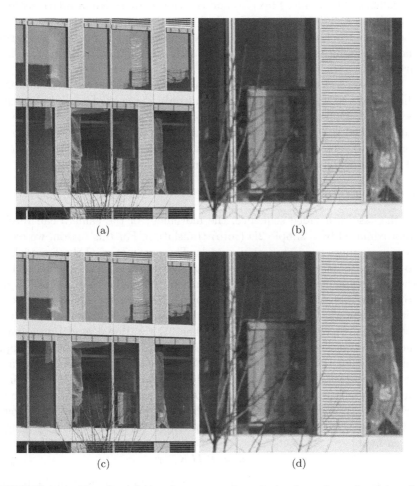

(a) (b)

(c) (d)

FIGURE 6.11: Results of the frequency domain registration algorithm for partially aliased images on the Sigma SD10 images of the outdoor scene. (a) (Part of) one of the 4 input images, with a detail in (b). (c) high-resolution output image, with a detail of the central part in (d) to show the differences better.

6.5.2 Super-Resolution from Low-Quality Videos

In this section, we explore the possibility of applying super-resolution techniques to low-resolution videos, such as those acquired with handheld devices. These devices typically acquire videos with low-resolution (such as CIF, i.e., 288x352, or QCIF, i.e., 144x176) and the videos are coded at relatively low bit rates (such as 128 Kb/s). Despite the low quality of these videos, we will show that super-resolution algorithms can be applied, under certain hypotheses, to increase resolution and obtain additionally a significant reduction of coding artifacts.

As discussed in Section 6.1, the limitations of the acquisition system for the considered application are such that only the case of partial aliasing can be applied. Moreover, the type of motion present in a video can rarely be accurately modeled by a simple 2D translation or even a 2D roto-translation. This forces us to consider more general motion models and prevents the application of algorithms based on the Fourier transform alone.

6.5.2.1 Motion Model

In the case of a video sequence, the movement of the camera can rarely be approximated by a simple 2D (roto)-translation. For this reason, we consider here some more general motion models. The parameters of these models have to be computed with high precision during the registration phase, such that the displacements in the image plane are much smaller than a pixel. In order to do this robustly, we restrict the choice of the motion model to those with few parameters. We used the 3D rotational and the planar model under the hypothesis of perspective projection. The 3D rotational model describes the motion in the scene as a 3D rotation around the camera. It is a good approximation when the objects present in the scene are distant from the camera or the translational component is negligible. The planar model approximates the scene with a planar surface in 3D space and the camera undergoes an arbitrary (3D) roto-translation. In both cases, the relation between a point at position p in the reference image and p' in one of the nonreference images is given by [6]

$$p' = KHK^{-1}p, \tag{6.25}$$

where $p = [x \ y \ f]^T$ and $p' = [x' \ y' \ f]^T$ are the homogeneous coordinates of the two points and f is the focal length. The matrix K is the camera calibration matrix and has the structure,

$$K = \begin{bmatrix} \alpha & 0 & u_0 \\ 0 & \beta & v_0 \\ 0 & 0 & 1 \end{bmatrix}, \tag{6.26}$$

where α and β are the magnification factors (i.e., the number of pixels per meter along the horizontal and vertical direction of the sensor) and $[u_0 \ v_0]$ is the position of the principal point, i.e., the position of the point where the optical axis intersects the sensor. The matrix H is a rotation matrix in the case of a 3D rotation centered at the focal point. In this case, no assumption is needed about the structure of the scene, since there is no parallax. In the case of a planar scene and an arbitrary roto-translation, H takes the structure of a homography matrix and has 8 degrees of freedom, since it is defined up to a scaling factor.

In the following, we assume that the parameters f, α, β, u_0, and v_0 are known. Their values can be determined precisely using a calibration procedure or they can be estimated based on the type of camera.

6.5.2.2 Image Registration

Aliasing typically perturbs registration and its influence should be reduced for accurate results. A first way to achieve this consists in applying a low-pass filter (or only considering low frequencies) as done in Section 6.5.1. This reduces the spectral components associated with a large aliasing amplitude. Another method is to operate in the spatial domain using robust estimators. The main idea is that the effect of aliasing is typically most visible along image edges, which are well localized in space. These regions give large registration errors and perturb algorithms that minimize the mean squared error (MSE). The problem is illustrated in Figure 6.12 for a simple one-dimensional case. The original continuous-time signal is represented in Figure 6.12a. It consists of a low frequency component (a sinusoid) and a step function, which is non-bandlimited and represents an image edge. Two sets of samples are taken from the continuous time signal with a relative shift of $\delta = 0.083$ and are used to obtain an approximation of the input signal. The shift between the two sets is unknown and should be determined by a registration algorithm. A way to estimate the shift is to interpolate the sets of samples using a low-pass filter, as in Figure 6.12b and Figure 6.12c, and determine the shift that minimizes the signal difference (MSE) between the two. However, this solution is not necessarily the correct one. For our example, such a minimization results in a shift $\delta = 0.25$ (while the correct shift was $\delta = 0.083$). Figure 6.12e shows the difference (error) between the interpolated signals in Figure 6.12b and Figure 6.12c for the estimated shift $\delta = 0.25$, while the difference for the correct shift is given in Figure 6.12d. The reason for this behavior is the presence of the discontinuity, which implies large registration errors. The minimum MSE method tends to minimize the error at all points, irrespectively of their amplitude. Instead, we see that the correct shift presents large errors in the region of the discontinuity, but much smaller errors for the remaining points. This type of problem has been previously addressed in statistics and in image registration leading to the development of robust algorithms [9, 16].

As shown in Figure 6.1, we call the reference image I_0 and the other images I_n, $n = \pm 1, \pm 2, \ldots, \pm(N-1)/2$, assuming that the size of the circular buffer is N. A way to implement a robust estimator is to use an M-estimator and

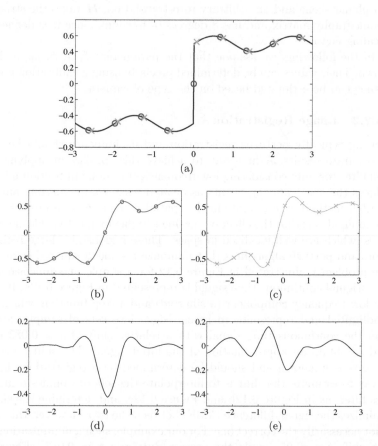

FIGURE 6.12: Example of signal registration for a 1D case. (a) A signal composed of a sinusoid and a step function is uniformly sampled twice with an unknown shift δ (in this example $\delta = 0.083$). The shift between the two sets of samples should be determined to reconstruct the original signal. (b) and (c) Low-pass interpolation of the two sets of samples. (d) Difference between the signals from (b) and (c) for the correct shift ($\delta = 0.083$). The error is large around the discontinuity (outlier) and small elsewhere. (e) Difference between the signals in (b) and (c) for the shift obtained using MSE minimization ($\delta = 0.25$). The error is minimized over the entire domain, but an incorrect shift value is obtained.

minimize the quantity

$$J_n = \sum_{\boldsymbol{p}} e(I_0(\boldsymbol{p}) - w(I_n(\boldsymbol{p}), \boldsymbol{h}_n)), \qquad (6.27)$$

for each non-reference image. The function $w(I_n(\boldsymbol{p}), \boldsymbol{h}_n)$ computes a warped version of $I_n(\boldsymbol{p})$, i.e., an image for which the movement of the camera has been compensated. The vector $\boldsymbol{h}_n$ is a parametric representation of the matrix $\boldsymbol{H}$ (for example, the three Euler angles in the case of a 3D rotation). The function e is used to measure how close the reference and the motion compensated image are. For example, in the case of the MSE, $e(x) = x^2$, i.e., the error on each pixel contributes to the total error with the square of its value. The optimal choice of the function e follows the Maximum Likelihood principle (ML) which determines e according to the Probability Distribution Function (PDF) of the residual error [9]. The MSE measure is optimal in the case of a Gaussian distribution of the residual error. Instead, when outliers are present, such as in the case of some forms of aliasing, large errors occur proportionally more often and a slowly decaying PDF is a more appropriate model. In our implementation, the cost functions J_n are minimized using a multiresolution Gauss-Newton descent method, similar to that described by Sawhney and Ayer [17].

Some registration error statistics are given in Table 6.1 for estimators designed for Cauchy and Gaussian distributions of the residual error. Additionally, a pre-filter can be applied to reduce aliasing at high-frequency components. The values are obtained by running 100 simulations with random motion parameters (for the 3D rotational model) corresponding to a pixel displacement in the range $[-0.5, 0.5]$.

The image used for the simulation is shown in Figure 6.13a. A part of the image is downsampled to obtain the image in Figure 6.13b, which is used as a reference image for the 100 runs of the registration algorithm. One of the 100 nonreference images is shown in Figure 6.13c. Aliasing is clearly visible in the region of the regular pattern of the car radiator.

The results in Table 6.1 show that the registration algorithm achieves sub-pixel precision. Both the low-pass filter and the robust estimator reduce the average registration error with respect to the Gaussian estimator. However, the low-pass filter is more effective than the robust estimator. This can be explained by the small amount of aliasing present in the images and the use of a Cauchy distribution for the residual error. Even if this is a better model than the Gaussian, it does not necessarily match the actual error distribution.

6.5.2.3 Image Reconstruction

When the input images are registered with respect to the reference I_0, it is possible to determine the super-resolution image I_0' corresponding to I_0. We follow an approach similar to Farsiu et al. [5]. The procedure consists in

(a)

(b)   (c)

FIGURE 6.13: Images used to test the motion estimation algorithm. (a) Original high-resolution (3072×2304) image. (b) low-resolution reference image (176×144). (c) One of the 100 low-resolution non reference images (176×144) registered with the reference image in (a).

minimizing, with respect to the unknown image I_0', the cost function

$$J_S = \sum_n \sum_p e(I_n(\boldsymbol{p}) - d \circ g \circ w(I_0'(\boldsymbol{p}, \boldsymbol{h}_n)) + \alpha T(I_0'), \qquad (6.28)$$

where I_n are the nonreference images, and the function $d \circ g \circ w$ represents the composition of down-sampling, low-pass filtering, and warping. These functions represent the transformations that one should apply to the super-resolution image I_0' to obtain each of the nonreference images. This is realized by warping I_0' according to the parameters computed in the registration step, and then low-pass filtering the result to simulate the behavior of camera optics, sensor, and motion blur. The last step is downsampling to reduce the number of pixels to that of the image I_n. The function e measures the error of the result with respect to the image I_n. When the image I_0' is correct, the error should be small. As in the case of registration, the function e should be chosen according to the distribution of the residual errors. Typical choices are the L_1, L_2, and L_p norms. The additional term $\alpha T(I_0')$ in equation (6.28) is needed to impose regularity on the image I_0', because in most cases, the input images are not sufficient to determine the solution unambiguously. The need for regularization depends on the type of motion and the low-pass filter g. For

Estimator type	Avg. abs. error (pixels)	Max abs. error (pixels)
Gaussian	0.0421	0.0756
Cauchy	0.0224	0.0488
Gaussian with prefilter	0.00556	0.0184
Cauchy with prefilter	0.00488	0.0127

TABLE 6.1: Registration errors for different estimators with and without prefilter. The estimation errors are computed on 100 simulations using a 3D rotational model with parameters corresponding to displacements in the range $[-0.5, \ 0.5]$ pixels. Both the prefilter and the use of robust estimators contribute to the reduction of the registration errors.

example, it is not possible to determine the components of I_0' that correspond to zeros of g. The typical choice for T is the Total Variation (TV) measure:

$$T(I_0') = \int \sqrt{\|\nabla I_0'\|^2 + \beta}, \qquad (6.29)$$

where β is a regularization term that makes T differentiable (we used $\beta = 0.01$ for frames with values in the range $[0, \ 255]$). The value of the Total Variation is related to the gradient magnitude of the image intensity, which is a measure of the image sharpness. In this way, the second term of equation (6.28) limits the amount of high frequencies added by the algorithm. In a software implementation of the Total Variation, the gradient operator in equation (6.29) is replaced by a difference operator. In our experiments, we approximated the derivatives with the average of the forward and backward differences along the x and the y coordinates. The constant α in (6.28) controls the trade-off between regularity and level of details in the output image.

The minimization of J_S is performed on the space of the possible images I_0', which has a dimension equal to the number of pixels. To limit the complexity of the algorithm, the steepest descent method is used. This consists in applying iterations

$$I_0'^{(i+1)} = I_0'^{(i)} + \mu \frac{dJ_S}{dI_0'}, \qquad (6.30)$$

to an initial guess $I_0'^{(0)}$. The step size μ determines the speed of convergence.

6.5.2.4 Results on Video Sequences

The proposed algorithm has been applied to a set of videos acquired using handheld devices and coded at low bit rates (around 128 Kb/s). The size of the circular buffer was $N = 9$. The motion parameters were computed assuming a 3D rotation, which is a good approximation when the distance of the scene is much larger than the translation of the camera. The large coding errors were the main source of noise during the frame registration step. This strongly reduced the advantage of the low-pass filter and robust estimation to minimize

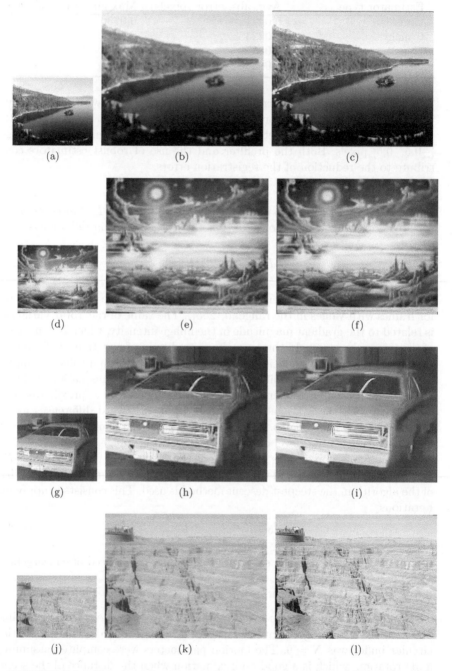

FIGURE 6.14: Results of super-resolution on low-rate encoded videos. The first column shows one video frame, the second column shows the result of bicubic interpolation, and the third column the result of the super-resolution algorithm applied to 9 consecutive video frames.

the effects of aliasing. However, robust estimators reduced the influence of regions not following the motion model, for example due to objects moving in the scene or model mismatch (such as optical distortion or non negligible translation). The descent method was applied to a multiscale representation of the frames. We found that 3 resolution levels and a maximum of 50 iterations were sufficient to obtain sub-pixel precision of the registrations. The super-resolution images were computed using the iterations from (6.30) with a step size $\mu = 0.07$. The term α in equation (6.28) determines the influence of the regularization term. A too small value will result in terms corresponding to the zeros of the filter g appearing in the solution. On the other hand, if α is too large, the improvement given by the algorithm is reduced. We found that $\alpha = 5$ was a good trade-off for our setup.

Some frames of the processed sequences are shown in Figure 6.14. The left column shows one of the original frames and the right column contains the super-resolution results. For comparison, the second column gives the result of bicubic interpolation. The increased level of details and sharpness is clearly visible on the super-resolution frames. In addition, we notice that some coding artifacts, like blockiness, are reduced in the output images. The reason is the temporal filtering introduced by the super-resolution algorithm, which tends to remove uncorrelated errors in the input images.

It is interesting to notice that, despite the high compression rate of the input videos, enough aliasing is still present in the video to be able to apply super-resolution. The reason is that the motion compensation step of a conventional video coder uses a block-based translational model, while the motion of the sequence is better modeled by a 3D rotation. Video encoders generally encode a subset of (intra) frames directly, and use motion compensation to predict the (inter) frames in between. If the motion model is accurate, the inter frames can be predicted correctly, and there is no need to encode any residual error. This would result in a single frame and a set of motion vectors, and it would be impossible to apply super-resolution. As the 3D rotational motion is not well modeled by the translational model, the encoder has to spend a large portion of its bit rate to represent the residual error, which includes errors due to both motion model mismatch and aliasing. Therefore, we can conclude that it is the inefficiency of (motion compensation in) current video coders that makes super-resolution possible.

6.6   Conclusions

We have presented a set of super-resolution algorithms, ranging from a more theoretical analysis of super-resolution as a multichannel sampling problem with unknown offsets to a practical algorithm for low-resolution videos captured with a mobile phone. A special emphasis was given to the registration part of such algorithms, as a precise sub-pixel registration is a necessary prerequisite for a good reconstruction of additional details. Through simulations and practical experiments, we have shown the good performance of our algorithms in the relevant use cases.

Bibliography

[1] C. B. Barber, D. P. Dobkin, and H. Huhdanpaa. The Quickhull algorithm for convex hulls. *ACM Transactions on Mathematical Software (TOMS)*, 22(4):469–483, December 1996.

[2] S. Borman and R. Stevenson. Spatial resolution enhancement of low-resolution image sequences - a comprehensive review with directions for future research. Technical report, University of Notre Dame, 1998.

[3] The Computer Journal, Special Issue on Super-Resolution, 2009.

[4] EURASIP Journal on Applied Signal Processing, Special Issue on Super-Resolution, 2006.

[5] S. Farsiu, M. Elad, and P. Milanfar. Multiframe demosaicing and super-resolution of color images. *IEEE Transactions on Image Processing*, 15(1):141–159, January 2006.

[6] D. A. Forsyth and J. Ponce. *Computer Vision: A Modern Approach*. Prentice Hall, August 2002.

[7] G. Golub and V. Pereyra. Separable nonlinear least squares: the variable projection method and its applications. *Inverse Problems*, 19(2):R1–R26, 2003.

[8] E. Hecht. *Optics*. Pearson - Addison Wesley, 2002.

[9] P. J. Huber. *Robust Statistics*. John Wiley and Sons, 1981.

[10] IEEE Signal Processing Magazine, Special Issue on Super-Resolution, May 2003.

[11] International Organization for Standardization. ISO 12233:2000 - Photography - Electronic still picture cameras - Resolution measurements, 2000.

[12] Mathworks (The). MATLAB function reference: griddata, 2009.

[13] A. Papoulis. Generalized sampling expansion. *IEEE Transactions on Circuits and Systems*, 24(11):652–654, November 1977.

[14] S. C. Park, M. K. Park, and M. G. Kang. Super-resolution image reconstruction: a technical overview. *IEEE Signal Processing Magazine*, 20(3):21–36, May 2003.

[15] D. Robinson, S. Farsiu, and P. Milanfar. Optimal registration of aliased images using variable projection with applications to super-resolution. *The Computer Journal*, 52(1):31–42, 2009.

[16] P. J. Rousseeuw and A. M. Leroy. *Robust Regression and Outlier Detection*. Wiley, 1986, 2003.

[17] H. Sawhney and S. Ayer. Compact representations of videos through dominant and multiple motion estimation. *IEEE Transactions on Pattern Analysis and Machine Intelligence*, 18:814–830, 1996.

[18] R. Y. Tsai and T. S. Huang. Multiframe image restoration and registration. In T. S. Huang, editor, *Advances in computer vision and image processing*, volume 1, pages 317–339. JAI Press, 1984.

[19] P. Vandewalle. *Super-Resolution from Unregistered Aliased Images*. PhD thesis, Ecole Polytechnique Fédérale de Lausanne, Switzerland, July 2006. No. 3591, [Reproducible] http://rr.epfl.ch/6.

[20] P. Vandewalle, L. Sbaiz, J. Vandewalle, and M. Vetterli. Super-Resolution from Unregistered and Totally Aliased Signals using Subspace Methods. *IEEE Transactions on Signal Processing*, 55(7, Part 2):3687–3703, 2007. [Reproducible] http://rr.epfl.ch/4.

[21] P. Vandewalle, S. Süsstrunk, and M. Vetterli. A Frequency Domain Approach to Registration of Aliased Images with Application to Super-Resolution. *EURASIP Journal on Applied Signal Processing, Special Issue on Super-Resolution Imaging*, 2006, 2006. Article ID 71459, 14 pages, [Reproducible] http://rr.epfl.ch/3.

7

Towards Super-Resolution in the Presence of
Spatially Varying Blur

Michal Šorel

Academy of Sciences of the Czech Republic

Filip Šroubek

Academy of Sciences of the Czech Republic

Jan Flusser

Academy of Sciences of the Czech Republic

CONTENTS

The effective resolution of an imaging system is limited not only by the physical resolution of an image sensor but also by blur. If the blur is present, super-resolution makes little sense without removing the blur. Some super-resolution methods considering space-invariant blur are described in other chapters of this book. The presence of a *spatially varying blur* makes the problem much more challenging and for the present, there are almost no algorithms designed specifically for this case. We argue that the critical part of such algorithms is precise estimation of the varying blur, which depends to a large extent on a specific application and type of blur.

In this chapter, we discuss possible sources of spatially varying blur, such as defocus, camera motion, or object motion. In each case we review known approaches to blur estimation, illustrate their performance on experiments with real data and indicate problems that must be solved to be applicable in super-resolution algorithms.

7.1   Introduction

At the very beginning, we should remark that in this chapter we consider only algorithms working with multiple acquisitions – situations where we fuse information from several images to get an image of better resolution. To our best knowledge, there are no true super-resolution algorithms working with an unknown space-variant blur. A first step in this direction is the algorithm [34], detailed in Section 7.5.1. On the other hand, considerable amount of literature exists on the deblurring of images degraded by space-variant blur. Our results [33, 32, 31] are described in Section 7.5, other relevant references [4, 22, 14, 8, 20] are commented in more detail at the beginning of Sections 7.4 and 7.5.3.

We do not treat super-resolution methods working with one image that needs a very strong prior knowledge – either in the form of shape priors describing whole objects or sets of possible local patches in the case of example based methods [11, 7, 13]. Nor do we consider approaches requiring hardware adjustments such as special shutters (coded-aperture camera [15]), camera actuators (motion-invariant photography [16]), or sensors (Penrose pixels [5]). However, these approaches can be considered in the same framework presented in this chapter.

We first introduce a general model of image acquisition that includes sampling, which we need for modeling resolution loss. This model is used for deriving a Bayesian solution to the problem of super-resolution. Next, a substantial part of the chapter discusses possible sources of spatially varying blur, such as defocus, camera motion, or object motion. Where possible, we included analytical expressions for the corresponding point-spread function (PSF). In each case we discuss possible approaches for blur estimation and illustrate their use in algorithms described in the second part of the chapter. Where the existing algorithms work only with deblurring, we indicate problems that must be solved to be applicable in true super-resolution.

All of the above-mentioned types of spatially varying blur can be described by a linear operator H acting on an image u in the form

$$[Hu](x,y) = \int u(x-s, y-t)h(s,t,x-s,y-t)\, dsdt, \qquad (7.1)$$

where h is a PSF. We can look at this formula as a convolution with a PSF that changes with its position in the image. The convolution is a special case thereof with the PSF independent of coordinates x and y, i.e., $h(s,t,x,y) = h(s,t)$ for an arbitrary x and y.

In practice, we work with a discrete representation of images and the same notation can be used with the following differences. Operator H in (7.1) corresponds to a matrix and u to a vector obtained by stacking columns of the image into one long vector. In the case of convolution, H is a block-Toeplitz matrix with Toeplitz blocks and each column of H contains the same PSF. In the space-variant case, each column may contain a different PSF that corresponds to the given position.

7.1.1 Representation of Spatially Varying PSF

An obvious problem of spatially varying blur is that the PSF is now a function of four variables. Except in trivial cases, it is hard to express it by an explicit formula. Even if the PSF is known, we must solve the problem of efficient representation.

If the PSF changes smoothly without discontinuities, we can store the PSF on a discrete set of positions and use interpolation to approximate the whole function h (see Figure 7.7). If the PSF is not known, as is usually the case, the local PSF's must be estimated as in the method described in Section 7.5.

Another type of representation is necessary if we consider for example moving objects, where the blur changes sharply at object boundaries. Then we usually assume that the blur is approximately space-invariant inside objects, and the PSF can be represented by a set of convolution kernels for each object and a corresponding set of object contours.

The final case occurs when the PSF depends on the depth. If the relation cannot be expressed by an explicit formula, as in the case of the ideal pillbox function for defocus, we must store a table of PSFs for every possible depth.

7.1.2 General Model of Resolution Loss

Let us represent the scene by two functions: intensity values of an ideal image $u(x,y)$ and a depth map $d(x,y)$. A full 3D representation is necessary only if occlusion is considered, which will not be our case.

Digital imaging devices have limited achievable resolution due to many theoretical and practical restrictions. In this section, we show a general model of image acquisition, which comprises commonly encountered degradations. Depending on the application, some of these degradations are known and some can be neglected.

First, light rays emanating from the scene come from different directions before they enter the lens as the camera orientation and position change, which can be modeled by a geometric transformation of the scene. Second, several external and internal phenomena degrade the perceived image. The external effects are, e.g., atmospheric turbulence and relative camera-scene motion. The internal effects include out-of-focus blur and all kinds of aberrations. As the light passes through the camera lens, warping due to lens distortion occurs. Finally, a camera digital sensor discretizes the image and produces a digitized noisy image $g(x, y)$. An acquisition model, which embraces all the above radiometric and geometric deformations, can be written as a composition of operators

$$g = DLHWu + n. \tag{7.2}$$

Operators W and L denote geometric deformation of the original scene and lens distortions, respectively. Blurring operator H describes the external and internal radiometric degradations. D is a decimation operator modeling the camera sensor and n stands for additive noise. Our goal is to solve an inverse problem, i.e., to estimate u from the observation g.

The decimation operator D consists of filtering followed by sampling. Filtering is a result of diffraction, shape of light sensitive elements and void spaces between them (fill factor), which cause the recorded signal to be band-limited. Sampling can be modeled by multiplication by a sum of delta functions placed on an evenly spaced grid. For principle reasons, D is not invertible but we will assume that its form is known.

Many restoration methods assume that the blurring operator H is known, which is only seldom true in practice. The first step towards more general cases is to assume that H is a traditional convolution with some unknown PSF. This model is true for some types of blurs (see, e.g., [23]) and narrow-angle lenses. In this chapter, we go one step further and assume spatially varying blur, which is the most general case that encompasses all the radiometric degradations if occlusion is not considered. Without additional constraints, the space-variant model is too complex. Various scenarios that are space-variant and allow solution are discussed in Section 7.5.

If lens parameters are known, one can remove lens distortions L from the observed image g without affecting blurring H, since H precedes L in (7.2). There is a considerable amount of literature on the estimation of distortion [36, 2]. In certain cases the distortion can be consider as a part of the estimated blurring operator as in the algorithm 7.5.2.

A more complicated situation materializes in the case of geometric deformation W. If a single acquisition is assumed, calculation of W is obsolete since we can only estimate Wu as a whole. In the case of multiple acquisitions in (7.3), the image u is generally deformed by different geometric transforms W_k's and one has to estimate each W_k by a proper image registration method [38]. By registering the images g_k's, we assume that the order of operators H_k and W_k is interchanged. In this case the blurring operator is $\tilde{H}_k = W_k^{-1} H_k W_k$ ($H_k W_k = W_k W_k^{-1} H_k W_k = W_k \tilde{H}_k$). If H_k is a standard convolution with some PSF h_k and W_k denotes a linear geometric transform, then by placing W_k in front of H_k, the new blurring operator $\tilde{H}_k$ remains a standard convolution but with h_k warped according to W_k. If W_k denotes a nonlinear geometric transform, then after interchanging the order, $\tilde{H}_k$ becomes a space-variant convolution operator in general. It is important to note that the blurring operator is unknown and instead of H_k we are estimating $\tilde{H}_k$, which is an equivalent problem as long as the nature of both blurring operators remains the same. Thus, to avoid extra symbols, we keep the symbol H_k for the blurring operator even if it would be more appropriate to write $\tilde{H}_k$ from now on.

As mentioned in the introduction, we need multiple acquisitions to have enough information to improve resolution. Hence, we write

$$g_k = DW_k H_k u + n_k = D_k H_k u + n_k \,, \tag{7.3}$$

where $k = 1, \ldots, K$, K is the number of input images, lens distortions L are not considered, D remains the same in all the acquisitions, and the order of operators H_k and W_k has been interchanged. We denote the combined operator of W_k and D as $D_k = DW_k$ and assume it is known.

In practice, there may be local degradations that are still not included in the model. A good example is a local motion that violates an assumption of global image degradation. If this is the case, restoration methods often fail. In order to increase flexibility of the above model, we introduce a masking operator M, which allows us to select regions that are in accordance with the model. The operator M multiplies the image with an indicator function (mask), which has ones in the valid regions and zeros elsewhere. The final acquisition model is then

$$g_k^v = M_k D_k H_k u + n_k = G_k u + n_k \,, \tag{7.4}$$

where g_k^v denotes the k-th acquired image with invalid regions masked out. The whole chain of degradations will be denoted as G_k. More about masking is in Section 7.5.1.

7.1.3 Bayesian View of Solution

There are a number of possible directions, from which we can approach the problem of super-resolution. One of the most frequent is the Bayesian approach, which we adopt here as well. Other approaches can be considered as approximations to the Bayesian solution.

An important fact is that if we know degradation operators G_k, the MAP (maximum a posteriori) solution under the assumption of Gaussian noise[1] corresponds to the minimum of a functional

$$E(u) = \sum_k \frac{1}{2\sigma_k^2} \|G_k u - g_k^v\|^2 + Q(u), \qquad (7.5)$$

where the first term describes an error of our model and the second term $Q(u)$ is a so-called regularization term that corresponds to the negative logarithm of the prior probability of the image u. Noise variance in the k-th image is denoted as σ_k.

The prior probability is difficult to obtain and it is often approximated by statistics of the image gradient distribution. A good approximation for common images is for example total variation regularization [21]

$$Q(u) = \lambda \int_\Omega |\nabla u|, \qquad (7.6)$$

which corresponds to an exponential decay of gradient magnitude. The total variation term can be replaced by an arbitrary suitable regularizer (Tikhonov, Mumford-Shah, etc.) [3, 29, 25]. The functional (7.5) can be extended to color images in quite a straightforward manner. The error term of the functional is summed over all three color channels (u_r, u_g, u_b) as in [28]:

$$Q(u) = \lambda \int \sqrt{|\nabla u_r|^2 + |\nabla u_g|^2 + |\nabla u_b|^2}. \qquad (7.7)$$

This approach has significant advantages as it suppresses noise effectively and prevents color artifacts at edges.

To minimize functional (7.5) we can use many existing algorithms, depending on a particular form of the regularization term. If it is quadratic (such as the classical Tikhonov regularization), we can use an arbitrary numerical method for the solution of systems of linear equations. In the case of total variation, the problem is usually solved by transforming the problem to a sequence of linear subproblems. In our implementations, we use the half-quadratic iterative approach as described for example in [32].

The derivative of functional (7.5) with the total variation regularizer (7.7) can be written as

$$\frac{\partial E(u)}{\partial u} = \sum_k \frac{G_k^*(G_k u - g_k^v)}{\sigma_k^2} - \lambda \operatorname{div}\left(\frac{\nabla u}{|\nabla u|}\right). \qquad (7.8)$$

$G_k^* = H_k^* D_k^* M_k^*$ is an operator adjoint to G_k and it is usually easy to construct. Adjoint masking M_k^* is equal to the original masking M_k. If D_k is

---

[1]Poisson noise can be considered by prescaling the operators G_k in equation (7.5) according to values of corresponding pixels in g_k.

downsampling, then D_k^* is upsampling. The operator adjoint to H_k defined in (7.1) can be written as

$$[H^*u](x,y) = \int u(x - s, y - t)h(-s, -t, x, y)\,dsdt. \qquad (7.9)$$

We can imagine this correlation-like operator as putting the PSF to all image positions and computing dot product. The gradient of any regularization functional of form $\int \kappa(|\nabla u|)$, where κ is an increasing smooth function, can be found in [28].

If we know the operators G_k, the solutions are in principle known, though the implementation of the above formulas can be quite complicated. In practice however, the operators G_k are not known and must be estimated.

Especially in the case of spatially varying blur, it turns out to be indispensable to have at least two observations of the same scene, which gives us additional information that makes the problem more tractable. Moreover, to solve such a complicated ill-posed problem, we must exploit the internal structure of the operator, according to the particular problem we solve. Some parts of the composition of sub-operators in (7.2) are known, some can be neglected or removed separately – for example, geometrical distortion. In certain cases we can remove the downsampling operator and solve only a deblurring problem, if we find out that we work at diffraction limit (read more about diffraction in 7.2.4). All the above cases are elaborated in the section on algorithms 7.5.

Without known PSFs it is in principle impossible to register precisely images blurred by motion. Consequently, it is important that image restoration does not necessarily require sub-pixel and even pixel precision of the registration. The registration error can be compensated in the algorithm by shifting the corresponding part of the space-variant PSF. Thus, the PSF estimation provides robustness to misalignment. As a side effect, misalignment due to lens distortion does not harm the algorithm as well.

In general, if each operator $G_k = G(\boldsymbol{\theta}_k)$ depends on a set of parameters $\boldsymbol{\theta}_k = \{\theta_k^1, \ldots, \theta_k^P\}$, we can again solve the problem in the MAP framework and maximize the joint probability over u and $\{\boldsymbol{\theta}_k\} = \{\boldsymbol{\theta}_1, \ldots, \boldsymbol{\theta}_K\}$. As the image and degradation parameters can be usually considered independent, the negative logarithm of probability gives a similar functional

$$E(u, \{\boldsymbol{\theta}_k\}) = \sum_{k=1}^{K} \frac{1}{2\sigma_k^2} \|G(\boldsymbol{\theta}_k)u - g_k^v\|^2 + Q(u) + R(\{\boldsymbol{\theta}_k\}), \qquad (7.10)$$

where the additional term $R(\{\boldsymbol{\theta}_k\})$ corresponds to a (negative logarithm of) prior probability of degradation parameters. The derivative of the error term in (7.10) with respect to the i-th parameter θ_k^i of $\boldsymbol{\theta}_k$, equals

$$\frac{\partial E(u, \{\boldsymbol{\theta}_k\})}{\partial \theta_k^i} = \frac{1}{\sigma_k^2} \langle \frac{\partial G(\boldsymbol{\theta}_k)}{\partial \theta_k^i} u, G(\boldsymbol{\theta}_k)u - g_k^v \rangle + \frac{\partial R(\{\boldsymbol{\theta}_k\})}{\partial \theta_k^i}, \qquad (7.11)$$

where $\langle . \rangle$ is the standard inner product in L_2. In discrete implementation,

$\frac{\partial G(\boldsymbol{\theta}_k)}{\partial \theta_k^i}$ is a matrix that is multiplied by the vector u before computing the dot product.

Each parameter vector $\boldsymbol{\theta}_k$ can contain registration parameters for images, PSFs, depth maps, masks for masking operators, etc., according to the type of degradation we consider.

Unfortunately in practice, it is by no means easy to minimize the functional (7.10). We must solve the following issues:

1. How to express the G_k as a function of parameters $\boldsymbol{\theta}_k$, which may be sometimes complex – for example, dependence of PSF on the depth of scene. We also need to be able to compute the corresponding derivatives.

2. How to design an efficient algorithm to minimize the nonconvex functional we derive. In particular, the algorithm should not get trapped in a local minimum.

All this turns out to be especially difficult in the case of spatially varying blur, which is also the reason why there are so few papers considering super-resolution or just deblurring in this framework.

An alternative to the MAP approach is to estimate the PSF in advance and then proceed with (nonblind) restoration by minimization over the possible images u. This can be regarded as an approximation to MAP. One such approach is demonstrated in Section 7.5.2.

To finalize this section, note that the MAP approach may not give optimal results, especially if we do not have enough information and the prior probability becomes more important. This is a typical situation for blind deconvolution of one image. It was documented (blind deconvolution method [10] and analysis [15]) that in these cases marginalization approaches can give better results. On the other hand, we are interested in the cases of multiple available images, where the MAP approach seems to be appropriate.

7.2   Defocus and Optical Aberrations

This section describes degradations produced by optical lens systems and the relation of the involved PSF to camera parameters and three-dimensional structure of an observed scene (depth).

We describe mainly the geometrical model of optical systems and corresponding PSFs, including the approximation by a Gaussian PSF. We mention also the case of general axially-symmetric optical system. Finally, we describe diffraction effects even though these can be considered space-invariant. The classical theory of Seidel aberrations [6] is not treated here as in practice the PSF is measured by an experiment and there is no need to express it in the form of the related decomposition. Also the geometrical distortion is omitted as it actually introduces no PSF and can be compensated by a geometrical transformation of images.

7.2.1 Geometrical Optics

Image processing applications widely use a simple model based on *geometrical (paraxial, Gaussian) optics* that follows the laws of ideal image formation. The name "paraxial" suggests that in reality it is valid only in a region close to the optical axis.

In real optical systems, there is also a roughly circular *aperture*, a hole formed by the blades that limit the pencils of rays propagating through the lens (rays emanate within solid angle subtended by the aperture). The aperture size is usually specified by f-number $F = f/2\rho$, where ρ is the radius of the aperture hole and f is a focal length. The aperture is usually assumed to be placed at the principal plane, i.e. somewhere inside the lens. It should be noted that this arrangement has an unpleasant property that magnification varies with the position of focal plane. If we work with more images of the same scene focused at different distances, it results in more complicated algorithms with precision deteriorated either by misregistration of corresponding points or by errors introduced by resampling and interpolation.[2]

If the aperture is assumed to be circular, the graph of the PSF has a cylindrical shape usually called a *pillbox* in literature. When we describe the appearance of the PSF in the image (or photograph), we speak about a *blur circle* or a *circle of confusion*.

It can be easily seen from the similarity of triangles that the blur circle radius for an arbitrary point at distance l is

$$r = \rho\zeta \left(\frac{1}{\zeta} + \frac{1}{l} - \frac{1}{f} \right) = \rho\zeta \left(\frac{1}{l} - \frac{1}{l_s} \right) , \qquad (7.12)$$

where ρ is the aperture radius, ζ is the distance of the image plane from the lens and l_s distance of the plane of focus (where objects are sharp) that can be computed from ζ using the relation $1/f = 1/l_s + 1/l$.

Notice the importance of inverse distances in these expressions. The expression (7.12) tells us that the radius r of the blur circle grows proportionally to the difference between inverse distances of the object and of the plane of focus. Other quantities, ρ, ζ, and f, depend only on the camera settings and are constant for one image.

---

[2]These problems can be eliminated using the so-called front telecentric optics, i.e. , optics with the aperture placed at the front focal plane. Then all principal rays (rays through principal point) become parallel to the optical axis behind the lens and consequently magnification remains constant as the sensor plane is displaced [35]. Unfortunately, most conventional lenses are not telecentric.

Thus, PSF can be written as

$$h(s,t,x,y) = \begin{cases} \frac{1}{\pi r^2(x,y)}, & \text{for } s^2 + t^2 \leq r^2(x,y), \\ 0, & \text{otherwise,} \end{cases} \qquad (7.13)$$

where $r(x,y)$ denotes the radius r of the blur circle corresponding to the distance of point (x,y) according to (7.12). Given camera parameters f, ζ and ρ, matrix r is only an alternative representation of depth map.

Now, suppose we have another image of the same scene, registered with the first image and taken with different camera settings. As the distance is the same for all pairs of points corresponding to the same part of the scene, inverse distance $1/l$ can be eliminated from (7.12) and we get a linear relation between the radii of blur circles in the first and the second image

$$r_2(x,y) = \frac{\rho_2}{\rho_1}\frac{\zeta_2}{\zeta_1}r_1(x,y) + \rho_2\zeta_2\left(\frac{1}{\zeta_2} - \frac{1}{\zeta_1} + \frac{1}{f_1} - \frac{1}{f_2}\right) \qquad (7.14)$$

Obviously, if we take both images with the same camera settings except for the aperture, i.e., $f_1 = f_2$ and $\zeta_1 = \zeta_2$, we get the right term zero and the left equal to the ratio of f-numbers.

In reality the aperture is not a circle but a polygonal shape with as many sides as there are blades. Note that at full aperture, where blades are completely released, the diaphragm plays no part and the PSF support is really circular. Still assuming geometrical optics, the aperture blur projects on the image plane with a scale changing the same way as for circular aperture, i.e. with a ratio

$$w = \frac{l' - \zeta}{l'} = \zeta\left(\frac{1}{l} - \frac{1}{l_s}\right) = \frac{1}{l}\zeta + \zeta\left(\frac{1}{\zeta} - \frac{1}{f}\right) \qquad (7.15)$$

and consequently

$$h(s,t,x,y) = \frac{1}{w^2(x,y)}\hat{h}\left(\frac{s}{w(x,y)}, \frac{t}{w(x,y)}\right), \qquad (7.16)$$

where $\hat{h}(s,t)$ is the shape of the aperture. The PSF keeps the unit integral thanks to the normalization factor $1/w^2$. Comparing (7.15) with (7.12), one can readily see that the blur circle (7.13) is a special case of (7.16) for $w(x,y) = r(x,y)/\rho$ and

$$\hat{h}(s,t) = \begin{cases} \frac{1}{\pi\rho^2}, & \text{for } s^2 + t^2 \leq \rho^2, \\ 0, & \text{otherwise.} \end{cases} \qquad (7.17)$$

Combining (7.15) for two images yields, analogously to (7.14),

$$w_2(x,y) = \frac{\zeta_2}{\zeta_1}w_1(x,y) + \zeta_2\left(\frac{1}{\zeta_2} - \frac{1}{\zeta_1} + \frac{1}{f_1} - \frac{1}{f_2}\right). \qquad (7.18)$$

Notice that if the two images differ only in the aperture, then the scale factors are the same, i.e., $w_2 = w_1$. The ratio ρ_2/ρ_1 from (7.14) is hidden in the different scale of the aperture hole.

7.2.2   Approximation of PSF by 2D Gaussian Function

In practice, due to lens aberrations and diffraction effects, PSF will be a circular blob, with brightness falling off gradually rather than sharply. Therefore, most algorithms use two-dimensional Gaussian function instead of pure pillbox shape. To map the variance σ to real depth, [26] proposes to use relation $\sigma = r/\sqrt{2}$ together with (7.12) with the exception of very small radii. Our experiments showed that it is often more precise to state the relation between σ and r more generally as $\sigma = \kappa r$, where κ is a constant found by camera calibration (for the lenses and settings we tested k varied around 1.2). Then analogously to (7.14) and (7.18)

$$\sigma_2 = \alpha\sigma_1 + \kappa\beta, \quad \alpha, \beta \in \mathbf{R}. \tag{7.19}$$

Again, if we change only the aperture then $\beta = 0$ and α equals the ratio of f-numbers.

The Corresponding PSF can be written as

$$h(s,t,x,y) = \frac{1}{2\pi\kappa^2 r^2(x,y)} e^{-\frac{s^2+t^2}{2\kappa^2 r^2(x,y)}} . \tag{7.20}$$

If possible we can calibrate the whole (as a rule monotonous) relation between σ and distance (or its representation) and consequently between σ_1 and σ_2.

In all cases, to use Gaussian efficiently, we need a reasonable size of its support. Fortunately Gaussian falls off quite quickly to zero and it is usually sufficient to truncate it by a circular window of radius 3σ or 4σ. Moreover, for common optical systems, an arbitrary real out-of-focus PSF has a finite support anyway.

7.2.3   General Form of PSF for Axially-Symmetric Optical Systems

In case of high-quality optics, pillbox and Gaussian shapes can give satisfactory results as the model fits the reality well. For poorly corrected optical systems, rays can be aberrated from their ideal paths to such an extent that it results in very irregular PSFs. In general, aberrations depend on the distance of the scene from the camera, position in the image and on the camera settings f, ζ and ρ. As a rule, the lenses are well corrected in the image center, but towards the edges of the image PSF may become completely asymmetrical.

Common lenses are usually axially-symmetric, i.e., they behave independently of its rotation about the optical axis. For such systems, it is easily seen (see Figure 7.1) that

1. in the image center, PSF is radially symmetric,

2. for the other points, PSF is bilaterally symmetric about the line passing

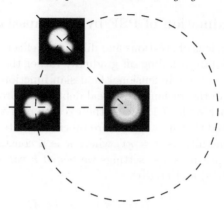

FIGURE 7.1: Three types of PSF symmetry in an optical system symmetrical about the optical axis.

through the center of the image and the respective point (two left PSFs in Figure 7.1),

3. for points of the same distance from the image center and corresponding to objects of the same depth, PSFs have the same shape, but they are rotated about the angle given by angular difference of their position with respect to the image center (again can be seen at two left PSFs in Figure 7.1).

The second and third property can be written as

$$h(s,t,x,y) = h\left(\frac{|(-t,s)(x,y)^T|}{|(x,y)|}, \frac{(s,t)(x,y)^T}{|(x,y)|}, 0, |(x,y)|\right). \qquad (7.21)$$

In most cases, it is impossible to derive an explicit expression for the PSF. On the other hand, it is relatively easy to get it by a raytracing algorithm. The above-mentioned properties of the axially-symmetric optical system can be used to save memory as we need not to store PSFs for all image coordinates but only for every distance from the image center. Naturally, it makes the algorithms more time-consuming as we need to rotate the PSFs every time they are used.

7.2.4 Diffraction

Diffraction is a wave phenomenon which makes a beam of parallel light passing through an aperture to spread out instead of converging to one point. For a circular aperture it shapes the well-known Airy disk (see Figure 7.2). The smaller the aperture, the larger the size of the disk and the signal is more blurry. Due to the diffraction the signal becomes band-limited, which defines

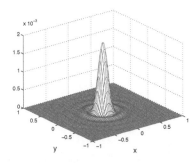 

FIGURE 7.2: Airy function: surface plot (left) and the corresponding grayscale image (right). The side lobes are very small and do not appear in the image plot. For this reason, we often talk about the Airy disk as only the central lobe is clearly visible.

a theoretical maximum spatial resolution and hence implies limits on super-resolution as will be shown later.

On a sensor array the signal is sampled by photosensitive devices (CCD/CMOS). Driven by marketing requirements of more and more megapixels, present day cameras were brought very close to this diffraction limit. Especially it is true for compacts with their small sensors. It means that we cannot neglect this phenomenon and should incorporate the corresponding PSF to deblurring algorithms.

To study the frequency response of a diffraction-limited optical system, we use transfer functions, i. e. , the Fourier transform of PSFs. If we assume an ideal circular aperture, neglect the defocus phenomena and other aberrations, the Optical Transfer Function (OTF) of the system due to diffraction is given [19] as

$$
\text{OTF}(\omega) = \begin{cases} \frac{2}{\pi}\left(\cos^{-1}\left(\frac{\omega}{\omega_c}\right) - \frac{\omega}{\omega_c}\sqrt{1 - \left(\frac{\omega}{\omega_c}\right)^2} \right) & \text{for } \omega < \omega_c \\ 0 & \text{otherwise,} \end{cases} \tag{7.22}
$$

where $\omega = \sqrt{\omega_x^2 + \omega_y^2}$ is the radial frequency in a 2D frequency space $[\omega_x, \omega_y]$, and $\omega_c = 1/(F\lambda)$ is the cutoff frequency of the lens (λ is the wavelength of incoming light). For example for aperture $F = 4$ and $\lambda = 500$ nm (in the middle of visible light), the cutoff frequency is $\omega_c = 0.5$ MHz and the corresponding OTF is plotted in Figure 7.3a as a solid line.

Assuming a square sensor without cross-talk, the Sensor Transfer Function (STF) is given by:

$$
\text{STF}(\omega_x, \omega_y) = \text{sinc}\left(\frac{\pi w \omega_x}{\omega_s}\right)\text{sinc}\left(\frac{\pi w \omega_y}{\omega_s}\right), \tag{7.23}
$$

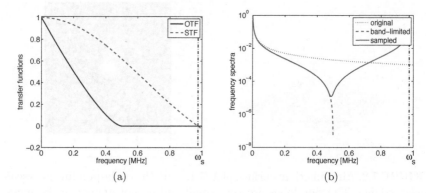

(a)　　　　　　　　　　　　(b)

FIGURE 7.3: Correctly sampled signal: (a) Optical transfer function and sensor transfer function; (b) Signal spectrum modified by diffraction and sensor sampling.

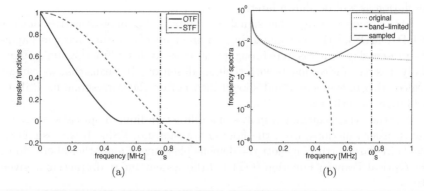

(a)　　　　　　　　　　　　(b)

FIGURE 7.4: Under-sampled signal: (a) Optical transfer function and sensor transfer function; (b) Signal spectrum modified by diffraction and sensor sampling.

where $\text{sinc}(x) = \sin(x)/x$ for $x \neq 0$ and $\text{sinc}(0) = 1$, ω_s is the sampling frequency, and w is the relative width of the square pixel ($w \leq 1$). For the fill-factor of 100% ($w = 1$) and if the signal is properly sampled ($\omega_s = 2\omega_c$), the corresponding STF is plotted in Figure 7.3a as a dashed line. As can be seen, the OTF is the main reason for a band-limited signal, since no information above its cutoff frequency passes through the optical system.

Figure 7.3b summarizes the effects of diffraction and sensor sampling on signal spectra. If the frequency spectrum of an original signal is modeled as a decaying dotted line, the spectrum of the band-limited signal is the attenuated dashed line, and the spectrum of the sampled signal is the solid line. The maximum frequency representable by the sampled signal is $\frac{1}{2}\omega_s$, which in this case is close to the cutoff frequency ω_c (proper sampling), and no aliasing is available, i.e. , the solid line matches the dashed line. It is clear that if super-resolution is applied to such data, no high-frequency information can be extracted and super-resolution merely interpolates.

On the other hand, if the optical system is undersampling the signal, the corresponding OTF and STF looks as in Fig. 7.4a. For the given aperture, wavelength and fill-factor, OTF is the same but STF shrinks. The sampled signal (solid line) has its high frequencies (around $\frac{1}{2}\omega_s$) disrupted due to aliasing as Fig. 7.4b illustrates. In this case, super-resolution can in principle unfold the signal spectra and recover the high-frequency information.

As mentioned above, the sampling of current consumer cameras approaches the diffraction limit that limits performance of any super-resolution algorithm. For example, a typical present day 10MP compact camera Canon PowerShot SX120 IS has its cut-off frequency about 2500 to 4000 per sensor width,[3] depending on the aperture, with maximum x-resolution of 3600 pixels. Especially with higher f-numbers it is very close to the theoretical limit. On the other hand, highly sensitive cameras (often near and mid-infrared) still undersample the images that leaves enough room for substantial resolution improvements.

If the decimation operator D is not considered in the acquisition model (7.2), the diffraction effect can be neglected as the degradation by H is far more important. Since the deconvolution algorithm estimates H, OTF and STF can be considered as part of H and thus estimated automatically as well. In the case of super-resolution, inclusion of D is essential as the goal is to increase sampling frequency. The diffraction phenomenon is irreversible for frequencies above the cutoff frequency ω_c and it is thus superfluous to try to increase image resolution beyond $2\omega_c$. (7.2). The diffraction phenomenon is irreversible and thus we will assume that the original image u is already band-limited. The decimation operator D will model only STF and sampling.

7.2.5 Summary

In this section, we described several shapes of PSF that can be used to model out-of-focus blur. Gaussian and pillbox shapes are adequate for good quality lenses or in the proximity of the image center, where the optical aberrations are usually well corrected. A more precise approach is to consider optical aberrations. However, an issue arises in this case that aberrations must be described for the whole range of possible focal lengths, apertures, and planes of focus. In practice, it is indispensable to take diffraction effects into account as many cameras are close to their diffraction limits.

---

[3]Aperture $f/2.8 - 4.3$, sensor size 1/2.5" (5.5 mm width), 3600×2700 maximum resolution, the diffraction limit (cut-off frequency), given by $\omega_c = 1/(F\lambda)$, is about 2500/sensor width (for $F = 4.3$) up to 4000/sensor width ($F = 2.8$). Light wavelength λ is taken as 500 nm.

7.3 Camera Motion Blur

In this section we analyze various types of camera motion for the classical pinhole camera model. We treat the case of a general motion in all six degrees of freedom and detail the special cases of camera rotation and translation in a plane.

To model camera motion blur by a PSF h from (7.1), we need to express the PSF as a function of the camera motion and a depth of the scene. In the case of a general camera motion, it can be computed from the formula for *velocity field* [12, 8] that gives apparent velocity of the scene for the point (x, y) of the image at time instant τ as

$$
v(x, y, \tau) = \frac{1}{d(x, y, \tau)} \begin{bmatrix} -1 & 0 & x \\ 0 & -1 & y \end{bmatrix} T(\tau) + \\
\begin{bmatrix} xy & -1 - x^2 & y \\ 1 + y^2 & -xy & -x \end{bmatrix} \Omega(\tau),
\tag{7.24}
$$

where $d(x, y, \tau)$ is the depth corresponding to point (x, y) and $\Omega(\tau)$ and $T(\tau) = [T_x(\tau), T_y(\tau), T_z(\tau)]^T$ are three-dimensional vectors of rotational and translational velocities of the camera at time τ. Both vectors are expressed with respect to the coordinate system originating in the optical center of the camera with axes parallel to x and y axes of the sensor and to the optical axis. All the quantities, except $\Omega(\tau)$, are in focal length units. The *depth* $d(x, y, \tau)$ is measured along the optical axis, the third axis of the coordinate system. The function d is called *depth map*.

The apparent curve $[\bar{x}(x, y, \tau), \bar{y}(x, y, \tau)]$ drawn by the given point (x, y) can be computed by the integration of the velocity field over the time when the shutter is open. Having the curves for all the points in the image, the two-dimensional space-variant PSF can be expressed as

$$
h(s, t, x, y) = \int \delta(s - \bar{x}(x, y, \tau), t - \bar{y}(x, y, \tau)) d\tau,
\tag{7.25}
$$

where δ is the two-dimensional Dirac delta function.

Complexity of derivation of an analytical form of (7.25) depends on the form of velocity vectors $\Omega(\tau)$ and $T(\tau)$. Meanwhile, most algorithms do not work directly with analytical forms and use a discrete representation extending standard convolution masks.

7.3.1 Rotation

Excessive complexity of a general camera movement can be overcome by imposing certain constraints. A good example is an approximation used in almost all[4] optical image stabilizers that they consider only rotational motion

---

[4]Recently Canon announced Hybrid IS that works with translational movements as well.

in two axes. What concerns ordinary photographs, it turns out that in most situations (landscapes and cityscapes without close objects, some portraits), translation can be neglected.

If we look at formula (7.24) with no translation, i.e. , $T(\tau) = 0$, we can see that the velocity field is independent of depth and changes slowly – realize that x and y are in focal length units that means the values are usually less then one (equals one for the border of an image taken with 35 mm equivalent lens). As a consequence, also the PSF has no discontinuities, the blur can be considered locally constant and can be locally approximated by convolution. This property can be used to efficiently estimate the space-variant PSF, as described in Section 7.5.2.

7.3.2 No Rotation

A more complicated special case it to disallow rotation and assume that the change of depth is negligible with an implication that also the velocity in the direction of view can be considered zero $(T(3) = 0)$. It can be easily seen [32] that in this special case, the PSF can be expressed explicitly using the knowledge of the PSF for one fixed depth of scene.

If the camera does not rotate, that is $\Omega = [0,0,0]^T$, and moves in only one plane perpendicular to the optical axis $(T_z(\tau) = 0)$, equation (7.24) becomes

$$v(x,y,\tau) = \frac{1}{d(x,y,\tau)} \begin{bmatrix} -T_x(\tau) \\ -T_y(\tau) \end{bmatrix}. \tag{7.26}$$

In other words, the velocity field has the direction opposite to camera velocity vector and the magnitudes of velocity vectors are proportional to inverse depth. Moreover, depth for the given part of the scene does not change during such a motion (depth is measured along the optical axis and the camera moves perpendicularly to it), $d(x,y,\tau)$ does not change in time, and consequently the PSF simply follows the (mirrored because of the minus sign) curve drawn by the camera in image plane. The curve only changes its scale proportionally to the inverse depth.

The same is true for the corresponding PSFs we get according to relation (7.25). Let us denote the PSF corresponding to an object of the depth equal to the focal length as h_0. Note that this "prototype" PSF also corresponds to the path covered by the camera. Recall that the depth is given in focal length units. After linear substitution in the integral (7.25) we get

$$h(s,t,x,y) = d^2(x,y)h_0(sd(x,y),td(x,y)). \tag{7.27}$$

Equation (7.27) implies that if we recover the PSF for an arbitrary fixed depth, we can compute it for any other depth by simple stretching proportionally to the ratio of the depths.

7.4 Scene Motion

The degradation models we have discussed so far resulted either in the camera motion or in the global scene motion. In many real scenarios, the observed scene is not static but contains moving objects. Local changes inflicted by moving objects are twofold. First, local motion creates additional varying blurring, and second, occlusion of the background may occur. To include these two phenomena in the acquisition model is complicated as it requires segmentation based on motion detection. Most restoration methods assume a rigid transform (e.g., homography) as the warping operator W in (7.3). If the registration parameters can be calculated, we can spatially align input images. If local motion occurs, the warping operator must implement a non-global transform, which is difficult to estimate. In addition, warping by itself cannot cope with occlusion. A reasonable approach is to segment the scene according to results obtained by local-motion estimation and deal with individual segments separately. Several attempts in this direction were explored in literature recently. Since PSFs may change abruptly, it is essential to precisely detect boundaries, where the PSFs change, and consider boundary effects. An attempt in this direction was for example proposed in [4], where level-sets were utilized. Another interesting approach is to identify blurs and segment the image accordingly by using local image statistics as proposed, e.g., in [14]. All these attempts consider only convolution degradation. If decimation is involved, then space-variant super-resolution was considered, e.g., in [22]. However, this technique assumes that PSFs are known or negligible. A method restoring scenes with local motion, which would perform blind deconvolution and super-resolution simultaneously, has not been proposed yet.

A natural way to avoid the extra burden implied by local motion is to introduce masking as in (7.4). Masking eliminates occluded, missing, or corrupted pixels. In the case of local motion, one can proceed in the following way. A rigid transform is first estimated between the input images and inserted in the warping operator. Then discrepancies in the registered images can be used for constructing masks. More details are provided in the next section on algorithms, 7.5.1.

7.5 Algorithms

This section outlines the deblurring and super-resolution algorithms that in a way consider spatially varying blur.

As we already mentioned, for the present, there are no super-resolution methods working with unknown spatially varying blur. Deblurring and super-resolution share the same problem of blur estimation and, as we saw in the introduction, it is useful to consider both in the same framework. This section describes deblurring algorithms based on the MAP framework explained in the introduction, where a similar approach could be used for true super-resolution as well.

As the number of blur parameters increases, so does the complexity of estimation algorithms. We will progress our review from simple to more complex scenarios. If the blur is space-invariant except relatively small areas, we can use a space-invariant method supplemented with masking described in the introduction. An algorithm of this type is described in Section 7.5.1. If the blur is caused by a more complex camera movement, it generally varies across the image but not randomly. The PSF is constrained by six degrees of freedom of a rigid body motion. Moreover, if we limit ourselves to only rotation, we not only get along with three degrees of freedom, but we also avoid the dependence on a depth map. This case is described in Section 7.5.2. If the PSF depends on the depth map, the problem becomes more complicated. Section 7.5.3 provides possible solutions for two such cases: defocus with a known optical system and blur caused by camera motion. In the latter case, the camera motion must be known or we must be able to estimate from the input images.

7.5.1 Super-Resolution of a Scene with Local Motion

We start with a super-resolution method [34] that works with space-invariant PSFs and treats possible discrepancies as an error of the convolutional model. This model can be used for super-resolution of a moving object on a stationary background. A similar approach with more elaborated treatment of object boundaries was applied for deblurring in a simplified case of unidirectional steady motion in [1].

We assume the K-channel acquisition model in (7.4) with H_k being convolution with an unknown PSF h_k of a small support. The corresponding functional to minimize is (7.10) where $\{\boldsymbol{\theta}_k\} = \{\boldsymbol{\theta}_1, \ldots, \boldsymbol{\theta}_K\}$ consists of registration parameters for images g_k's, PSFs h_k's, and masks for masking operators M_k's. Due to the decimation operators D_k's, the acquired images g_k's are of lower resolution than the sought-after image u. Minimization of the functional provides estimates of the PSFs and original image. As the PSFs are estimated in the scale of the original image, positions of PSFs centroids correspond to sub-pixel shifts in the scale of the acquired images. Therefore, by estimating PSFs, we automatically estimate shifts with sub-pixel accuracy, which is essential for a good performance of super-resolution. One image from the input sequence is selected as a reference image g_r $(r \in 1, \ldots, K)$ and registration is performed with respect to this image. If the camera position changes slightly between acquisitions, which is typically a case of video sequences, we can assume homography model. However, homography cannot compensate for local motion, whereas masking can to some extent. Discrepancies in preregistered (with homography) images give us regions where local motion is highly probable. Masking out such regions and performing simultaneously blind deconvolution and super-resolution, produces naturally looking high-resolution images.

The algorithm runs in two steps:

1. Initialize parameters $\{\boldsymbol{\theta}_k\}$: Estimate homography between the reference frame g_r and each g_k for $k \in 1, \ldots, K$. Calculate masks M_k's and construct decimation operators D_k's. Initialize $\{h_k\}$ with delta functions.

2. Minimization of $E(u, \{\boldsymbol{\theta}_k\})$ in (7.10): alternate between minimization with respect to u and with respect to $\{\boldsymbol{\theta}_k\}$. Run this step for a predefined number of iterations or until a convergence criterion is met.

To determine M_k, we take the difference between the registered image g_k and the reference image g_r and threshold its magnitude. Values below 10% of the intensity range of input images are considered as correctly registered and the mask is set to one in these regions; remaining areas are zeroed. In order to attenuate the effect of misregistration errors, the morphological operator "closing" is then applied to the mask. Note that M_r will be always identity and therefore high-resolution pixels of u in regions of local motion will be at least mapped to low-resolution pixels of g_r. Depending on how many input images map to the original image, the restoration algorithm performs any task from simple interpolation to well-posed super-resolution.

The regularization term $R(\{\boldsymbol{\theta}_k\})$ is a function of h_k's and utilizes relations between all the input images g_k's. An exact derivation is given in [23]. Here, we leave the discussion by stating that the regularization term is of the form

$$R(\{h_k\}) \propto \sum_{\substack{1 \le i,j \le K \\ i \ne j}} \|h_i * g_j - h_j * g_i\|^2, \qquad (7.28)$$

which is convex.

We use a standard web camera to capture a short video sequence of a child waving a hand with following setting: 30 FPS, shutter speed 1/30s, and resolution 320×200. An example of 5 low-resolution frames is in the top row in Figure 7.5. The position of the waving hand slightly differs from frame to frame. Registering the frames in the first step of the algorithm removes homography. Estimated masks in the middle row in Figure 7.5 show that most of the erroneous pixels are around the waving hand. Note that only the middle frame, which is the reference one and does not have any mask, provides information about the pixels in the region of the waving hand. Comparison of estimating the high-resolution frame with and without masking together with simple interpolation is in the bottom row. Ignoring masks results in heavy artifacts in the region of local motion. On the contrary, masking produces smooth results with the masked-out regions properly interpolated. Remaining artifacts are the result of imprecise masking. Small intensity differences between the images, which set the mask to one, do not always imply that

FIGURE 7.5: Super-resolution of a scene with local motion. The first row shows five consecutive input frames acquired by a web camera. The second row shows masks (white areas), which indicate regions with possible local motion. The third row shows the estimated original image using simple interpolation (left), super-resolution without masking (central), and proposed super-resolution with masking (right).

the corresponding areas in the image are properly registered. Such a situation may occur for example in regions with a small variance or periodic texture.

7.5.2 Smoothly Changing Blur

This section demonstrates space-variant restoration in situations where the PSF changes gradually without sharp discontinuities, which means that the blur can be locally approximated by convolution. A typical case is the blur caused by camera shake, when taking photos of a static scene without too close objects from hand. Under these conditions, the rotational component of camera motion is dominant and, as was shown in Section 7.3.1, the blur caused by camera rotation does not depend on the depth map.

In principle, in this case, the super-resolution methods that use convolution could be applied locally and the results of deconvolution/super-resolution could be fused together. Unfortunately, it is not easy to sew the patches together without artifacts on the seams. An alternative way is first to use the estimated PSFs to approximate the spatially varying PSF by interpolation of adjacent kernels (see Figure 7.7) and then compute the image of improved

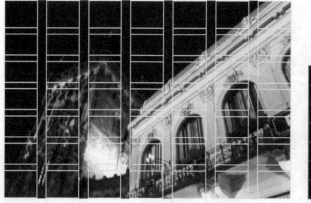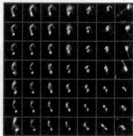

FIGURE 7.6: A night photo taken from hand with shutter speed 1.3s. The right image shows PSFs computed within white squares on the left using the algorithm described in Section 7.5.2. Short focal length (36 mm equivalent) accents spatial variance of the PSF.

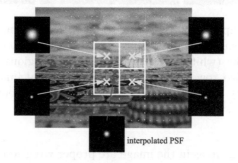

interpolated PSF

FIGURE 7.7: If the blur changes gradually, we can estimate convolution kernels on a grid of positions and approximate the PSF in the rest of the image (bottom kernel) by interpolation from four adjacent kernels.

resolution by minimization of the functional (7.5). The main problem of these naive procedures is that they are relatively slow, especially if applied on too many positions. A partial speed up of the latter can be achieved at the expense of precision by estimating the PSF based solely on blind deconvolution and then upscaling to the desired resolution. This algorithm has not been tested yet.

To see, whether the interpolation of the PSF can work in practice and what is the necessary density of the PSFs, we applied this approach for the purpose of image stabilization in [33].

We worked with a special setup that simplifies the involved computations and makes them more stable. It considers the possibility to set the exposure

FIGURE 7.8: Details of restoration. From left to right – the blurred image, noisy image and the result of the algorithm combining them to get a low-noise sharp photo.

time of the involved camera, which is an acceptable assumption as we can alway balance noise with motion blur by setting a suitable shutter speed. In particular, we set the exposure time of one of the images to be so short, that the image is sharp, of course at the expense of noise amplification. The whole idea was explored relatively recently [27, 17, 37].

In Figure 7.6, we can see a night photo of a historical building taken at ISO 100 with shutter speed 1.3s. The same photo was taken once more at ISO 1600 with 2 stops underexposure to achieve a hand-holdable shutter time 1/50s. The following algorithm fuses them to get one sharp photo.

The algorithm works in three phases:

1. Robust image registration;

2. Estimation of convolution kernels (Figure 7.6 right) on a grid of windows (white squares in Figure 7.6 left) followed by an adjustment at places where the estimation failed;

3. Restoration of the sharp image by minimizing the functional (7.5). The PSF described by the operator H for the blurred image is approximated by interpolation from the kernels estimated in the previous step.

We do not describe in detail the image registration here. Just note that the ambiguous registration discussed in Section 7.1.3 does not harm the procedure because the registration error is compensated by the shift of the corresponding part of the PSF.

The second step is a critical part of the algorithm and we describe it here in more detail. In the example in Figure 7.6, we took 49 square subwindows (white squares), in which we estimated kernels $h_{i,j}$ ($i, j = 1..7$). The estimated kernels are assigned to centers of the windows where they were computed. In the rest of the image, the PSF h is approximated by bilinear interpolation from blur kernels in the four adjacent sub-windows.

The blur kernel corresponding to each white square is calculated as

$$h_{i,j} = \arg\min_c \|d_{i,j} * c - z_{i,j}\|^2 + \alpha \|\nabla c\|^2, \quad c(\boldsymbol{x}) \geq 0, \qquad (7.29)$$

where $h_{i,j}(s,t)$ is an estimate of $h(x_0, y_0; s, t)$, x_0, y_0 being the center of the current window $z_{i,j}$, $d_{i,j}$ the corresponding part of the noisy image, and c the locally valid convolution kernel.

The kernel estimation procedure (7.29) can naturally fail. In a robust system, such kernels must be identified, removed, and replaced by for example an average of adjacent (valid) kernels. There are basically two reasons why kernel estimation fails – a lack of texture and pixel saturation. Two simple measures, sum of the kernel values and its entropy turned out to be sufficient to identify such failures.

For minimization of the functional (7.5), we used a variant of the half-quadratic iterative approach, solving iteratively a sequence of linear subproblems, as described for example in [32]. In this case, the decimation operator D and masking operator M are identities for both images. Blurring operator H is the identity for the noisy image. The geometric deformation is removed in the registration step. Note that the blurring operator can be speeded up by Fourier transform computed separately on each square corresponding to the neighborhood of four adjacent PSFs [18].

To help reader recognize differences in quite a large photograph (1154 × 1736 pixels), we show details of the result in Figure 7.8. Details of the algorithm can be found in [33].

7.5.3　Depth-Dependent Blur

In this section, we demonstrate algorithms working for PSFs that depend on the depth, which implies that besides the restored image we must estimate

(a) Two motion blurred images with small depth of focus

(b) Result of the algorithm (left) and ground truth (right)

FIGURE 7.9: Removing motion blur from images degraded simultaneously by motion blur and defocus by the algorithm described in Section 7.5.3.

 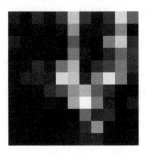

FIGURE 7.10: Depth map corresponding to images in Figure 7.9 and the PSF estimated locally around the flowers close to the center of the left input image.

also an unknown depth map. This includes the blur caused by a camera motion and defocus. Similarly to the previous section, there are no published algorithms that actually increase the physical resolution. On the other hand, a considerable work has been devoted to deblurring.

In the case of scenes with significant depth variations, the methods requiring PSFs without discontinuities are not suitable. Artifacts would appear especially at the edges of objects. For this case, so far, the only approach that seems to give relatively precise results is based on the MAP approach, which estimates simultaneously an unknown image and depth map by minimization of a functional in the form (7.10). The main assumption of these algorithms is that the relation between the PSF and the depth is known. One exception is [32], where this relation is estimated for a camera motion constrained to movement in one plane and without rotation. This result is described later in this section.

First this approach appeared in the context of out-of-focus images in [20] proposing to use simulated annealing to minimize the corresponding cost functional. This guarantees global convergence, but in practice, it is prohibitively slow. Later, this approach was adopted by Favaro et al. [8] who modeled the camera motion blur by a Gaussian PSF, locally deformed according to the direction and extent of blur. To make the minimization feasible, they take advantage of special properties of Gaussian PSFs as to view the corresponding blur as an anisotropical diffusion. This model can be appropriate for small blurs corresponding to short locally linear translations. An extension of [8] proposed in [9] segments moving objects but it keeps the limitations of the original paper concerning the shape of the PSF. Other papers related to this type of variational problem can be found also in the context of optical flow estimation, such as [30].

We start our discussion with a difficult case of the blur caused by an unconstrained camera motion. If the cameras' motion and parameters (focal length, resolution of the sensor, initial relative position of cameras) are known, we can, at least in theory, compute the PSF as a function of depth map and solve the MAP problem (7.10) for an unknown image u and a parameter set $\{\boldsymbol{\theta}_k\}$ corresponding now to a depth map for one of observed images g_k. An issue arises from the fact that the PSF is a function of not only depth but also of coordinates (x, y). In other words, different points of the scene draw different apparent curves during the motion even if they are of the same depth. In addition, the depth map is no longer common for all the images and must be transformed to a common coordinate system before computing H_k using (7.24) and (7.25). The numerical integration of the velocity field is unfortunately quite time-consuming. A solution could be to precompute the PSF for every possible combination of coordinates (x, y) and depth values. As it is hardly possible, a reasonable solution seems to store them at least on a grid of positions and compute the rest by interpolation. The density of this grid would depend on application.

In [32] we show that obstacles of the general case described above can be avoided by constraining camera motion to only one plane without rotations. This corresponds to vibrations of a camera fixed for example to an engine or machine tool.

A nice property of this case is that the PSF actually changes only its scale proportionally to inverse depth (see Section 7.3.2). As a consequence, if we

(a) Two out-of-focus images taken with aperture $F/5.0$ and $F/6.3$

(b) Results of the algorithm (left) and ground truth (right).

FIGURE 7.11: Removing out-of-focus blur by the algorithm described in Section 7.5.3. The extent of blur increases from front to back.

estimate the PSF for one depth, we know the whole relation between the PSF and depth (7.27). In addition, the depth map is common for all images.

The algorithm works in three steps:

1. PSF estimation at a fixed depth using the blind deconvolution algorithm [24]. A region where the PSF is estimated is specified by user (depth variations must be negligible). This region must be in focus, otherwise we would not be able to separate motion and out-of-focus blur.

2. Rough depth map estimation using a simpler method assuming that the blur is space-invariant in a neighborhood of each pixel (also described in [32]).

3. Minimization of the functional (7.10) to get a sharp image and a more precise depth map.

What concerns the degradation operators G_k in the functional (7.10), the operators D_k and M_k are identities and we work only with a blurring. The minimization proceeds alternately by conjugate gradients in the image subspace and steepest descent in the depth map subspace. We chose total variation for image regularization and Tikhonov regularization for the depth map. Note that the depth map we estimate is relative to the distance of the object on which we estimated the PSF in the first step.

An example in Figure 7.9 illustrates the performance of the algorithm compared to ground truth. Besides the motion blur the photographs contain also defocus but the defocus is common for both images and is not to be removed. Figure 7.10 shows the convolution kernel estimated in the first step of the algorithm (right) and the recovered depth map (left).

Figure 7.11 shows a result of the above described algorithm [32] modified to remove defocus (there is no motion blur in the images). It assumes that the PSF of the lens can be modeled by a pillbox function as a function of depth according to relation (7.13). For minimization of the corresponding functional, we use the same method as in [32]. Details are given in [31].

7.6 Conclusion

Bringing this all together, for the present, the restoration of images blurred by spatially varying blur is not resolved satisfactorily for most cases. In this chapter, we went through the special cases where at least a partial solution is known and we explained the basic principles on which published algorithms are based. We showed that from the Bayesian perspective it is usefull to consider deblurring and super-resolution in one framework.

Many open questions and unresolved problems remain. A large number of blur parameters we need to estimate brings significant errors to the solution and for the present there is no analysis of super-resolution limits for these cases. It may turn out that in many cases super-resolution does not bring much more than mere deblurring. We have shown several algorithms that estimated space-variant blur considering only the deblurring problem. It will be interesting to see if the extension to true super-resolution really works. Especially difficult is the situation when the changes in the PSF are not continuous, e.g., several independently moving objects (motion blur) or even worse, if the PSF depends on the depth of scene (defocus, camera motion).

7.7 Acknowledgments

Financial support of this research was provided by Czech Ministry of Education under the project 1M0572 (Research Center DAR) and the Grant Agency of the Czech Republic under the project 102/08/1593.

Bibliography

[1] Amit Agrawal, Yi Xu, and Ramesh Raskar. Invertible motion blur in video. *ACM Trans. Graph.*, 28(3):1–8, 2009.

[2] M. Ahmed and A. Farag. Non-metric calibration of camera lens distortion. In *Image Processing, 2001. Proceedings. 2001 International Conference on*, volume 2, pages 157–160 vol.2, Oct 2001.

[3] Mark R. Banham and Aggelos K. Katsaggelos. Digital image restoration. *IEEE Signal Process. Mag.*, 14(2):24–41, March 1997.

[4] Leah Bar, Nir A. Sochen, and Nahum Kiryati. Restoration of images with piecewise space-variant blur. In *SSVM*, pages 533–544, 2007.

[5] Moshe Ben-Ezra, Zhouchen Lin, and Bennett Wilburn. Penrose pixels super-resolution in the detector layout domain. In *Proc. IEEE Int. Conf. Computer Vision*, pages 1–8, 2007.

[6] M. Born and E. Wolf. *Principles of Optics*. Pergamon Press, Oxford, 1980.

[7] Dmitry Datsenko and Michael Elad. Example-based single document image super-resolution: a global map approach with outlier rejection. *Multidimensional Syst. Signal Process.*, 18(2-3):103–121, 2007.

[8] Paolo Favaro, Martin Burger, and Stefano Soatto. Scene and motion reconstruction from defocus and motion-blurred images via anisothropic diffusion. In Tomáš Pajdla and Jiří Matas, editors, *ECCV 2004, LNCS 3021, Springer Verlag, Berlin Heidelberg*, pages 257–269, 2004.

[9] Paolo Favaro and Stefano Soatto. A variational approach to scene reconstruction and image segmentation from motion-blur cues. In *Proc. IEEE Conf. Computer Vision and Pattern Recognition*, volume 1, pages 631–637, 2004.

[10] Rob Fergus, Barun Singh, Aaron Hertzmann, Sam T. Roweis, and William T. Freeman. Removing camera shake from a single photograph. *ACM Trans. Graph.*, 25(3):787–794, 2006.

[11] William T. Freeman, Thouis R. Jones, and Egon C Pasztor. Example-based super-resolution. *IEEE Comput. Graph. Appl.*, 22(2):56–65, 2002.

[12] D. J. Heeger and A. D. Jepson. Subspace methods for recovering rigid motion. *International Journal of Computer Vision*, 7(2):95–117, 1992.

[13] Kwang In Kim and Younghee Kwon. Example-based learning for single-image super-resolution. In *Proceedings of the 30th DAGM symposium on Pattern Recognition*, pages 456–465, Berlin, Heidelberg, 2008. Springer-Verlag.

[14] A. Levin. Blind motion deblurring using image statistics. In *NIPS*, pages 841–848, 2006.

[15] Anat Levin, Robert Fergus, Frédo Durand, and William T. Freeman. Image and depth from a conventional camera with a coded aperture. *ACM Trans. Graph.*, 26(3):70, 2007.

[16] Anat Levin, Peter Sand, Taeg Sang Cho, Frédo Durand, and William T. Freeman. Motion-invariant photography. In *SIGGRAPH '08: ACM SIGGRAPH 2008 papers*, pages 1–9, New York, NY, USA, 2008. ACM.

[17] S. H. Lim and D. A. Silverstein. Method for deblurring an image. US Patent Application, Pub. No. US2006/0187308 A1, Aug 24 2006.

[18] James G. Nagy and Dianne P. O'Leary. Restoring images degraded by spatially variant blur. *SIAM J. Sci. Comput.*, 19(4):1063–1082, 1998.

[19] T. Q. Pham. *Spatiotonal adaptivity in super-resolution of under-sampled image sequences*. PhD thesis, TU Delft, 2006.

[20] A. N. Rajagopalan and S. Chaudhuri. An MRF model-based approach to simultaneous recovery of depth and restoration from defocused images. *IEEE Trans. Pattern Anal. Mach. Intell.*, 21(7), July 1999.

[21] L. I. Rudin, S. Osher, and E. Fatemi. Nonlinear total variation based noise removal algorithms. *Physica D*, 60:259–268, 1992.

[22] Huanfeng Shen, Liangpei Zhang, Bo Huang, and Pingxiang Li. A MAP approach for joint motion estimation, segmentation, and super resolution. *IEEE Trans. Image Process.*, 16(2):479–490, February 2007.

[23] F. Šroubek, G. Cristobal, and J. Flusser. A Unified Approach to Super-resolution and Multichannel Blind Deconvolution. *IEEE Transactions on Image Processing*, 16:2322–2332, September 2007.

[24] Filip Šroubek and Jan Flusser. Multichannel blind iterative image restoration. *IEEE Trans. Image Process.*, 12(9):1094–1106, September 2003.

[25] Filip Šroubek and Jan Flusser. Multichannel blind deconvolution of spatially misaligned images. *IEEE Trans. Image Process.*, 14(7):874–883, July 2005.

[26] Muralidhara Subbarao and G. Surya. Depth from defocus: a spatial domain approach. *International Journal of Computer Vision*, 3(13):271–294, 1994.

[27] M. Tico, M. Trimeche, and M. Vehvilainen. Motion blur identification based on differently exposed images. In *Proc. IEEE Int. Conf. Image Processing*, pages 2021–2024, 2006.

[28] D. Tschumperlé and R. Deriche. Diffusion PDE's on vector-valued images. *IEEE Signal Process. Mag.*, 19(5):16–25, September 2002.

[29] David Tschumperlé and Rachid Deriche. Vector-valued image regularization with PDEs: A common framework for different applications. *IEEE Transactions on Pattern Analysis and Machine Intelligence*, 27(4):506–517, 2005.

[30] Damon L. Tull and Aggelos K. Katsaggelos. Regularized blur-assisted displacement field estimation. In *Proc. Int. Conf. Image Processing*, volume 3, pages 85–88, September 1996.

[31] Michal Šorel. *Multichannel Blind Restoration of Images with Space-Variant Degradations*. PhD thesis, Charles University in Prague, 2007.

[32] Michal Šorel and Jan Flusser. Space-variant restoration of images degraded by camera motion blur. *IEEE Trans. Image Process.*, 17(2):105–116, February 2008.

[33] Michal Šorel and Filip Šroubek. Space-variant deblurring using one blurred and one underexposed image. In *Proc. Int. Conf. Image Processing*, 2009.

[34] Filip Šroubek, Jan Flusser, and Michal Šorel. Superresolution and blind deconvolution of video. In *Proc. Int. Conf. on Pattern Recognition*, pages 1–4, 2008.

[35] Masahiro Watanabe and Shree K. Nayar. Rational filters for passive depth from defocus. *International Journal of Computer Vision*, 3(27):203–225, 1998.

[36] Wonpil Yu. An embedded camera lens distortion correction method for mobile computing applications. *Consumer Electronics, IEEE Transactions on*, 49(4):894–901, Nov. 2003.

[37] Lu Yuan, Jian Sun, Long Quan, and Heung-Yeung Shum. Image deblurring with blurred/noisy image pairs. In *SIGGRAPH '07: ACM SIGGRAPH 2007 papers*, page 1, New York, NY, USA, 2007. ACM.

[38] Barbara Zitová and Jan Flusser. Image registration methods: a survey. *Image and Vision Computing*, 11(21):977–1000, 2003.

8

Toward Robust Reconstruction-Based Super-Resolution

Masayuki Tanaka

Tokyo Institute of Technology

Masatoshi Okutomi

Tokyo Institute of Technology

CONTENTS

8.1 Introduction

Super-resolution is a technique to reconstruct a high-resolution image from low-resolution images. Super-resolution algorithms are roughly divisible into two main categories. One category includes reconstruction-based algorithms. The other includes learning-based algorithms. Learning-based algorithms estimate high-frequency image details based on the training database or the model learned with the training database. Learning-based algorithms are usually applied to enhance the resolution of single low-resolution images. Reconstruction-based algorithms reconstruct the high-resolution image using overlapped multiple low-resolution images. This chapter presents discussion of reconstruction-based algorithms. Reconstruction-based algorithms consist of two phases: high-accuracy image registration and super-resolution (SR) reconstruction. In image registration, multiple input images are registered with subpixel accuracy. Then SR reconstruction is performed based on the registration information and the multiple input images. Accurate registration for real scenes sometimes fails because of nonrigid motions, luminance changes, occlusions, and multiple motions, which all must be addressed adequately. For practical super-resolution, robustness against them is a key factor. Since reconstruction-based super-resolution consists of the registration and reconstruction, we must improve the robustness of both the registration and the super-resolution. A common key technique for both robust registration and robust reconstruction is outlier rejection, or to select appropriate pixels that can correctly contribute the high-resolution image reconstruction. The robustness of the registration and the super-resolution could be improved if we were able to select appropriate pixels. We introduce a novel outlier rejection method based on the similarity measure and displacement estimation. In this chapter, we briefly overview existing robust registration and robust super-resolution techniques. Then, we address the novel robust and accurate registration method consisting of feature-based registration, region-based registration, and outlier rejection.

Section 8.2 presents a brief description of existing robust SR reconstruction and robust registration. The proposed SR reconstruction, using pixel selection based on the similarity measure and displacement estimation, is discussed in Section 8.3. We demonstrate that the proposed SR reconstruction can improve the image quality using the motion vectors encoded in the MPEG data in Section 8.4. We also propose, in Section 8.5, robust and accurate image registration algorithms: a feature-based algorithm, a region-based algorithm, and a region extraction algorithm using the proposed pixel selection.

8.2 Overviews

8.2.1 Super-Resolution Reconstruction

The reconstruction process of the reconstruction-based super-resolution can be formulated as an inverse problem that estimates the source of the high-resolution image from observed low-resolution images assuming an image generative model. Several image generative models have been proposed for SR reconstruction [5, 12, 20, 13]. A widely used image generative model is

$$\boldsymbol{y}_i \;=\; \boldsymbol{D}\,\boldsymbol{H}_i\,\boldsymbol{F}_i\,\boldsymbol{x} + \boldsymbol{v}_i \,, \tag{8.1}$$

where $\boldsymbol{y}_i$ is the vectorized i-th observed image, $\boldsymbol{x}$ is the vectorized high-resolution image, $\boldsymbol{F}_i$ is a matrix representing the motion of the i-th image, $\boldsymbol{H}_i$ is a matrix representing the point-spread function (PSF) of the i-th image, $\boldsymbol{D}$ is a matrix representing the downsampling, and $\boldsymbol{v}_i$ represents the noise of the i-th image.

A maximum a posteriori (MAP) estimation is the popular solution of the inverse problem associated with the generative model in Eq. (8.1). The MAP estimation is performed by minimizing the cost function:

$$E \;=\; \sum_{i=1}^{N} ||\boldsymbol{y}_i - \boldsymbol{D}\,\boldsymbol{H}_i\,\boldsymbol{F}_i\,\boldsymbol{x}||_2^2 + \alpha\,\lambda\,(\boldsymbol{x}) \,, \tag{8.2}$$

where $\lambda(\boldsymbol{x})$ is a constraining function derived from the prior distribution of the high-resolution image, α is a hyper-parameter, and N is the number of observed images. The cost function is the negative logarithm of the posterior distribution of the high-resolution image when the observed low-resolution images are given, wherein the noise model is assumed to be independent Gaussian distribution.

Tikhonov regularization, $\lambda(\boldsymbol{x}) = ||\boldsymbol{Q}\,\boldsymbol{x}||_2^2$, is a traditional constraint function in which the matrix $\boldsymbol{Q}$ represents a high-pass filter so that the regularization represents the general smoothness constraint. Tikhonov regularization is known to tend to be oversmooth. Therefore, several edge-preserving smoothness constraints have been proposed [22, 13, 15].

Herein, we specifically examine the first term of the cost function, or the fidelity term, to improve the SR reconstruction robustness. The fidelity term is related directly to the image generative model in Eq. (8.1). However, this image generative model is often insufficient for practical applications.

In the real application of the super-resolution, a set of low-resolution images is usually given, meaning that the PSF and the motion should be assumed and/or estimated to construct the image generative model. For this reason, accurate registration between low-resolution images is important. However, accurate subpixel registration is well known to be a difficult problem. The estimated motion is usually inaccurate and includes registration error. That is true also for PSF estimation. In addition, the image generative model in Eq. (8.1) does not express illumination changes and/or occlusions.

All of the registration error, the inaccurate PSF, the illumination changes, and the occlusions violate the image generative model in Eq. 8.1. It is possible to incorporate these errors into a generative model, if we know the statistics of these errors. However, the statistics of these errors are usually unknown and complicated. These errors of the image generative model must be addressed to improve the robustness of SR reconstruction. In other words, outlier handling is the key to robust SR reconstruction.

8.2.2 Robust SR Reconstruction

First, we show the pixel-wise cost function to express a general robust cost function. The pixel-wise generative model can be derived from Eq. (8.1) as

$$y_{i,j} \;=\; \boldsymbol{b}_{i,j}^T \cdot \boldsymbol{x} + v_{i,j} \,, \tag{8.3}$$

where $y_{i,j}$ is the j-th pixel value of the i-th observed image, $v_{i,j}$ represents the noise of the j-th pixel value of the i-th observed image, and $\boldsymbol{b}_{i,j}^T$ is defined as

$$\boldsymbol{D}\,\boldsymbol{H}_i\,\boldsymbol{F}_i \;=\; \begin{pmatrix} \boldsymbol{b}_{i,1}^T \\ \boldsymbol{b}_{i,2}^T \\ \vdots \end{pmatrix}. \tag{8.4}$$

The cost function of the MAP estimation in Eq. 8.2 can also be rewritten as a pixel-wise expression,

$$E \;=\; \sum_{i=1}^{N}\sum_{j=1}^{M_i} \left[y_{i,j} - \boldsymbol{b}_{i,j}^T \cdot \boldsymbol{x}\right]^2 + \alpha\,\lambda\,(\boldsymbol{x}) \,, \tag{8.5}$$

where M_i is the pixel number of the i-th observed image.

The occlusion and/or the registration error are not involved in the pixel-wise generative model. Therefore, we must reject the occluded pixels and the pixels with registration error as outliers for SR reconstruction. This is the general concept of robust SR reconstruction.

The mathematical expression to reject the certain pixel is to assign it zero weight. The other approach is to use a robust error function or M-estimator instead of the L_2 norm.

The general expression of the cost function of the robust SR reconstruction is

$$I \;=\; \sum_{i=1}^{N}\sum_{j=1}^{M_i} w_{i,j}\,\rho\left(\boldsymbol{b}_{i,j}^T \cdot \boldsymbol{x} - y_{i,j}\right) + \alpha\,\lambda\,(\boldsymbol{x}) \,, \tag{8.6}$$

where, $w_{i,j}$ is a real value weight or a binary mask for the j-th pixel of the

i-th observed image, and function $\rho(x)$ is a robust error function. The key of robust super-resolution is how to design the weight and robust error function adaptively. The existing SR reconstruction algorithms are also classifiable into the pixel-weight approach [26, 17, 16, 18] and the robust error function approach [10, 28, 13].

First, we overview the pixel-weight approach. In the pixel-weight approach, zero weights or small weights are used to reject the outlier pixels, where the error function is the L_2 norm.

Zhao and Sawhney [26] have proposed robust SR reconstruction, which assigns zero weight pixel-by-pixel based on the similarity between the reference image and registered input images, where the reference image is the image to be super-resolved. They use cross-correlation to measure the similarity and apply a simple thresholding approach. They concluded that super-resolution with unstable registration such as optical flow is possible if outlier pixels that are occluded and/or those of large registration error can be rejected.

Ivanovski et al. [17] proposed the outlier rejection algorithm based on differences between pixel values. They compared several criteria for pixel rejection: the pixels of registered input image are compared to the median of the registered pixels (MDM), the initially interpolated high-resolution image (MDIM), and the current estimated high-resolution image (MDSRE). They reported that MDM is slightly better than MDIM and MDSRE. They also mentioned that MDIM and MDSRE yield similar results.

Lee and Kang [18] proposed an algorithm that assigns real-value weights instead of binary weights frame-by-frame. They design the weight to satisfy the following properties: the weight is inversely proportional to $||\boldsymbol{y}_i - \boldsymbol{D}\,\boldsymbol{H}_i\,\boldsymbol{F}_i\,\boldsymbol{x}||_2^2$, and the weight is proportional to the constraint term. Their algorithm, which assigns a lower weight to a frame that has the larger fidelity term, can adaptively remove an input frame with large registration error. Their cost function is not a quadratic form with respect to the high-resolution image because the weight is a function of the high-resolution image. Although they experimentally described that the algorithm converges, that convergence is not theoretically guaranteed.

He and Kondi [16] extended Lee's algorithm with mathematical analysis of the convergence. Although the basic idea is similar to that of Lee's algorithm, they mathematically guarantee the convergence of their algorithm.

The second approach is to use the robust error function. Although the L_2 norm is derived from the Gaussian distribution, the L_2 norm is too sensitive to outliers. In other words, the outlier pixels have a dominant effect for the L_2 norm case. Therefore, the robust error function is designed to have less effect by the outlier pixels.

Farsiu et al. [11] proposed the L_1 norm to use for the fidelity term instead of the L_2 norm. The error function of the L_1 norm case is $\rho(x) = |x|$. The L_1 norm has a smaller effect for large-error pixels than that of the L_2 norm. In this regard, the L_1 norm is more robust than the L_2 norm.

Other famous robust error functions are Lorentzian, Huber, and Geman–

McClure. Actually, SR reconstruction algorithms using the Lorentzian function have been proposed [21, 10]. A Lorentzian function is defined as

$$\rho(x;\tau) \;=\; \log\left(1 + \frac{1}{2}\frac{x^2}{\tau^2}\right), \tag{8.7}$$

where τ is a thresholding parameter.

Patanavijit and Jitapunkul [21] use the Lorentzian function for the data fidelity term. They manually set the thresholding parameter τ. Then they reported that their algorithm is robust and better than either the L1 or L2 norm.

The different thresholding parameter yields different results. Therefore, El-Yamany and Papamichalis [10] proposed an adaptive design of the thresholding parameter for each frame based on similarities of the frame between the reference frame and the registered input frames. They assign a smaller thresholding parameter for the frame that has a lower similarity to the reference frame.

Both the pixel-weight and the robust error function approaches have a common framework: outlier handling is performed based on similarities. However, similarities have strong correlation with the texture, as described in Section 8.3.1. Therefore, we propose a novel pixel-weighting algorithm based on similarities and registration error estimations.

8.2.3 Robust Registration

General image registration includes photometric and geometric registration [5]. In this chapter, we specifically examine geometric registration, which is a process to obtain a dense correspondence between multiple images. In a general sense, the dense correspondence must be defined for every image point. However, if we can assume a motion model such as planar projective transformation, then we can obtain a dense correspondence by estimating the motion model parameters. Registration is generally complex and includes many challenging problems, especially for super-resolution, which requires subpixel accuracy registration.

Registration algorithms are classifiable into two major categories: region-based algorithms and feature-based algorithms. Region-based algorithms estimate the motion parameters to minimize the intensity difference between the reference image and the warped input image for a given region [3]. Region-based algorithms can estimate with subpixel accuracy. Region-based algorithms are known to be very sensitive to illumination changes, specular effects, occlusions, etc. Black and Anandan [4] proposed a registration algorithm to improve the robustness using M-estimator.

One other challenge is to specify the region that includes only a single motion. Region-based algorithms can estimate the parameters of the single motion in the given region. However, the region in real scenes often includes multiple motions, such as the foreground and the background. In that case, we must specify the region that includes only a single motion before the registration process. However, region specification is not an easy task.

The second category of registration includes feature-based algorithms. Feature-based algorithms perform registration based on point matching instead of the difference of pixel intensities. In feature-based algorithms, the feature points are first extracted from the reference image and the input image. Subsequently, feature point matching is performed based on feature descriptors. SIFT [19], which is robust and applicable for large motions, is a powerful tool for feature point extraction and matching. The motion parameters are estimated based on corresponding feature points with the elimination of outliers. Several algorithms for outlier elimination have been proposed [8, 14, 7]. Feature-based algorithms are robust compared to region-based algorithms because feature-based algorithms can easily include outlier elimination mechanisms. Super-resolution requires information related to the motion and the region associated with the motion. However, feature-based algorithms merely provide the motion parameters and the inlier feature point pairs. The associated region should be estimated after feature-based registration. Therefore, in Section 8.5, we propose a novel algorithm that sequentially estimates the multiple motions and the associated regions.

8.3   Robust SR Reconstruction with Pixel Selection

In this section, we propose a novel SR reconstruction with pixel selection based on both the similarity measure and displacement estimation. Existing pixel selection algorithms only use the similarity measure. First, we show a strong dependence of the similarity measure on the image texture. The similarity measure is insufficient to reject a pixel with the registration error. Then, pixel selection based on both the similarity measure and displacement estimation is proposed. Experiments are done to assess whether the proposed algorithm can correctly reject both the occluded pixels and the pixels with the registration error. Results show that the proposed SR reconstruction can reconstruct the high-resolution images correctly, even if the registration is not accurate.

8.3.1   Displacement and Similarity Measure

For outlier rejection, the existing SR reconstruction algorithms fundamentally use similarities between the reference image and the warped input image. These algorithms reject pixels that have low similarities. This is a common idea for both the pixel weighting approach and the robust error function approach.

However, outlier rejection based on the similarity measure has limitation. This outlier rejection can remove the occluded pixels and the pixels whose luminance is changed because the similarities of these pixels present a large

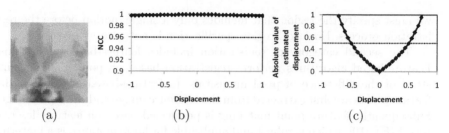

(a)   (b)   (c)

FIGURE 8.1: Displacement and NCC of the poor texture image: (a) Image,
(b) NCC, and (c) Absolute value of the estimated displacement using NCC.

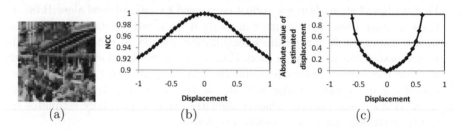

(a)   (b)   (c)

FIGURE 8.2: Displacement and NCC of the rich texture image: (a) Image,
(b) NCC, and (c) Absolute value of the estimated displacement using NCC.

difference between the reference image and the warped input image. Furthermore, the difference of the similarities depends strongly on the texture, especially when the pixels include registration error, which cannot be evaluated
using the similarity measure alone. For this reason, it is difficult to remove the
pixels with the registration error for outlier rejection based on the similarity
measure alone.

We show, experimentally, the texture dependency of the similarity when
the registration error is included. We evaluate similarities of the poor and
the rich texture images shifting one image to other. The Normalized Cross-
Correlation (NCC) is used for the similarity evaluation. The calculated NCC
for the poor texture image and the rich texture image are portrayed respectively in Figures 8.1 and 8.2. Images with subpixel displacement are synthesized by down-sampling from the high-resolution image.

The NCCs of the poor texture image are almost identical even if subpixel
displacements exist, although the NCCs of the rich texture image decrease
for the subpixel displacements. The pixels, whose registration error is greater
than 0.5, can be removed using the outlier rejection algorithm based on the
similarity measure if we set the threshold of the NCC 0.96 for the rich texture
image. However, for images with poor texture, the outlier rejection algorithm
based on the similarity measure judges inlier pixels, even if those pixels have a

large registration error. The subpixel level registration error must be handled because the super-resolution requires subpixel accuracy registration.

Our idea is to evaluate the registration error directly instead of the similarity measure. A parabola fitting algorithm is used to estimate the subpixel displacement [23]. In a one-dimensional case, the parabola-fitting algorithm gives subpixel displacement from three pixel-level similarities as

$$\Delta = \frac{R(-1) - R(1)}{2R(-1) - 4R(0) + 2R(1)} \tag{8.8}$$

where Δ is the estimated subpixel displacement, $R(0)$ is the pixel-level maximum similarity, and $R(1)$ and $R(-1)$ respectively signify similarities for which one image is shifted one pixel from the pixel-level maximum position. Figures 8.1(c) and 8.2(c) show the estimated subpixel displacement for each given displacement, where we show the absolute value of the displacement because only the absolute value is important for outlier rejection. These results show that the displacement estimation is independent of the texture, in contrast to the similarity measure, which depends strongly on the texture. For instance, when the estimated displacement at a position is larger than 0.5 pixels, we can reject the pixel at the position for the poor and rich texture images. Therefore, we propose novel pixel selection algorithms using the similarity measure and the displacement estimation.

8.3.2   Proposed Pixel Selection Algorithm

First, we specify what kind of pixels should be used for SR reconstruction. To reconstruct the high-resolution image robustly, we should use pixels that satisfy the following properties:

1. no occlusions,

2. registration with subpixel accuracy, and

3. no luminance change.

In the following section, we present our pixel selection algorithm in consideration of these properties. The proposed algorithm selects pixels based on the similarity measure and the displacement estimation, so that the pixels that are not occluded are registered correctly with subpixel accuracy. In the proposed robust SR reconstruction, luminance correction is also performed for selected pixels. Then, the high-resolution image is reconstructed using these pixels.

8.3.2.1   Pixel Selection Based on Similarity Measure and Displacement Estimation

We use the similarity measure and the displacement estimation to select pixels for robust SR reconstruction, as discussed earlier. The NCC is used for the similarity measure. Luminance correction is performed after pixel selection.

Therefore, the NCC becomes a suitable similarity measure with the proposed algorithms. The NCC of the region that includes aliasing effects tends to be a small value. Therefore, a low-pass filter is applied to the image before evaluating the NCC to reduce the aliasing effects. For subpixel displacement estimation, we apply two-dimensional parabola fitting using eight-neighbor similarities.

Suppose that the reference image $I_b(x, y)$, the input image $I_k(x, y)$, and the estimated warp parameters are given, where (x, y) is the image coordinate in the low-resolution image. The NCC for each pixel between the input image and the reference image that is warped using the estimated warp parameter is calculated. The local NCC at the position (x, y) in the input image is

$$R(\xi, \eta; x, y) = \frac{\displaystyle\sum_{(u,v)\in C(x,y)} [I_k(u, v) \times I_b'(u + \xi, v + \eta)]}{\sqrt{\displaystyle\sum_{(u,v)\in C(x,y)} I_k(u, v)^2 \times \displaystyle\sum_{(u,v)\in C(x,y)} I_b'(u + \xi, v + \eta)^2}}, \qquad (8.9)$$

where $I_b'(x, y)$ is the warped reference image, (ξ, η) represents the translational displacement, $C(x, y)$ is the neighbor pixel at the position (x, y), and (u, v) is the image coordinate to represent the neighbor pixel.

Let consider the input image pixel at the position (x, y). When the position (x, y) is given, the NCC $R(\xi, \eta; x, y)$ can be considered as a function of the displacement (ξ, η). Then, the NCCs are calculated for the nine displacements: (-1,-1), (-1,0), (-1,1), (0,-1), (0,0), (0,1), (1,-1), (1,0), and (1,1). To estimate subpixel displacement from these nine points of NCCs, we fit these data to the two-dimensional parabola function, $z(\xi, \eta) = a\xi^2 + b\xi\eta + c\eta^2 + d\xi + e\eta + f$, by least squares [24]. The translational displacement is estimated by finding the maximum point of the fitted parabola. The fitted parabola must be concave upward to have the maximum point. The necessary and sufficient condition for being concave upward is

$$a < 0, \quad c < 0, \quad D < 0, \qquad (8.10)$$

where $D = b^2 - 4ac$. In the case where the fitted parabola function is not concave upward, it can be considered that registration of the associated pixel has failed. Then, the associated pixel is rejected. The translational displacement $(\hat{\xi}, \hat{\eta})$ is estimated as

$$\begin{cases} \hat{\xi} = \dfrac{2cd - be}{D} \\ \hat{\eta} = \dfrac{2ae - bd}{D} \end{cases}. \qquad (8.11)$$

That result reflects that the registration of the associated pixel is inaccurate if the estimated translational displacement is greater than 0.5 pixels. These pixels are also rejected.

The maximum NCC should be corrected if the subpixel translational displacement exists. The corrected maximum NCC is obtainable as

$$z_0 = \frac{ae^2 + cd^2}{D} + f.$$ (8.12)

The associated pixel is rejected if the corrected maximum NCC is a small value for some reason, such as an occlusion.

In summary, the proposed pixel selection algorithm selects pixels that satisfy

$$z_0 \geq \kappa, \ |\hat{\xi}| \leq 0.5, \ |\hat{\eta}| \leq 0.5,$$ (8.13)

where κ is the threshold for the similarity measure. The main difference from the existing robust SR reconstruction algorithm is that the proposed algorithm selects pixels based on the similarity measure and the displacement estimation, while existing algorithms use only the similarity measure.

8.3.3 Luminance Correction

The luminance changes also degrade super-resolution results. An approach to handle the luminance changes is to reject the pixels if the luminance of the corresponding pixel is changed. However, this approach also reduces the number of pixels that contribute to reconstruction of the high-resolution image. Another approach is to correct the luminance. The number of pixels that contribute to reconstruction of the high-resolution image increases if we can correct the luminance. Therefore, we propose a luminance-correction approach.

The luminance can be assumed as spatially smooth [9]. We simply estimate the average luminance using low-pass filtering of the value image represented in the HSV color space. The estimated luminance image is

$$L(x, y) = G(x, y) * V(x, y),$$ (8.14)

where $G(x, y)$ is the low-pass filtering kernel that is typical Gaussian kernel, $*$ represents the convolution operator, and $V(x, y)$ is the pixel value in HSV color space for the position (x, y). The pixel value in HSV color space is given as

$$V = \max\{R, G, B\}.$$ (8.15)

Using the estimated luminance of the reference image and the input image, the pixel value can be corrected as

$$I'_k(x, y) = I_k(x, y) \times \frac{L_b(x, y)}{L_k(x, y)},$$ (8.16)

where $I'_k(x, y)$ is the corrected pixel value, $L_b(x, y)$ is the estimated luminance of the reference image, and $L_k(x, y)$ is the estimated luminance of the input image.

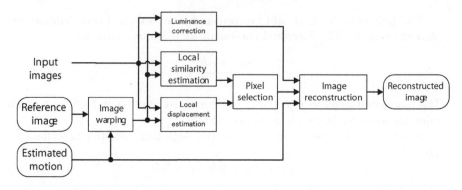

FIGURE 8.3: Process pipeline of the proposed robust SR reconstruction.

8.3.4   Experiments

The proposed robust super-resolution with pixel selection based on the similarity measure and the displacement estimation is compared experimentally to those with the pixel selection based solely on the similarity measure. Figure 8.3 shows the process pipeline of the proposed robust SR reconstruction. We capture an input image sequence, rotating a nonplanar complex shaped object. Three input frames are portrayed in Figure 8.4.

Motion parameters are then estimated for the region, as presented in Figure 8.4(b), assuming that the motion is the planar projective transformation, where the region size is 180×180. The inverse compositional image alignment (ICIA) algorithm [3] is used for motion parameter estimation. Although the planar projective transformation is a theoretically perfect model for the planar object, the object of this experiment is a nonplanar object. Therefore, the estimated motions include subpixel registration error, although motion can be estimated.

For the SR reconstruction, we apply the fast MAP reconstruction with simple smooth prior [25]. Figure 8.5 presents comparisons of the super-resolution results, where the 40×40 sized region is super-resolved to the 120×120 sized image using 60 frames.

The SR reconstruction without the pixel selection yields a blurry result because all pixels are used for reconstruction even if those pixels have large registration error. The result of the SR reconstruction with the pixel selection based on the similarity measure is improved compared to that of the SR reconstruction without the pixel selection. However, the pixel selection based on subpixel registration error is so difficult that only the similarity measure is insufficient to suppress all the blurry effect. The proposed SR reconstruction with the pixel selection based on the similarity measure and the displacement estimation yields the best result among the three algorithms because the proposed pixel selection based on the similarity measure and the displacement estimation can correctly reject the pixels even if these pixels have subpixel

(a) Frame 0   (b) Frame 20   (b) Frame 40

FIGURE 8.4: Input images.

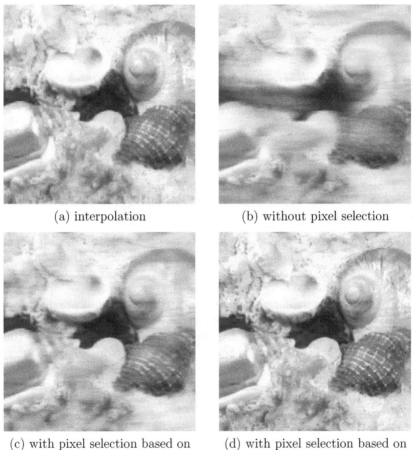

(a) interpolation

(b) without pixel selection

(c) with pixel selection based on
the similarity measure only

(d) with pixel selection based on
the similarity measure and
displacement estimation
(proposed)

FIGURE 8.5: Comparison of super-resolution results.

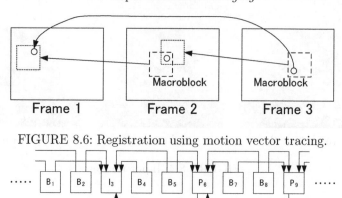

FIGURE 8.6: Registration using motion vector tracing.

FIGURE 8.7: I, P, and B pictures.

registration error. This comparison tells that evaluating the displacement estimation by adding to the similarity measure for the pixel selection can improve the robustness, especially against the subpixel registration error.

8.4 Robust Super-Resolution Using MPEG Motion Vectors

In the previous section, we proposed the robust SR algorithm with pixel selection based on the similarity measure and displacement estimation. The proposed algorithm can correctly reject pixels with subpixel registration error.

In this section, we demonstrate that the proposed robust super-resolution can improve the resolution using the registration which is obtained from the motion compensation vector embedded in compressed video data. Although the embedded motion vectors are often in 1/8 or 1/4 pixel accuracy (or even worse) that is not accurate enough for the SR reconstruction, our robust algorithm is capable of producing higher-resolution images because it can effectively reject the pixels registered at a wrong position. The experimental results show that our SR algorithm is suitable for improving the visual quality of MPEG videos.

8.4.1 Registration Using MPEG Motion Vectors

Motion compensation is usually used in video compression such as the Moving Picture Experts Group (MPEG) [1, 2] standard. A motion vector of each frame is encoded in MPEG data. The motion vector represents the translational motion of the block between two images. The correspondence between any frame distance can be obtained by tracing the motion vectors. Figure 8.6 illustrates an example that is useful to trace the motion vectors from frame 3 to frame 1. This tracing process is designated as motion vector tracing. Motion vector tracing enables us to correspond between frames, even if the motion vectors are not encoded directly between these frames.

The actual motion vector tracing is not as simple as that shown in Figure 8.6 because MPEG data usually include an Intra picture (I-picture), a Predictive picture (P-picture), and a Bidirectionially predictive picture (B-picture). A schematic drawing of the motion vector tracing for typical Group of Pictures (GOP) is illustrated in Figure 8.7. The I-picture does not include the motion vector. The P-picture includes the motion vectors to the previous I- or P- picture. The B-picture includes the motion vectors to the previous and the future I- or P- pictures. The P- and/or B- picture sometimes include I-blocks that do not include the motion vector. Motion estimation is then applied for I-pictures and I-blocks.

Motion vector tracing is the computationally effective algorithm for image registration. However, motion vector tracing is inaccurate for super-resolution. The registration error is also accumulated through the motion vector tracing. The proposed SR algorithm is robust against registration errors and it is still able to super-resolve MPEG.

8.4.2 Experiments of Robust SR Reconstruction

Figure 8.8 shows an image sequence that is synthesized by rotating the Lena image clockwise and occluding the lower right region. This image sequence is encoded using MEPG4 with the Advanced Simple profile. The motion vector tracing algorithm is applied to register all frames to the initial frame (frame 0). Figure 8.9 shows the registration results, where the white region represents the correctly registered region within 0.5 pixel error, and the black region represents the pixels registered incorrectly. We classify registered pixels into three categories: correctly registered, incorrectly registered, or occluded. Pixel quantities of these three categories are counted for each frame and are presented in Figure 8.10. The number of correctly registered pixels decreases as the frame distance between the reference image and the input image increases. That is because, although the motion of the synthesized image sequence is rotation, the translational motion is estimated in the MPEG encoding process. Thus, there are possibly many pixels that are incorrectly registered and the pixel selection algorithms or the robust SR reconstruction algorithms are necessary to reconstruct the high-resolution image by using the motion vector tracing of the MPEG motion vector.

Non-pixel-selection, pixel selection based on similarity measure only, and the proposed pixel selection based on the similarity measure and displacement estimation are applied for motion vector tracing results. In the proposed algorithm, we re-estimate the subpixel displacement for the pixel selection after the vector tracing. Figures 8.11 and 8.12 depict the number of correctly se-

TABLE 8.1: Categories of selected pixels.

	Correctly registered	Falsely registered	Occluded
Non-pixel-selection	63030	288262	86039
Similarity only	50221	138276	0
proposed	32659	11055	0

lected pixels and the number of incorrectly selected pixels in the non-occluded regions, respectively. Although the number of the selected pixels by the proposed pixel selection algorithm is slightly lower than the number of the pixels selected by the pixel selection with the similarity measure only as shown in Figure 8.11, Figure 8.12 shows that the proposed pixel selection rejects incorrectly registered pixels more effectively than the pixel selection with the similarity measure only. Also, the occluded pixels and the pixels with large registration error severely degrade the reconstructed image. Therefore, the proposed pixel selection, which can effectively reject the occluded pixels and the pixels with large registration error, is suitable for super-resolution. Selected pixels are classifiable into three categories: correctly registered pixels, falsely registered pixels, and occluded pixels. Table 8.1 presents the numbers of pixels of respective categories. This result demonstrates that pixel selection based solely on similarity measures can reject the occluded pixels. However, pixel selection based on similarity measures only selects many falsely registered pixels or pixels with large registration error. Therefore, similarity measures can accommodate the occlusion, but it is difficult to reject pixels with registration error using only the similarity measure.

Next, Figure 8.13 shows the comparison of the SR reconstruction by the three algorithms using a cropped section from the synthesized MPEG sequence shown in Figure 8.8. The magnification factor of this example is three. Figures 8.13(b) and 8.13(c) show that the non-pixel-selection approach and the pixel selection algorithm with the similarity measure only produce undesired artifacts and a deformed eye shape. On the other hand, the result of the proposed pixel selection, Figure 8.13(d), successfully suppresses all the artifacts.

| Frame 1 | Frame 9 | Frame 17 | Frame 25 |

FIGURE 8.8: The synthetic image sequence is generated by rotating the Lena image clockwise. The occluding object also moves in from the bottom-right. The resolution of the image sequence is 128×128.

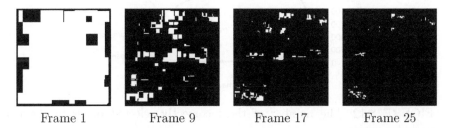

| Frame 1 | Frame 9 | Frame 17 | Frame 25 |

FIGURE 8.9: The results of registering Frames 1, 9, 17, and 25 to Frame 0 by the motion vector tracing of the embedded motion vector in the MPEG data. The white regions represent that correctly registered pixel by the motion vector tracing.

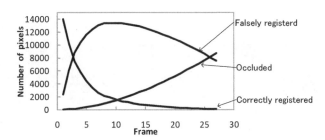

FIGURE 8.10: The number of the pixels that are correctly registered, incorrectly registered and occluded.

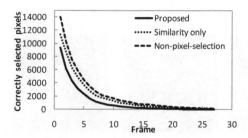

FIGURE 8.11: The number of the correctly selected pixels.

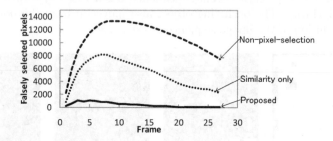

FIGURE 8.12: The number of the incorrectly selected pixels in the non-occluded region.

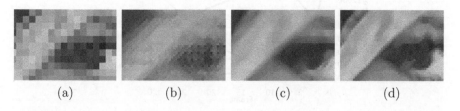

FIGURE 8.13: Comparison of super-resolution results: (a) nearest neighbor interpolation, (b) non-pixel-selection, (c) pixel selection based on similarity measure only, and (d) proposed pixel selection based on the similarity measure and displacement estimation.

Motion vector tracing is applied to the well-known Mobile Calendar sequence. First, the image sequence is downsampled by a factor of two. Then the down-sampled image sequence is encoded using MPEG4 with the Advanced Simple profile. Motion vector tracing is applied to this encoded image sequence. We generate stabilized images for the initial frame to evaluate the performance of the motion vector tracing. The generated stabilized images are portrayed in Figure 8.14, where the black region represents the region in which the motion vector tracing fails to track.

Although motion vector tracing is a very simple algorithm, Figure 8.14 shows that motion vector tracing is sufficient for rough registration. However, registration by the motion vector tracing includes estimation error.

Next, similar to the previous example, we apply the three SR algorithm for the registered pixels and reconstruct a high resolution image with the magnification factor three and the same parameter that we used for the results in Figure 8.13. The results are shown in Figure 8.15.

The result of the SR reconstruction without the pixel selection includes particle noise. It is blurred because the occluded pixels and pixels with registration error are used for the reconstruction. The occluded pixels and the pixels with registration error yield the particle noise and the blur. The result of the SR reconstruction with the pixel selection only with the similarity measure is better than that of the SR reconstruction without the pixel selection. However, several portions are still deformed, especially the letters "5" and "6". The pixels with small registration error yield visible deformation. As discussed above, the pixel selection with only the similarity measure can only reject the occluded pixels and the pixels with large registration error; the pixels with a small registration error are difficult to reject. The two reconstruction examples in Figures 8.13 and 8.15 show that our robust SR algorithm using the proposed pixel selection can reconstruct higher-quality images than the other two algorithms by detecting pixels with subpixel registration errors and reject them effectively.

8.5 Robust Registration for Super-Resolution

As demonstrated in Section 8.4.2, our robust SR algorithm is capable of producing high-resolution images even in the presence of the registration errors. Even though we apply the robust SR reconstruction, we need many pixels to reconstruct high-quality image. When the registration is inaccurate, the pixel selection process rejects a lot of pixels with registration error. As a result, the quality of the reconstruction image cannot be improved with inaccurate registration. An optical flow based registration and the motion vector tracing using the motion vector embedded in the MPEG data as discussed in Section 8.4.1 can register images in the presence of complex motions. However, these two

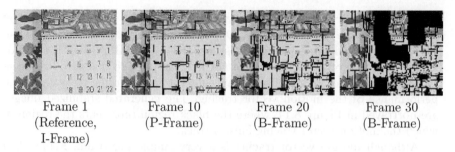

| Frame 1 | Frame 10 | Frame 20 | Frame 30 |
| (Reference, I-Frame) | (P-Frame) | (B-Frame) | (B-Frame) |

FIGURE 8.14: Reference image and stabilized images (Mobile Calendar).

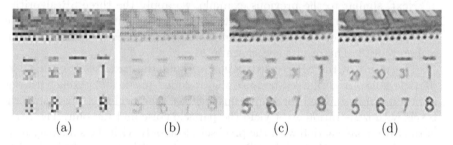

   (a)   (b)   (c)   (d)

FIGURE 8.15: Comparison of super-resolution results (Mobile Calendar): (a) nearest neighbor interpolation, (b) non-pixel-selection, (c) pixel selection based on similarity measure only, and (d) proposed pixel selection based on similarity measure and displacement estimation.

algorithms often register inaccurately. Robust and accurate registration is still preferable to improve the visual quality of the output images further. Robust and accurate registration makes more pixels available for the pixel estimation at every position of the high-resolution image. Hence, the quality of the reconstructed image strongly depends on the accuracy of the registration. In this section, we discuss robust and accurate registration for multiple motions.

8.5.1 Proposed Multiple Motion Estimation

Feature-based algorithms are robust because they can include outlier elimination mechanisms as mentioned in Section 8.2.3. Feature-based algorithms merely provide the motion parameters and the inlier feature point pairs. The region associated to the estimated motion remains unknown. Region-based algorithms can estimate motion parameters with higher accuracy than that provided by feature-based algorithms. However, region-based algorithms require a single motion within the region and initial motion parameters that are close to true parameters. We use these two algorithms complementarily to achieve robust and high-accuracy motion estimation required for super-resolution. First, we describe estimation of the motion parameters and the associated region for a single object. Then, we extend the proposed algorithm to multiple motion estimation.

8.5.1.1 Motion Estimation and Region Extraction for a Single Object

Figure 8.16 shows the motion estimation and the region extraction algorithm for the single object. In general, the region that includes only a single motion is also unknown. Therefore, we need to extract the region associated to the estimated motion.

First, the feature-based algorithm estimates motion parameters between the reference image and the input image, so that the robustness of the proposed registration is improved. Then, the region-based algorithm is applied for accurate registration. The motion parameters estimated by the feature-based algorithm is used for the initial motion parameters of the region-based algorithm. Although the region-based algorithm also requires the region that includes the single motion, the region is not provided by the feature-based algorithm. We extract the single motion region associated with the estimated motion by the same manner of the proposed pixel selection described in Section 8.3.2. We call this process single-motion region extraction. The region-based algorithm refines the motion parameters with the extracted region from the motion parameters estimated by the feature-based algorithm. Finally, we obtain the refined motion parameters and the region.

For feature point extraction and matching, we use SIFT [19]. We apply the PROSAC algorithm [8] to estimate the motion parameters from corresponding points, having eliminated outliers. The region associated with the estimated motion parameters is extracted using the algorithm proposed in Section 8.3.2. The pixel selection algorithm is also used to extract the region. The selected pixels are considered to represent the region. Region extraction using the pixel selection algorithm is simply called region extraction. For the region-based algorithm, we use the ICIA algorithm [3], which is known as a fast and accurate region-based algorithm.

The accuracy of the feature-based algorithm is roughly equivalent to the low-resolution image (LRI) pixel accuracy, whereas super-resolution requires subpixel accuracy. Although the region-based algorithm can estimate motions in subpixel accuracy, the single-motion region and the initial motion parameters are required. Fortunately, region extraction can be done with LRI pixel accuracy estimation. Then region extraction using the motion parameters estimated with the feature-based algorithm generates a single-motion region that is associated with the estimated motion. The motion parameters estimated using feature-based algorithms are useful as initial motion parameters. Once we obtain the region and the initial motion parameters, we can apply the region-based algorithm. The region-based algorithm refines motion parameters so that super-resolution can reconstruct the high-resolution image (HRI).

In summary, our motion estimation algorithm first computes large motions

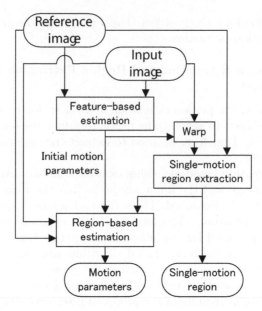

FIGURE 8.16: The block diagram of our robust and accurate estimation algorithm of the single motion.

roughly by the feature-based algorithm, and then refines the motions in sub-pixel accuracy by the region-based algorithm. In addition, unlike the region-based algorithm [3] require a single-motion region, which is usually manually selected, our algorithm selects the single-motion regions automatically.

8.5.1.2 Multiple Motion Estimation

Practical algorithms to classify the feature point pairs based on their motions have been reported [8, 7]. We perform multiple motion estimation in the same manner of the feature point classification algorithm. First, we apply the feature-based algorithm using all extracted and matched feature point pairs. The outputs of the feature-based algorithm are the estimated motion parameters and inlier feature point pairs. These motion parameters represent a dominant motion. The dominant motion is the motion that is associated with the most numerous inlier feature point pairs. Then, the inlier feature point pairs are removed. The feature-based algorithm is applied for the remainder of the feature point pairs. This scheme can sequentially estimate multiple motion parameters and associated feature point pairs.

The difference between the proposed algorithm and the existing algorithms [8, 7] is that the proposed algorithm can estimate the region associated with the dominant motion, whereas existing algorithms merely classify the feature point pairs.

8.5.2 Super-Resolution for Multiple Motions

The MAP-based super-resolution is formulated as the optimization problem [13]. The cost function for the single motion is same Equation (8.2).

The proposed multiple motion estimation algorithm provides multiple motion parameters and associated regions. The estimated region can be considered as a set of inlier pixels associated with motion parameters. In this sense, the proposed multiple motion estimation algorithm is closely related to the robust super-resolution using outlier rejection [17, 26]. Super-resolution with outlier rejection is formulated using the weighted cost function. The weight, or mask, is generated from outlier rejection results. The cost function of the multiple motion super-resolution can be derived using the weight representing the extracted motion regions. The cost function for the multiple motions is

$$E = \sum_{i=1}^{N} \sum_{\ell=1}^{K_i} (A_{i\ell} x - y_i)^T \operatorname{diag}(w_{i\ell}) (A_{i\ell} x - y_i) + \alpha \lambda (x) , \quad (8.17)$$

where the suffix $i\ell$ represents the ℓ-th single motion region of the i-th frame, w is the vectorized weight image, which represents the single-motion region, $\operatorname{diag}(w)$ is the diagonal matrix whose diagonal elements are elements of w, and K_ℓ is the number of motions in the i-th frame. In addition, the matrix $A_{i\ell}$ is

$$A_{i\ell} = D H_i F_{i\ell} , \quad (8.18)$$

where $F_{i\ell}$ is a matrix representing the ℓ-th motion of the i-th frame. The pixel value of the image w is 1 if the pixel is inside the region and 0 if the pixel is outside.

8.5.3 Experiments

We apply the proposed algorithm to a real-image sequence. Figure 8.17 shows that the captured sequence includes two planar objects. Planar homography is assumed for the motion model. These targets move independently. A single planar homography cannot represent the motions of two planes.

Figure 8.18 illustrates the estimated regions, where the first and the second rows show the extracted regions for the left and the right objects, respectively. The results show that the regions of the planar objects are extracted correctly, although the extraction is not perfect, especially the object's boundaries. The correctness of the extracted region is important because super-resolution results are degraded in comparison to the original reference image if the extracted region includes the wrong region. Super-resolution requires highly accurate motion parameters, but perfect regions of the objects are not necessary. Figure 8.19 illustrates the images warped to the reference image for the left and right objects. These regions can also be improved by our SR algorithm.

Next, we apply the proposed robust super-resolution described in Section 8.3 using the motion parameters estimated using the proposed algorithm.

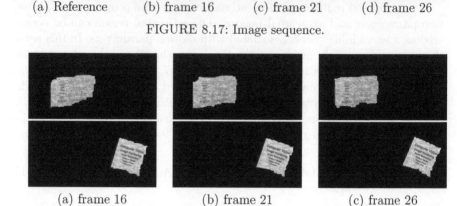

| (a) Reference | (b) frame 16 | (c) frame 21 | (d) frame 26 |

FIGURE 8.17: Image sequence.

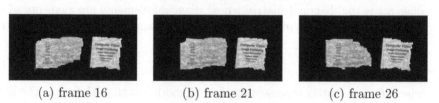

| (a) frame 16 | (b) frame 21 | (c) frame 26 |

FIGURE 8.18: Estimated regions.

| (a) frame 16 | (b) frame 21 | (c) frame 26 |

FIGURE 8.19: Images warped to the reference image.

For comparison, we also generate super-resolution results obtained using the motion parameters estimated using the ICIA algorithm [3]. As regions of interest for the ICIA algorithm, we manually set three different regions: whole image, left object, and right object. For robust super-resolution, we use the outlier rejection algorithm, which is based on local similarity and local motion estimation error [6]. The reconstruction algorithm proposed in [25] is applied, where the zooming factor is 3×3 and the number of frames is 30. Figure 8.20 presents the super-resolution results. We apply the proposed robust super-resolution algorithms, and none of them produce any undesired artifacts. In this regard, the proposed robust super-resolution algorithm is powerful. However, the robust super-resolution cannot improve the image quality of the region that includes registration error. In left images of Figure 8.20, Figures 8.20 (c) and (e) show improved image quality compared to Figures 8.20 (b) and (d). In right images of Figure 8.20, image qualities of Figures 8.20 (d) and (e) are better than those of Figures 8.20 (b) and (c). These differences are caused by the motion estimation errors. The experimental results described herein

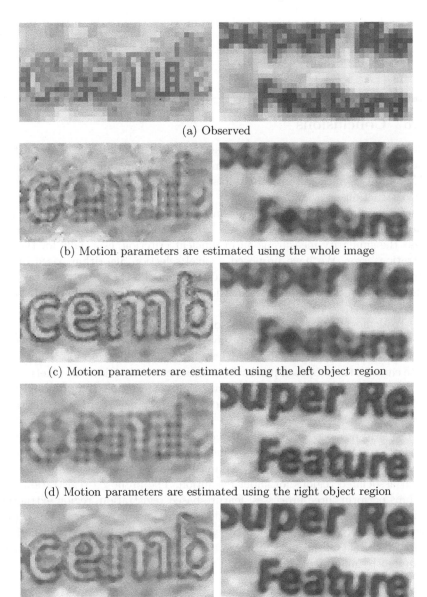

(a) Observed

(b) Motion parameters are estimated using the whole image

(c) Motion parameters are estimated using the left object region

(d) Motion parameters are estimated using the right object region

(e) Motion parameters are estimated using the proposed algorithm

FIGURE 8.20: Super-resolution results using 30 frames, where (a) portrays the observed image (size: 40×20), and (b), (c), (d), and (e) presents super-resolution results (size: 120×60) with different motion estimation. The whole image, the region of the left object, and the region of the right object are used, respectively, for motion estimations of (b), (c), and (d); (e) depicts the result obtained using the proposed algorithm.

demonstrate that the proposed algorithm can estimate the motion parameters of multiple targets.

8.6 Conclusions

Robustness is a key factor of the practical super-resolution. The reconstruction based SR approach consists of image registration and SR reconstruction. Therefore, we must improve the robustness of both image registration and SR reconstruction.

Although several robust SR reconstruction algorithms based on the similarity measure have been proposed [26, 17, 16, 18], we have demonstrated that the similarity measure is an insufficient criterion, especially for the registration error. We have proposed a pixel selection algorithm based on the similarity measure and displacement estimation. The proposed pixel selection algorithm uses displacement estimation directly in order to effectively reject pixels registered incorrectly.

To demonstrate the robustness of the SR reconstruction with the proposed pixel selection, we super-resolved MPEG videos using the embedded motion vectors in MPEG data. Although the embedded motion is not accurate enough, the proposed pixel selection can correctly reject the pixels placed at wrong positions. The experimental results attest that the proposed pixel selection correctly rejects the occluded pixels and pixels that have registration error.

We also proposed a robust and accurate registration for multiple motions in Section 8.5. Real scenes usually include multiple motions such as those of the background and foreground. We must estimate the multiple motions and their associated regions for super-resolution. The proposed registration algorithms perform three processes: feature-based registration, region extraction, and region-based registration. For region extraction, we use the proposed pixel selection algorithm. It can robustly estimate multiple motions with high accuracy without initial motion parameters or an associated region, which are required by many motion estimation algorithms [3, 27].

Bibliography

[1] Generic coding of moving pictures and associated audio: Video. *ISO/IEC*, pages 13818–2, 1996.

[2] Coding of audio-visual objects: Visual. *ISO/IEC*, pages 14496–2, 1998.

[3] S. Baker and I. Matthews. Lucas-Kanade 20 years on: a unifying framework. *International Journal of Computer Vision*, 56(3):221–255, 2004.

[4] M.J. Black and P. Anandan. The robust estimation of multiple motions: Parametric and piecewise-smooth flow fields. *Computer Vision and Image Understanding*, 63(1):75–104, 1996.

[5] D. Capel. *Image mosaicing and super-resolution*. Springer-Verlag New York Inc, 2004.

[6] S Chang, M Shimizu, and M Okutomi. Multi-frame super-resolution with multiple motion regions. *Korean Japan Joint Workshop on Pattern Recognition (KJPR)*, 107(281):57–62, 2007.

[7] O. Choi, H. Kim, and I.S. Kweon. Simultaneous plane extraction and 2D homography estimation using local feature transformations. *Asian Conference on Computer Vision (ACCV)*, 4844:269–278, 207.

[8] O. Chum and J. Matas. Matching with PROSAC-progressive sample consensus. *IEEE Computer Society Conference on Computer Vision and Pattern Recognition (CVPR)*, 1:220–226, 2005.

[9] F. Durand and J. Dorsey. Fast bilateral filtering for the display of high-dynamic-range images. *Proc. the Annual Conference on Computer Graphics and Interactive Techniques*, pages 257–266, 2002.

[10] N.A. El-Yamany and P.E. Papamichalis. Robust color image super-resolution: an adaptive M-estimation framework. *Journal on Image and Video Processing*, 8(2), 2008.

[11] S. Farsiu, M. Elad, P. Milanfar, et al. Multiframe demosaicing and super-resolution of color images. *IEEE Transactions on Image Processing*, 15(1):141–159, 2006.

[12] S. Farsiu, D. Robinson, M. Elad, and P. Milanfar. Advances and challenges in super-resolution. *International Journal of Imaging Systems and Technology*, 14(2):47–57, 2004.

[13] S. Farsiu, MD Robinson, M. Elad, and P. Milanfar. Fast and robust multiframe super resolution. *IEEE Transactions on Image Processing*, 13(10):1327–1344, 2004.

[14] M.A. Fischler and R.C. Bolles. Random sample consensus: a paradigm for model fitting with applications to image analysis and automated cartography. *Communications of the ACM*, 24(6):381–395, 1981.

[15] T. Gotoh and M. Okutomi. Direct super-resolution and registration using raw CFA images. In *IEEE Computer Society Conference on Computer Vision and Pattern Recognition (CVPR)*, volume 2. IEEE Computer Society; 1999, 2004.

[16] H. He and LP Kondi. An image super-resolution algorithm for different error levels per frame. *IEEE Transactions on image processing*, 15(3):592–603, 2006.

[17] Z.A. Ivanovski, L. Panovski, and L.J. Karam. Robust super-resolution based on pixel-level selectivity. In *Proceedings of SPIE*, volume 6077, 2006.

[18] E.S. Lee and M.G. Kang. Regularized adaptive high-resolution image reconstruction considering inaccurate subpixel registration. *IEEE Transactions on Image Processing*, 12(7):826–837, 2003.

[19] D.G. Lowe. Distinctive image features from scale-invariant keypoints. *International Journal of Computer Vision*, 60(2):91–110, 2004.

[20] S.C. Park, M.K. Park, and M.G. Kang. Super-resolution image reconstruction: a technical overview. *IEEE Signal Processing Magazine*, 20(3):21–36, 2003.

[21] V. Patanavijit and S. Jitapunkul. A Lorentzian stochastic estimation for a robust iterative multiframe superresolution reconstruction with Lorentzian-Tikhonov regularization. *EURASIP Journal on Advances in Signal Processing*, 2007(2):21–21, 2007.

[22] R.R. Schultz and R.L. Stevenson. Extraction of high-resolution frames from video sequences. *IEEE Transactions on Image Processing*, 5(6):996–1011, 1996.

[23] M. Shimizu and M. Okutomi. Subpixel estimation error cancellation on area-based matching. *International Journal of Computer Vision*, 63(3):207–224, 2005.

[24] C. Sun. Fast algorithms for stereo matching and motion estimation. *Proc. of Australia-Japan Advanced Workshop on Computer Vision*, pages 38–48, 2003.

[25] M. Tanaka and M. Okutomi. A fast MAP-based super-resolution algorithm for general motion. *Electronic Imaging Computational Imaging IV*, 6065:1–12, 2006.

[26] W.Y. Zhao and H.S. Sawhney. Is super-resolution with optical flow feasible? *Lecture Notes in Computer Science*, pages 599–613, 2002.

[27] B. Zitová and J. Flusser. Image registration methods: a survey. *Image and Vision Computing*, 21(11):977–1000, 2003.

[28] A. Zomet, A. Rav-Acha, and S. Peleg. Robust super-resolution. In *IEEE Computer Society Conference on Computer Vision and Pattern Recognition (CVPR)*, volume 1, 2001.

9

Multiframe Super-Resolution from a
Bayesian Perspective

Lyndsey Pickup, Stephen Roberts, Andrew Zisserman
University of Oxford

David Capel
2d3 Inc.

CONTENTS

This chapter examines multiframe image super-resolution in a probabilistic framework. Many multiframe super-resolution algorithms begin by a point estimate of the unknown *latent* parameters, such as those describing the motion or the blur function. The focus of this chapter is on alternatives to this practice that can yield superior super-resolution results.

We begin with the generative model and simple Maximum Likelihood (ML) and Maximum a Posteriori (MAP) solutions for the super-resolution image. Then, in the second section, we illustrate the consequences of inaccurate point estimates of the latent parameters on the simple MAP algorithm.

The third section introduces the simultaneous MAP algorithm, which estimates the super-resolution image together with parameters such as the image registration, allowing the high-resolution information to influence and improve the estimates of the latent parameters. In the fourth section, we show how Bayesian marginalization can integrate the latent parameters out of the problem, leading to a cost function in terms of the low-resolution images that can be optimized with respect to the high-resolution pixels directly. We conclude with a brief discussion of the benefits of these two Bayesian approaches to super-resolution.

9.1 The Generative Model

A generative model is a parameterized, probabilistic model of data generation, which attempts to capture the forward process by which observed data (in this case low-resolution images) is generated by an underlying system (the scene and imaging parameters), and corrupted by various noise processes. This translates to a top-down view of the super-resolution problem, starting with the scene or high-resolution image, and resulting in the low-resolution images, via the physical imaging and noise processes.

For super-resolution, the generative model approach is intuitive, since the goal is to recover the initial scene, and an understanding of the way it has influenced the observed low-resolution images is crucial. The generative model's advantage over classical descriptive models is that it allows us to express a probability distribution directly over the "hidden" high-resolution image *given* the low-resolution inputs, while handling the uncertainty introduced by the noise.

A high-resolution scene x, with N pixels (represented as an $N \times 1$ vector), is assumed to have generated a set of K low-resolution images, where the kth such image is $y^{(k)}$, and has M pixels. The warping (e.g., to account for motion), blurring, and subsampling of the scene is modelled by an $M \times N$ sparse matrix $W^{(k)}$ [4, 15], and a global affine photometric correction results from multiplication and addition across all pixels by scalars $\lambda_1^{(k)}$ and $\lambda_2^{(k)}$ respectively [4]. Thus, the generative model for one of the low-resolution images

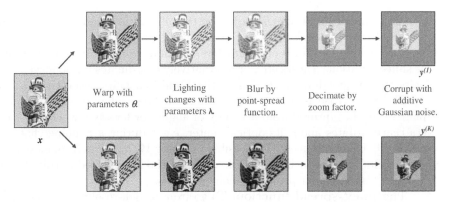

FIGURE 9.1: The generative model for two typical low-resolution images.

is

$$y^{(k)} = \lambda_1^{(k)} W^{(k)} x + \lambda_2^{(k)} 1 + \mathcal{N}\left(0, \beta^{-1} I\right), \tag{9.1}$$

where λ_2 is the scalar λ_2 multiplied by a vector of ones, and the final term on the right is a noise term consisting of *i.i.d.* samples from a zero-mean Gaussian with precision β, or alternatively with standard deviation σ_N, where $\beta^{-1} = \sigma_N^2$.

Figure 9.1 shows the generative model for two typical greyscale low-resolution images in terms of the images at each step in the procedure. On the left is the single ground truth scene, and on the extreme right are the two images (in this case $y^{(1)}$ and $y^{(K)}$, which might represent the first and last images in a K-image sequence) as they are observed by the camera sensors. Given a set of low-resolution images like this, $\left\{y^{(k)}\right\}$, the goal is to recover x, without knowing the values associated with $\left\{W^{(k)}, \lambda^{(k)}, \sigma_N\right\}$.

9.1.1 Considerations in the Forward Model

While specific elements of W are unknown, it is still highly structured, and generally can be parameterized by relatively few values compared to its overall number of nonzero elements, though this depends upon the type of motion assumed to exist between the input images, and on the form of the point-spread function.

Motion Models: Early super-resolution research was predominantly concerned with simple motion models where the registration typically had only two or three degrees of freedom (DoF) per image, e.g., from datasets acquired using a flatbed scanner and an image target. Some models are even more restrictive, and in addition to the 2DoF shift-only registration, the low-resolution image pixel centers are assumed to lie on a fixed integer grid on the super-resolution image plane [7, 13].

Affine (6DoF) and planar projective (8DoF) motion models are generally applicable to a much wider range of common scenes. These are typically parameterized by a vector $\boldsymbol{\theta}$ of parameters, which may either be quantities such as shifts, rotations, shears etc., or alternatively coordinates for nondegenerate pairs of corresponding image points (three and four pairs, respectively).

The 8DoF case is most suitable for modelling planar or approximately planar objects captured from a variety of angles, or for cases where the camera centre rotates about its optical center, e.g., during a panning shot in a movie [8]. Though it is equally possible to create $\boldsymbol{W}$ matrices from more complex motion models such as optic flow, the methods described in this chapter are based upon planar projective homographies.

The point-spread function: To go from a high-resolution image (or a continuous scene), to a low-resolution image, the function representing the light levels reaching the image plane of the low-resolution image is convolved with a *point-spread function* (PSF) and sampled at discrete intervals to represent the low-resolution image pixels. This point-spread function can be decomposed into factors representing the blurring caused by camera optics and the spatial integration performed by a CCD sensor [1].

Generally, the PSF associated with each low-resolution pixel is approximated by a simple parametric function centered on the corresponding location in the high-resolution: the two most common are an isotropic 2D Gaussian with a covariance $\sigma_{PSF}^2 \boldsymbol{I}$, or a circular disk (top-hat function) with a radius r_{PSF}.

The noise model: The image noise is assumed to be *i.i.d.* Gaussian, which leads to an L_2 data error measure. Several common types of image noise contain more structure than this model would predict, e.g., noise levels which depend on observed pixel intensities, or errors introduced by quantization artifacts in the JPEG or MPEG compression process. However, it can be shown that even for nonlinear noise such as JPEG quantization, the Gaussianity assumption performs well [14], and has the benefit that the problem remains convex. Some authors assume other families of noise, e.g., Laplacian or salt-and-pepper noise, for which the L_1 norm is more appropriate [6]. It is worth noting that in most of the algorithms presented in this chapter the L_1 norm can be substituted into the derived objective functions if appropriate, though gradient expressions must then be adjusted accordingly.

Constructing $\boldsymbol{W}^{(k)}$: Each low-resolution image pixel can be created by dropping a blur kernel into the high-resolution scene and taking the corresponding weighted sum of the pixel intensity values. The center of the blur kernel is given by the location of the center of low-resolution pixel when its location is mapped into the frame of the high-resolution image. This means that the ith row in $\boldsymbol{W}^{(k)}$ represents the kernel for the ith low-resolution image pixel over the whole of the high-resolution image, and $\boldsymbol{W}^{(k)}$ is therefore sparse, because pixels far from the kernel center should not have significantly nonzero weights.

9.1.2 A Probabilistic Setting

From the generative model given in (9.1) and the assumption that the noise model is Gaussian, the likelihood of a low-resolution image $\boldsymbol{y}^{(k)}$, given the high-resolution image $\boldsymbol{x}$, geometric registration parameters $\boldsymbol{\theta}^{(k)}$ and photometric registration parameters $\boldsymbol{\lambda}^{(k)}$, may be expressed

$$p\left(\boldsymbol{y}^{(k)}\,\Big|\,\boldsymbol{x},\boldsymbol{\theta}^{(k)},\boldsymbol{\lambda}^{(k)}\right) \;=\; \left(\frac{\beta}{2\pi}\right)^{\frac{M}{2}}\exp\left\{-\frac{\beta}{2}\left\|\boldsymbol{y}^{(k)}-\lambda_1^{(k)}\boldsymbol{W}^{(k)}\boldsymbol{x}-\lambda_2^{(k)}\right\|_2^2\right\},$$

where $\boldsymbol{W}^{(k)}$ is a function of the PSF and of $\boldsymbol{\theta}^{(k)}$.

It can be helpful to think in terms of the *residual* errors, where the residual refers to the parts of the data (in this case our low-resolution images), which are not explained by the model (i.e., the high-resolution estimate), given values for all the imaging parameters. We define the kth residual, $\boldsymbol{r}^{(k)}$, to be

$$\boldsymbol{r}^{(k)} \;=\; \boldsymbol{y}^{(k)}-\lambda_1^{(k)}\boldsymbol{W}^{(k)}\boldsymbol{x}-\lambda_2^{(k)}. \tag{9.2}$$

Using this notation, the compact form of the data likelihood for the whole low-resolution dataset may be written

$$p\left(\left\{\boldsymbol{y}^{(k)}\right\}\Big|\boldsymbol{x},\left\{\boldsymbol{\theta}^{(k)},\boldsymbol{\lambda}^{(k)}\right\}\right) \;=\; \left(\frac{\beta}{2\pi}\right)^{\frac{KM}{2}}\exp\left\{-\frac{\beta}{2}\sum_{k=1}^{K}\left\|\boldsymbol{r}^{(k)}\right\|_2^2\right\}. \tag{9.3}$$

9.1.2.1 The Maximum Likelihood Solution

The Maximum Likelihood (ML) solution to the super-resolution problem is simply the super-resolution image which maximizes the probability of having observed the dataset,

$$\hat{\boldsymbol{x}}_{\text{ML}} \;=\; \arg\max_{\boldsymbol{x}}\left(p\left(\{\boldsymbol{y}^{(k)}\}|\boldsymbol{x},\{\boldsymbol{\theta}^{(k)}\boldsymbol{\lambda}^{(k)}\}\right)\right). \tag{9.4}$$

If all other parameters are known, $\hat{\boldsymbol{x}}_{\text{ML}}$ can be computed directly as the *pseudoinverse* of the problem. Neglecting the photometric parameters for the moment, if $\boldsymbol{y}^{(k)}=\boldsymbol{W}^{(k)}\boldsymbol{x}+\mathcal{N}\left(\boldsymbol{0},\beta^{-1}\boldsymbol{I}\right)$, then the pseudoinverse would be

$$\hat{\boldsymbol{x}}_{\text{ML}} \;=\; \left(\boldsymbol{W}^T\boldsymbol{W}\right)^{-1}\boldsymbol{W}^T\boldsymbol{y}, \tag{9.5}$$

where $\boldsymbol{W}$ is the $KM \times N$ stack of all K of the $\boldsymbol{W}^{(k)}$ matrices, and $\boldsymbol{y}$ is the $KM \times 1$ stack of all the vectorized low-resolution images. Re-introducing the photometric components gives

$$\hat{\boldsymbol{x}}_{\text{ML}} \;=\; \left(\sum_{k=1}^{K}\lambda_1^{(k)2}\boldsymbol{W}^{(k)T}\boldsymbol{W}^{(k)}\right)^{-1}\left[\sum_{k=1}^{K}\lambda_1^{(k)}\boldsymbol{W}^{(k)T}\left(\boldsymbol{y}^{(k)}-\lambda_2^{(k)}\right)\right]. \tag{9.6}$$

Thus we can solve for $\hat{\boldsymbol{x}}_{\text{ML}}$ directly if we know $\left\{\boldsymbol{W}^{(k)},\boldsymbol{\lambda}^{(k)}\right\}$. This can be a time-consuming process if the $\boldsymbol{W}$ matrices are large or have many nonzero elements, and if the matrix $\boldsymbol{W}^T\boldsymbol{W}$ is singular (e.g., when $KM < N$) the direct inversion is problematic. Instead, the ML solution can be found efficiently using a gradient descent algorithm like Scaled Conjugate Gradients (SCG) [12]. Such schemes take an objective function $\mathcal{L}$, and its derivative with respect to the current estimate of the high-resolution image, $\frac{\partial \mathcal{L}}{\partial \boldsymbol{x}}$, and find the optimal $\boldsymbol{x}$ using an iterative scheme. For the ML solution, the expressions of interest are therefore

$$\mathcal{L} = \frac{1}{2}\sum_{k=1}^{K}\left\|\boldsymbol{y}^{(k)} - \lambda_1^{(k)}\boldsymbol{W}^{(k)}\boldsymbol{x} - \boldsymbol{\lambda}_2^{(k)}\right\|_2^2 \tag{9.7}$$

$$= \frac{1}{2}\sum_{k=1}^{K}\left\|\boldsymbol{r}^{(k)}\right\|_2^2 \tag{9.8}$$

$$\frac{\partial \mathcal{L}}{\partial \boldsymbol{x}} = \sum_{k=1}^{K} -\lambda_1^{(k)}\boldsymbol{W}^{(k)T}\boldsymbol{r}^{(k)}. \tag{9.9}$$

When this is initialized with a reasonable estimate of the super-resolution image, this scheme can be used to improve the super-resolution estimate iteratively, even when $KM < N$.

Note that $\mathcal{L}$ is essentially a quadratic function of $\boldsymbol{x}$, so this problem is *convex*. A unique global minimum exists, and gradient-descent methods (which include SCG) can find it given enough steps. Typically one might need up to N steps (where there are N pixels in the super-resolution image) to solve exactly for $\boldsymbol{x}$, but generally far fewer iterations are required to obtain a good image. Using SCG, small super-resolution images (under 200×200 pixels) tend to require fewer than 50 iterations before the super-resolution image intensity values change by less than a gray level per iteration. This requirement roughly scales with the number of pixels under consideration.

9.1.2.2 The ML Solution in Practice

Unfortunately, ML super-resolution is an ill-conditioned problem whose solution is prone to corruption by very strong high-frequency oscillations. To illustrate this, a set of synthetic *Graffiti* datasets is introduced in Figure 9.2, where the number of images and the amplitude of the additive Gaussian noise is varied, and the registration parameters are determined randomly under a planar projective motion model. The noise amplitude is measured here in *gray levels*, with one gray level being one 255th of the intensity range (e.g., assuming 8-bit images, which have 256 possible intensity values per pixel).

The ML super-resolutions for some of these are shown in Figure 9.3. There are four input images in the datasets in the left column, going up in powers of two to 64 images for each output in the right column. The standard deviation of the noise goes from zero for the top row to 2, 5, and 10 gray levels proceeding down the figure. The more images there are in the dataset, the

FIGURE 9.2: The synthetic graffiti dataset. Left: ground truth graffiti wall image. Right: four of the low-resolution images generated according to the forward model, with a Gaussian PSF of *std* 0.4 low-resolution pixels, and a zoom factor of 2.

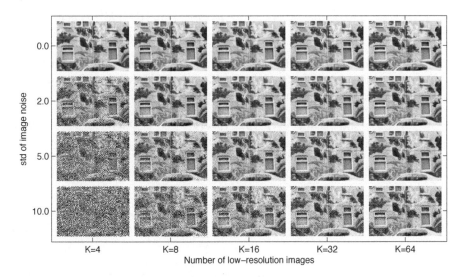

FIGURE 9.3: The ML super-resolution estimate. Synthetic datasets with varying numbers of images and varying levels of additive Gaussian noise were super-resolved using the ML algorithm.

less the ocillations dominate the output, though even with 64 input images, visible degradation is still evident in the output for cases with 5 and 10 gray levels of noise.

9.1.2.3 The Maximum a Posteriori Solution

A prior over x is usually introduced into the super-resolution model to avoid solutions which are subjectively very implausible to the human viewer. The Maximum a Posteriori (MAP) approach is explained here in terms of the generative model and its probabilistic interpretation. We go on to cover a few of the general image priors commonly selected for image super-resolution.

The MAP estimate of the super-resolution image comes about by an application of Bayes' theorem,

$$p(x|d) = \frac{p(d|x)p(x)}{p(d)}. \tag{9.10}$$

The left hand side is known as the *posterior* distribution over x, and if d (which in this case might represent our observed data) is held constant, then $p(d)$ may be considered as a normalization constant.

Applying these identities to the super-resolution model, we have

$$p\left(x\middle|\left\{y^{(k)},\theta^{(k)},\lambda^{(k)}\right\}\right) = \frac{p\left(\{y^{(k)}\}\,|x,\{\theta^{(k)},\lambda^{(k)}\}\right)p(x)}{p\left(\{y^{(k)}\}\,|\,\{\theta^{(k)},\lambda^{(k)}\}\right)} \tag{9.11}$$

If we again assume that the denominator is a normalization constant in this case — it is not a function of x — then the MAP solution, $\widehat{x}_{\text{MAP}}$, can be found by maximizing the numerator with respect to x, giving

$$\widehat{x}_{\text{MAP}} = \arg\max_{x} p(\{y^{(k)}\}|x,\{\theta^{(k)},\lambda^{(k)}\})p(x). \tag{9.12}$$

We take the objective function $\mathcal{L}$ to be the negative log of the numerator of (9.11), and minimize $\mathcal{L}$ with respect to x. The objective function and its gradient are

$$\mathcal{L} = -\log\left(p(x)\right) + \frac{\beta}{2}\sum_{k=1}^{K}\left\|r^{(k)}\right\|_{2}^{2} \tag{9.13}$$

$$\frac{\partial\mathcal{L}}{\partial x} = \frac{\partial}{\partial x}\left[-\log\left(p(x)\right)\right] - \sum_{k=1}^{K}\lambda_{1}^{(k)}W^{(k)T}r^{(k)}. \tag{9.14}$$

In order to solve this, one requires a form for the image prior $p(x)$.

In general the prior should favor smoother solutions than the ML approach typically yields, so it is usual to promote smoothness by penalizing excessive gradients or higher derivatives. Log priors that are *convex* and *continuous* are desirable, so that gradient-descent methods like SCG [12] can be used along with (9.13) and (9.14) to solve for x efficiently. A least-squares-style penalty term for image gradient values leads to a Gaussian image prior that gives a closed-form solution for the super-resolution image. However, natural images *do* contain edges where there are locally high image gradients, which it is undesirable to smooth out.

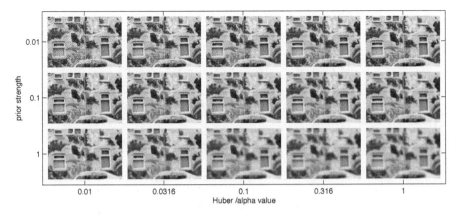

FIGURE 9.4: Maximum a Posteriori super-resolution images. This figure uses the same input data as 9.3, but uses the Maximum a Posteriori method to infer the super-resolution images using an array of prior parameter settings. With only 9 images and 5 gray levels of noise, the corresponding ML output would be swamped with noise (not shown). However, in all but the top left case (weakest prior), the MAP images clearly show the details of the underlying scene.

Figure 9.4 shows the improvement in super-resolution image estimates that can be achieved using a very simple prior on the super-resolution image, x. The super-resolution images were reconstructed using exactly the same input datasets as Figure 9.3, with 9 images and 5 gray levels of noise, but this time a Huber function [9] prior was used on image gradients, and all of the noise present in the ML solutions is gone. A few simple forms of prior will be considered next.

9.1.3 Selected Priors Used in MAP Super-Resolution

While ostensibly the prior is merely required to steer the objective function away from the "bad" solutions, in practice the exact selection of image prior does have an impact on the image reconstruction accuracy and on the computational cost of the algorithm, since some priors are much more expensive to evaluate than others.

This section introduces a few families of image priors commonly used in super-resolution, examines their structure, and derives the relevant objective functions to be optimized in order to make a MAP super-resolution image estimate in each case.

GMRF Image Priors

Gaussian Markov Random Field (GMRF) priors arise from a formulation where the gradient of the super-resolution solution is penalized, and correspond to specifying a Gaussian distribution over x:

$$p(x) = (2\pi)^{-\frac{N}{2}} |Z_x|^{-\frac{1}{2}} \exp\left\{-\frac{1}{2}x^T Z_x^{-1} x\right\}, \qquad (9.15)$$

where N is the size of the vector x, and Z_x is the covariance of a zero-mean Gaussian distribution:

$$p(x) \sim \mathcal{N}(0, Z_x). \qquad (9.16)$$

For super-resolution using any zero-mean GMRF prior, we have:

$$\mathcal{L} = \beta \|r\|^2 + x^T Z_x^{-1} x \qquad (9.17)$$

$$\frac{\partial \mathcal{L}}{\partial x} = -2\beta\lambda_1 W^T r + 2Z_x^{-1} x, \qquad (9.18)$$

where $\mathcal{L}$ and its derivative can be used in a gradient-descent scheme to find the MAP estimate for x.

Because the data error term and this prior are both Gaussian, it follows that the posterior distribution over x will also be Gaussian. It is possible to derive a closed-form solution in this case:

$$\widehat{x}_{\text{GMRF}} = \beta \Sigma \left(\sum_{k=1}^{K} \lambda_1^{(k)} W^{(k)T} \left(y^{(k)} - \lambda_2^{(k)}\right)\right) \qquad (9.19)$$

$$\Sigma = \left[Z_x^{-1} + \beta \left(\sum_{k=1}^{K} \lambda_1^{(k)2} W^{(k)T} W^{(k)}\right)\right]^{-1}, \qquad (9.20)$$

where Σ here is the covariance of the posterior distribution. However, the size of the matrices involved means that the iterative approach using SCG is far more practical for all but the very smallest of super-resolution problems.

Depending on the construction of the matrix Z_x, the GMRF may have several different interpretations. They are often expressed as the square of some linear operation on the pixels of x, such as an approximation to the image gradient or image Laplacian. This gives a general form

$$p(x) = \frac{1}{Z} \exp\left\{-\frac{\varsigma}{2}\|Dx\|_2^2\right\}, \qquad (9.21)$$

so that

$$Z_x^{-1} = \varsigma D^T D. \qquad (9.22)$$

In [4], D is a matrix which premultiplies x to give a vector of first-order approximations to the magnitude of the image gradient in horizontal, vertical and two perpendicular diagonal directions, giving a $4N \times N$ sparse D matrix with two nonzero elements per row. In [7], D is a small discrete approximation to the Laplacian-of-Gaussian filter, so the result for each pixel is the difference between its own value and the average of its four cardinal neighbors. Thus D is an $N \times N$ sparse matrix where the ith row has an entry of 1 at position i, and four entries of $-\frac{1}{4}$ corresponding to the four cardinal neighbors of pixel i. In neither of these cases do Z and Z^{-1} need to be computed explicitly for the iterative solution to be found.

In their *Bayesian Image Super-resolution* work [15], Tipping and Bishop treat the high-resolution image as a Gaussian Process, and suggest a form of Gaussian image prior where Z_x is calculated directly according to

$$Z_x(i,j) = A \exp \left\{ -\frac{||\boldsymbol{v}_i - \boldsymbol{v}_j||^2}{r^2} \right\}, \qquad (9.23)$$

where $\boldsymbol{v}_i$ is the two-dimensional position of the pixel that is lexicographically i^{th} in super-resolution image $\boldsymbol{x}$, r defines the distance scale for the correlations on the MRF, and A determines their strength.

This differs from the other two GMRF priors described above because here the long-range correlations in the high-resolution space are described explicitly, rather than resulting from the short-range weights prescribed in a difference matrix.

Image Priors with Heavier Tails than a Gaussian

The *Bilateral Total Variation* (BTV) prior is used by Farsiu *et al.* [5]. It compares the high-resolution image to versions of itself shifted by an integer number of pixels in various directions, and weights the resulting absolute image differences to form a penalty function. This leads again to a prior that penalizes high spatial frequency signals, but is less harsh than a Gaussian because the norm chosen is L_1 rather than L_2.

Huber Prior

The Huber function is used as a simple prior for image super-resolution which benefits from penalizing edges less severely than any of the Gaussian image priors. The form of the prior is

$$p(\boldsymbol{x}) = \frac{1}{Z} \exp \left\{ -\nu \sum_{g \in \mathcal{D}(\boldsymbol{x})} \rho(g, \alpha) \right\}, \qquad (9.24)$$

where $\mathcal{D}$ is the same set of gradient estimates as (9.21), given by $\boldsymbol{Dx}$. The parameter ν is a prior strength somewhat similar to a variance term, Z is the normalization constant, and α is a parameter of the Huber function specifying the gradient value at which the penalty switches from being quadratic to being linear:

$$\rho(x, \alpha) = \left\{ \begin{array}{ll} x^2, & \text{if } |x| \leq \alpha \\ 2\alpha|x| - \alpha^2, & \text{otherwise.} \end{array} \right. \qquad (9.25)$$

For a very simple 1D case, the probability function corresponding to the Huber function is

$$p(x) \;=\; \frac{1}{Z} \exp\left\{-\nu\rho(x,\alpha)\right\}, \qquad (9.26)$$

and one can verify by integration that

$$Z \;=\; \frac{1}{\nu\alpha} \exp\left\{-2\nu\alpha^2\right\} + \left(\frac{\pi}{\nu}\right)^{\frac{1}{2}} \mathrm{erf}\left\{\alpha\nu\right\}. \qquad (9.27)$$

Super-resolving with the Huber-MRF prior is very straightforward. The objective function and its gradient with respect to the high-resolution image pixels are

$$\mathcal{L} \;=\; \sum_{k=1}^{K} \beta \|\boldsymbol{r}^{(k)}\|_2^2 + \nu \sum_{g \in \mathcal{D}\boldsymbol{x}} \rho(g,\alpha) \qquad (9.28)$$

$$\frac{\partial \mathcal{L}}{\partial \boldsymbol{x}} \;=\; -2\beta \sum_{k=1}^{K} \lambda_1^{(k)} \boldsymbol{W}^{(k)T} \boldsymbol{r}^{(k)} + \nu \boldsymbol{D}^T \rho'(\boldsymbol{D}\boldsymbol{x}, \alpha) \qquad (9.29)$$

where

$$\rho'(x) \;=\; \left\{ \begin{array}{ll} 2x, & \text{if } |x| \le \alpha \\ 2\alpha\,\mathrm{sign}(x), & \text{otherwise.} \end{array} \right. \qquad (9.30)$$

and $\boldsymbol{D}$ is again a $4N \times N$ matrix giving first order approximations of the gradients in four directions in the image space. This has the advantage over the TV prior that unless $\alpha \to 0$, the function and its gradient with respect to $\boldsymbol{x}$ are continuous as well as convex, and can be solved easily using gradient-descent methods like SCG.

9.2 Where Super-Resolution Algorithms Go Wrong

There are many reasons why simply applying the MAP super-resolution algorithm to a collection of low-resolution images may not yield a perfect result immediately. This section consists of a few brief examples to highlight the causes of poor super-resolution results from what should be good and well-behaved low-resolution image sets. In particular, the super-resolution problem involves several closely-interrelated components: geometric registration, photometric registration, parameters for the prior, noise estimates and the point-spread function, not to mention the estimates of the values of the high-resolution image pixels themselves. If one component is estimated badly, there can be a knock-on effect on the values of the other parameters needed in order to produce best super-resolution estimate for a given low-resolution dataset.

FIGURE 9.5: Keble and Frog datasets. Ground truth images for the synthetic Keble and Frog datasets, along with two low-resolution images of each (not to scale).

9.2.1   Point-Spread Function Example

A bad estimate of the size and shape of the point-spread function kernel leads to a poor super-resolution image, because the weights in the system matrix do not accurately reflect the responsibility each high-resolution pixel (or scene element) takes for the measurement at any given low-resolution image pixel. The solution that minimizes the error in the objective function does not necessarily represent a good super-resolution image in this case, and it is common to see "ringing" around edges in the scene. These ringing artifacts can be attenuated by the prior on x, so in general, a dataset with an incorrectly-estimated PSF parameter will require a stronger image prior to create a reasonable-looking super-resolution image than a dataset where the PSF size and shape are known accurately.

To illustrate this, 16 images from the synthetic "Keble" dataset (see Figure 9.5) with a noise standard deviation of 2.5 gray levels are super-resolve at a zoom factor of 4 using the known geometric (8Dof) and photometric (2DoF) registration values, and a Huber prior with $\alpha = 0.05$. The PSF standard deviation, γ, is varied, and for each value the the prior strength ratio (ν/β in (9.28)), which minimizes the RMS error with respect to the ground-truth image is found.

The results are plotted in Figure 9.6, which clearly shows that as γ moves away from its true value of 0.4 low-resolution image pixels, the error increases, and the prior strength ratio needed to achieve the minimal error also increases. Figure 9.7 shows three images from the results. The first is the image reconstructed with the true γ, showing a very good super-resolution result. The next is the image reconstructed with $\gamma = 0.7$, and is a very poor super-resolution result. The final image of the three shows the super-resolution image reconstructed with the same poor value of γ, but with a much stronger prior; while the result is smoother than our ideal result, the quality is definitely superior to the middle case.

The important point to note is that all these images are constructed using

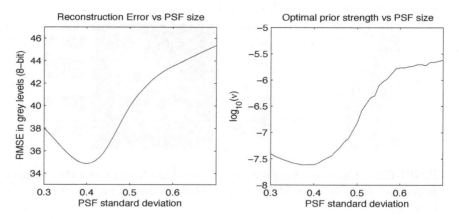

FIGURE 9.6: Reconstructing the Keble dataset with various PSF estimates. When the PSF standard deviation, γ, is overestimated by a factor of 50%, the prior needs to be almost two orders of magnitude larger to reconstruct the image as well as possible.

(a) γ=0.4, v_1: err=34.8   (b) γ=0.7, v_1: err=78.9   (c) γ=0.7, v_2: err=45.4

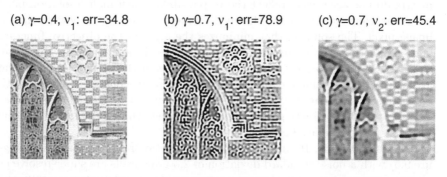

FIGURE 9.7: Reconstructing the Keble dataset with various PSF estimates: images. (a) The best reconstruction achieved at the correct value, $\gamma = 0.4$. (b) The super-resolution image obtained with the same prior as the left-hand image, but with $\gamma = 0.7$. Heavy ringing is induced by the bad PSF estimate. (c) The best possible reconstruction using $\gamma = 0.7$. This time the prior strength ratio is almost 100 times stronger than for the first image, even though the input images themselves are identical.

exactly the same input data, and only γ and ν were varied. The consequence of this kind of relationship is that even when it is possible to make a *reasonably* accurate estimate of each of the hyperparameters needed for super-resolution, the values themselves must be selected together in order to guarantee a good-looking super-resolution image.

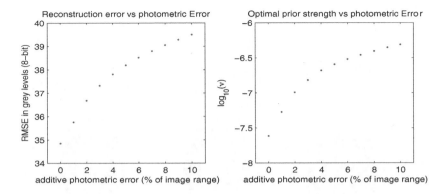

FIGURE 9.8: Reconstructing the Keble dataset with photometric error. Left: RMSE of the reconstruction increases with the uncertainty in the photometric shift parameter (0 to 10%). Right: the prior strength setting necessary to achieve the best reconstruction for each setting of the photometric parameters. The prior strength increases by well over an order of magnitude between the no-error case and the 10% case.

9.2.2   Photometric Registration Example

The photometric part of the model accounts for relative changes in illumination between images, either due to changes in the incident lighting in the scene, or due to camera settings such as automatic white balance and exposure time. When the photometric registration has been calculated using pixel correspondences resulting from the geometric registration step and bilinear interpolation onto a common frame, some errors may be expected because both sets of images are noisy, and because such interpolation does not agree with the generative model of how the low-resolution images relate to the original scene.

To understand the effect of errors in the photometric estimates, several super-resolution reconstructions of the Keble dataset are made with a noise standard deviation of 2.5 gray levels, using the ground truth point-spread function (a Gaussian with std 0.4 low-resolution pixels), the true geometric registration, and a set of photometric shift parameters that are gradually perturbed by random amounts, meaning that each image is assumed to be globally very slightly brighter or darker than it really is relative to the ground truth image.

For each setting of the photometric parameters, a set of super-resolution images was recovered using the Huber-MAP algorithm with different strengths of Huber prior. The plots in Figure 9.8 show the lowest error (left) and prior strength ratio, $\log_{10}(\nu/\beta)$ (right) for each case. Figure 9.9 shows the deterioration of the quality of the super-resolution image for the cases where the sixteen photometric shift parameters (one per image) were perturbed by an

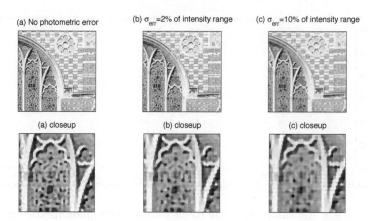

FIGURE 9.9: Reconstructing the Keble dataset with photometric error: images. Top: full super-resolution image; Bottom: close-up of the main window where the different levels of detail are very noticeable. Left-to-right: reconstruction with additive errors of 0%, 2%, and 10% on the photometric shift parameters. As the error increases, the edges still remain well-localized, but finer details are smoothed out due to the necessary increase in prior strength.

amount whose standard deviation was equal to 2% and 10% of the image range, respectively.

The edges are still very well localized even in the 10% case, because the geometric parameters are perfect. However, the ill conditioning caused in the linear system of equations solved in the Huber-MAP algorithm means that the optimal solutions require stronger and stronger image priors as the photometric error increases, and this results in the loss of some of the high-frequency detail, like the brick pattern and stained-glass window leading, which are visible in the error-free solution.

9.2.3   Geometric Registration Example

Errors from two different sources can also be very closely-coupled in the super-resolution problem. In this example, we show that errors in some of the geometric parameters $\theta^{(k)}$, can to a small extent, be mitigated by a small increase in the size of the blur kernel.

Sixteen images from the synthetic Frog dataset of Figure 9.5 with a zoom factor of 4, a PSF width of 0.4 low-resolution pixels, and various levels of *i.i.d.* Gaussian noise are taken as a starting point. Because the ground-truth image is much smoother than the Keble image, with fewer high-frequency details, it is in general easier to super-resolve, and leads to reconstructions with much lower RMS error than the Keble image dataset.

Errors are applied to the θ parameters governing the horizontal and verti-

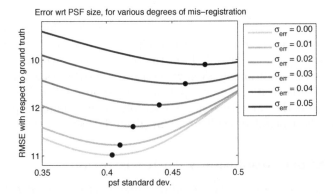

FIGURE 9.10: Reconstructing the Frog image with small errors in geometric registration and point-spread function size. The six colors represent six levels of additive random noise added to the shift parameters in the geometric registration. The curves represent the optimal error as the PSF parameter γ was varied about its ground truth value of 0.4. The larger the registration error, the bigger the error in γ is in order to optimize the result.

cal shifts of each image, with standard deviations of 0, 0.01, 0.02, 0.03, 0.04, and 0.05 low-resolution pixels, i.e., a very small amount. For each of these six registrations of the input data, a set of super-resolution images is recovered as the PSF standard deviation, γ is varied, and for each setting of γ, the prior strength ratio giving the best reconstruction is found.

Figure 9.10 shows how the best error for each of the six registration cases varies with the point-spread function size. When the geometric registration is known exceedingly accurately, the minimum falls at the true value of γ. However, as the geometric registration parameters drift more with respect to one another, the point at which the lowest error is found for any given geometric registration increases. This can be explained intuitively because a single high-resolution pixel is related by W to several low-resolution image regions, and if mis-registration causes these regions not to be aligned with one another exactly, then the super-resolution image pixel may end up being required to explain a mixture of values from that neighborhood in the high-resolution space, and thus more blur will help.

9.3 Simultaneous Super-Resolution

In the preceding section, we saw that the the problems of determining image registration or motion estimation, low-resolution image blur estimation, selection of a suitable prior, and super-resolution image estimation are seldom truly independent. In addition, it is also expected that each scene will have different underlying image statistics, and a super-resolution user generally has to hand-tune these in order to obtain an output that preserves as much richness and detail as possible from the original scene without encountering problems with conditioning, even when using a GMRF. Taken together, these observations motivate the development of an approach capable of registering the images at the same time as super-resolving and tuning a prior, in order to get the best possible result from any given dataset.

9.3.1   Super-Resolution with Registration

Standard approaches to super-resolution *first* determine the registration, *then* fix it and optimize a function like the MAP objective function of (9.13) with respect only to x to obtain the final super-resolution estimate. However, if the set of input images is assumed to be noisy, it is reasonable to expect the registration to be adversely affected by the noise.

In contrast, we make use of the high-resolution image estimate common to all the low-resolution images, and aim to find a solution in terms of the high-resolution image x, the set of geometric registration parameters, θ (which parameterize W), and the photometric parameters λ (composed of the λ_1 and λ_2 values), at the same time, i.e., we determine the point at which

$$\frac{\partial \mathcal{L}}{\partial x} = \frac{\partial \mathcal{L}}{\partial \theta} = \frac{\partial \mathcal{L}}{\partial \lambda} = 0 \qquad (9.31)$$

where $\mathcal{L}$ is defined e.g., as in 9.28.

The registration problem itself is not convex – for example, repeating textures can cause naïve intensity-based registration algorithms to fall into local minima, though when initialized sensibly, very accurate results are obtained. The pathological case where the footprints of the low-resolution images fail to overlap in the high-resolution frame can be avoided by adding an extra term to $\mathcal{L}$ to penalize large deviations in the registration parameters from the initial registration estimate, e.g., by assuming a very broad Gaussian prior distribution over relevant components of the geometric registration.

The simultaneous super-resolution and image registration problem closely resembles the well-studied problem of Bundle Adjustment [16], in that the camera parameters and image features (which are 3D points in Bundle Adjustment) are found simultaneously. Because most high-resolution pixels are observed in most frames, the super-resolution problem is closest to the "strongly convergent camera geometry" setup, and conjugate gradient methods are expected to converge rapidly [16].

The objective function for simultaneous registration and super-resolution is very similar to the regular MAP negative log likelihood, except that it is optimized with respect to the registration parameters as well as the super-resolution image estimate, e.g.,

$$\mathcal{L} = \sum_{k=1}^{K} \beta \|\boldsymbol{r}^{(k)}\|_2^2 + \nu \sum_{g \in \mathcal{D}\boldsymbol{x}} \rho(g, \alpha) \qquad (9.32)$$

$$[\boldsymbol{x}_{\text{MAP}}, \boldsymbol{\theta}_{\text{MAP}}, \boldsymbol{\lambda}_{\text{MAP}}] = \arg \max_{\boldsymbol{x}, \boldsymbol{\theta}, \boldsymbol{\lambda}} \mathcal{L}, \qquad (9.33)$$

where (9.32) is the same as (9.28), repeated here for convenience, though any reasonable image prior can be used in place of the Huber-MRF here.

The gradient with respect to $\boldsymbol{x}$ is given by (9.29), and the gradients with respect to $\boldsymbol{W}^{(k)}$ and the photometric registration parameters are

$$\frac{\partial \mathcal{L}}{\partial \boldsymbol{W}^{(k)}} = -2\beta \lambda_1^{(k)} \boldsymbol{r}^{(k)} \boldsymbol{x}^T \qquad (9.34)$$

$$\frac{\partial \mathcal{L}}{\partial \lambda_1^{(k)}} = -2\beta \boldsymbol{x}^T \boldsymbol{W}^{(k)T} \boldsymbol{r}^{(k)} \qquad (9.35)$$

$$\frac{\partial \mathcal{L}}{\partial \lambda_2^{(k)}} = -2\beta \sum_{i=1}^{M} r_i^{(k)}. \qquad (9.36)$$

The gradient of the elements of $\boldsymbol{W}^{(k)}$ with respect to $\boldsymbol{\theta}^{(k)}$ could be found analytically for simple parametric motion models, but for projective homographies it is simpler to use a finite difference approximation, because as well as the location, shape and size of the PSF kernel's footprint in the high-resolution image frame, each parameter also affects the entire normalization of $\boldsymbol{W}^{(k)}$, requiring a great deal of computation to find the exact derivatives of each individual matrix element.

Initializations for this scheme will be outlined in Section 9.3.4, but first the extensions to learning parameters for the prior distributions are considered.

9.3.2 Learning Prior Strength Parameters from Data

For most forms of MAP super-resolution, one must determine values for free parameters like the prior strength, and prior-specific additional parameters like the Huber-MRF's α value. In order to learn parameter values in a usual ML or MAP framework, it would be necessary to be able to evaluate the partition function (normalization constant), which is a function of ν and α. For example, the expression for the Huber-MRF is

$$p(\boldsymbol{x}) = \frac{1}{Z(\nu, \alpha)} \exp \left\{ -\nu \sum_{g \in \mathcal{D}\boldsymbol{x}} \rho(g, \alpha) \right\}, \qquad (9.37)$$

and so the full negative log likelihood function is

$$\mathcal{L} \;=\; \frac{1}{2} \sum_{k=1}^{K} \beta \| r^{(k)} \|_2^2 - \log p(x) \tag{9.38}$$

$$=\; \frac{1}{2} \sum_{k=1}^{K} \beta \| r^{(k)} \|_2^2 + \nu \sum_{g \in \mathcal{D}x} \rho(g, \alpha) - \log Z(\nu, \alpha). \tag{9.39}$$

For a ML solution to ν and α, this should be optimized with respect to those two variables, and in fact the entire data-error term could be neglected, since it does not depend on these variables. However, the partition function $Z(\nu, \alpha)$ for these sorts of edge-based priors is not generally easy to compute because the structure of the image means that each pair of pixel differences cannot be assumed independent.

Rather than setting the prior parameters using an ML or MAP technique, therefore, cross-validation is chosen for parameter-fitting. However, it is necessary to determine these parameters while still in the process of converging on the estimates of x, θ, and λ. This is done by removing some *individual low-resolution pixels* from the problem to create a *validation set*, then solving for x using the remaining pixels. This solution for x is projected back into the original low-resolution image frames, giving pixel intensity estimates for all low-resolution pixels (including those in the validation set). The error in the super-resolution estimate is determined by examining the L_1 norm of the difference between the predicted and measured intensity values over the validation set, though the L_2 norm or the Huber potential are also suitable measures, and give comparable results in practice. The selected α and ν should minimize this cross-validation error.

To evaluate a proposed pair of α and ν values, we optimize $\mathcal{L}$ w.r.t. x, starting with the current x estimate, for *just a few steps* to determine whether the parameter combination improves the estimate of x, as determined by cross-validation. This optimization to re-estimate x does not need to run to convergence in order to determine whether the proposed values ν and α make an improvement over the previous values, and are therefore worthy of consideration. A gradient-descent scheme can then be used on (ν, α)-space to select prior parameter values that give better high-resolution image estimates.

This scheme is much faster than the usual approach of running a complete optimization for a number of parameter combinations. An arbitrary 5% of pixels are used for validation, ignoring regions within a few pixels of edges, to avoid boundary complications.

9.3.3 Scaling and Convergence

The elements of x are scaled to lie in the range $[-\frac{1}{2}, \frac{1}{2}]$, and the geometric registration is decomposed into a "fixed" component, which is the initial mapping from $y^{(k)}$ to x, and a projective correction term, which is itself decomposed into constituent shifts, rotations, axis scalings and projective parameters, which are the geometric registration parameters, $\boldsymbol{\theta}$. The registration vector for a low-resolution image k, $\boldsymbol{\theta}^{(k)}$, is combined with the photometric registration parameter vector, $\boldsymbol{\lambda}^{(k)}$, to give one ten-element parameter vector per low-resolution image. These registration vectors are concatenated with the image vector $\boldsymbol{x}$ to form the full vector of free parameters over which the algorithm optimizes. Further details of parameter scaling are given in [14].

Using the Scaled Conjugate Gradients (SCG) implementation from Netlab [12], rapid convergence is observed up to a point, beyond which a slow steady decrease in the negative log likelihood gives no subjective improvement in the solution, but this extra computation can be avoided by specifying sensible convergence criteria.

Convergence for the *simultaneous* algorithm is defined to be the point at which all parameters change by less than a preset threshold in successive iterations. The outer loop is repeated till this point, typically taking 3-10 iterations. Thresholds are defined differently depending on the nature of the parameter. For $\boldsymbol{x}$, the point at which the iteration has failed to change any pixel value in $\boldsymbol{x}$ by more than 0.3 gray levels (e.g., 1/850 of the image range) is chosen. For ν and α, we work with the log values of the parameters (since neither should take a negative value), and look for a change of less than 10^{-4} between subsequent iterations.

A description of the overall structure of the simultaneous super-resolution algorithm described here is summarized in Table 9.1. In the inner-loop iterations use the same convergence criteria as the outer-loop, but additionally the number of steps for the update of $\boldsymbol{x}$ and $\boldsymbol{\theta}$ (algorithm part 2c) is limited to 20, and the number if steps for the prior update (algorithm part 2b) is limited to ten, so that the optimization is divided more effectively between the two groups or parameters.

9.3.4   Initialization

In our experiments, input images are assumed to be pre-registered by a standard algorithm [8] (e.g., RANSAC on features detected in the images) such that points at the image centers correspond to within a small number of low-resolution pixels. This takes us comfortably into the region of convergence for the global optimum even in cases with considerable repeating texture.

The Huber image prior parameters are initialized to around $\alpha = 0.01$ and $\nu = \beta/10$; as these are both strictly positive quantities, they are represented as log values throughout. A candidate PSF is selected in order to compute the *average image*, $\boldsymbol{a}$, which is a stable though excessively smooth approximation to $\boldsymbol{x}$. Each pixel in $\boldsymbol{a}$ is a weighted combination of pixels in $\boldsymbol{y}$, such that a_i depends strongly on y_j if y_j depends strongly on x_i according to the weights in $\boldsymbol{W}$. Lighting changes must also be taken into consideration, so

$$\boldsymbol{a} = \boldsymbol{D}^{-1}\boldsymbol{W}^T\boldsymbol{\Lambda}_1^{-1}(\boldsymbol{y} - \boldsymbol{\lambda}_2), \tag{9.40}$$

1. **Initialize PSF, image registrations, high-resolution image, and prior parameters.**

2. (a) **(Re)-sample the set of validation pixels**, selecting them from across all low-resolution images.

   (b) **Update α and ν (prior parameters).** Perform gradient descent on the cross-validation error to improve the values of α and ν.

   (c) **Update the super-resolution image and the registration parameters.** Optimize $\mathcal{L}$ (equation 9.32) jointly with respect to x (super-resolution image), λ (photometric transform), and θ (geometric transform).

3. If the maximum absolute change in α, ν, or any element of x, λ or ϕ is above preset convergence thresholds, return to 2.

TABLE 9.1: Basic structure of the simultaneous algorithm.

where W, y, and λ_2 are the stacks of the K groups of $W^{(k)}$, $y^{(k)}$, and $\lambda_2^{(k)}\mathbf{1}$ respectively, Λ_1 is a diagonal matrix whose entries are a stack of $\lambda_1^{(k)}\mathbf{1}$ values, and D is a diagonal matrix whose elements are the column sums of W. Notice that both inverted matrices are diagonal, so a is simple to compute. The resulting images are generally much smoother than the corresponding high-resolution frame calculated from the same input images will be, but are very robust to noise on the low-resolution images.

In order to get a registration estimate, it is possible to optimize $\mathcal{L}$ of (9.32) with respect to θ and λ only, by using a in place of x to estimate the high-resolution pixels. This provides a good estimate for the registration parameters, without requiring x or the prior parameters, and we refer to the output from this step as the *average image registration*. We find empirically that this out-performs popular alternatives such as mapping images to a common reference frame by bilinear interpolation and setting the registration parameters to minimize the resulting pixel-wise error.

To initialize x, we begin with a, and use the SCG algorithm with the ML solution equations as in (9.8) and (9.9) to improve the result. The optimization is terminated after around $\frac{K}{4}$ steps (where K is the number of low-resolution images), before the instabilities dominate the solution. This gives a sharper starting point than initializing with a as in [4]. When only a few images are available, a more stable ML solution can be found by using a constrained optimization to bound the pixel values so they must lie in the permitted image intensity range.

(a) ground truth image   (b) 1st input, interpolated

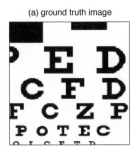
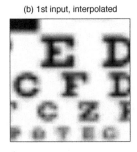

FIGURE 9.11: Synthetic "eyechart" dataset. Left: Ground truth "eyechart" image; Right: one of 16 low-resolution images, upsampled using bilinear interpolation.

9.3.5   Evaluation on Synthetic Data

Before applying the simultaneous algorithm to real low-resolution image data, an evaluation is performed on synthetic data, generated using the generative model (9.1) applied to ground truth images. The first experiment uses only the simultaneous registration and super-resolution part of the algorithm, and the second covers the cross-validation. Before presenting these experiments, it is worth considering the problem of making comparisons with ground-truth images when the registration itself is part of the algorithm.

Joint Registration and Super-Resolution

The simple "eyechart" image of Figure 9.11 was used to generate 16 low-resolution images at a zoom factor of 4, with each pixel being corrupted by additive Gaussian noise to give a SNR of $30dB$. The image is of text and has 256 gray levels (scaled to lie in $\left[-\frac{1}{2}, \frac{1}{2}\right]$ for the experiments), though the majority of the pixels are black or white, i.e., the original image is effectively binary-valued. The low-resolution images are 30×30 pixels in size. Values for a shift-only geometric registration, θ, and a 2D photometric registration λ are sampled independently from uniform distributions.

An initial registration was then carried out using the *average image registration* technique described above. This is taken as the "fixed" registration for comparison with the joint MAP algorithm, and it differs from the ground truth by an average of 0.0142 pixels, and 1.00 gray levels for the photometric shift.

Two sets of super-resolution images were computed: one set with a fixed registration, and one set using the simultaneous approach. Within each set, the value for the prior strength parameter ν was varied while keeping the Huber parameter α set to 0.04. The noise precision parameter β is chosen so that the noise in every case is assumed to have a standard deviation of 5 gray levels, so a single "prior strength" quantity equal to $\log_{10}(\nu/\beta)$ is used. For

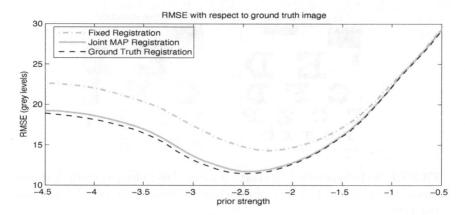

FIGURE 9.12: Synthetic data results. RMSE compared to ground truth for the "eyechart" text image, plotted for the fixed and joint MAP algorithms, and for the Huber super-resolution image found using the ground truth registration.

each prior strength, both the fixed-registration and the joint MAP methods are applied to the data, and the *root mean square error* (RMSE) compared to the ground truth image is calculated.

The RMSE compared to the ground truth image for both the fixed registration and the joint MAP approach are plotted, in Figure 9.12, along with a curve representing the performance if the ground truth registration is known. The joint MAP curve is very close to that of the true registrations, showing that allowing the super-resolution image to inform an update to the registrations does indeed yield an improvement in super-resolution performance.

Cross-Validation Example

A second synthetic-data experiment shows that the cross-validation-based prior learning phase is effective. The cross-validation error is measured by holding a percentage of the low-resolution pixels in each image back, and performing Huber-MAP super-resolution using the remaining pixels. The super-resolution image is then projected back down into the withheld set, and the mean absolute error is recorded. This is done for three different validation set sizes (2%, 5%, and 10%), and at a range of prior strengths, where all the prior strength values are given as $\log(\nu/\beta)$.

The low-resolution images were generated from the "Keble" image, and the results are plotted in Figure 9.13, along with a plot of the error measured from reconstructing the image using all low-resolution pixels and comparing the results with the ground truth high-resolution image. The best ground-truth comparison occurs when the log prior strength ratio is -2.92. In the cross-validation plots shown in the center of the figure, the curves' minima are at -2.90, -2.90, and -2.96, respectively, which is very close indeed. The

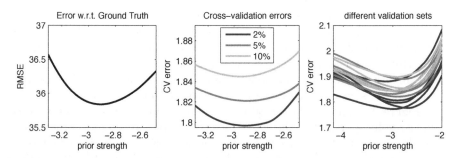

FIGURE 9.13: Cross-validation errors on synthetic data. Left: Error with respect to ground truth on the Keble dataset for this noise level. center: Three cross-validation curves, corresponding to 2%, 5%, and 10% of pixels being selected. Right: More cross-validation curves at each ratio.

final plot shows that for a variety of different random choices of validation pixel sets, the minima are all very close. All the curves are also very smooth and continuous, meaning that we expect optimization using the cross-validation error to be straightforward to achieve.

9.3.6 Experiments on Real Data

The Surrey library sequence: An area of interest is highlighted in the 30-frame Surrey Library sequence from http://www.robots.ox.ac.uk/~vgg/data/, shown in Figure 9.14. The camera motion is a slow pan through a small angle, and the sign on a wall is illegible given any one of the inputs alone. Gaussian PSFs with $std = 0.375, 0.45, 0.525$ are selected, and super-resolution images estimated are made with both the fixed-registration or simultaneous super-resolution algorithms. There are 77003 elements in y, and x has 45936 elements with a zoom factor of 4. W has around 3.5×10^9 elements, of which around 0.26% are nonzero with the smallest of these PSF kernels, and 0.49% with the largest. Most instances of the simultaneous algorithm converge in 2 to 5 iterations. Results in Figure 9.15 show that while both algorithms perform well with the middle PSF size, the simultaneous-registration algorithm handles the worse PSF estimates more gracefully.

The Colored Book Sequence: Finally, the simultaneous super-resolution method is demonstrated on a publicly-available color super-resolution dataset[1]. The dataset consists of 30 low-resolution images captured with a color camera following a global translation model. The PSF standard deviation was chosen to be 0.4 low-resolution pixels, the desired zoom factor was set to π, and the simultaneous algorithm was run as described above.

---

[1] Data from http://www.soe.ucsc.edu/~milanfar/software/sr-datasets.html

In generalizing the basic algorithm to handle color images, a very simple approach is taken. The three color channels are treated separately, though the geometric and photometric components are shared between the three channels. The Huber prior and data error terms are summed over the three channels, and all other algorithm components remain the same. Note that no color demosaicing step has been introduced for the purpose of this example.

Figure 9.16 shows 9 of the 30 low-resolution input images along with the super-resolution result in this case. This result compares well to other results for algorithms that do not employ a color demosaicing step in their models; the "jagging" artifacts visible at book edges, especially on the left-hand side

FIGURE 9.14: The "Surrey library" real dataset: close-ups of text from 9 of the 30 images across the low-resolution sequence, recorded with a real camera panning over a scene including the external wall of a library.

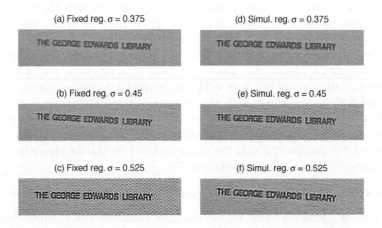

FIGURE 9.15: Results on the Surrey Library sequence. Left column (a,b,c) Super-resolution found using fixed registrations. Right column (d,e,f) Super-resolution images using our algorithm. While both algorithms perform well with the middle PSF size, the simultaneous-registration algorithm handles the worse PSF estimates more gracefully.

FIGURE 9.16: Results on the colored book sequence. Left: nine of the 30 color low-resolution images. Right: the super-resolution image when the zoom factor was chosen arbitrarily to be π.

of the image, result from this color channel alignment, but overall the prior term has adapted effectively to keep the overall image natural-looking at most of the edges, and book titles such as *Kalman Filtering* and *French Review and Practice* are clearly legible, which is not the case in the input images. Other than specifying the PSF and the desired zoom factor, this is achieved entirely automatically without any parameter-tuning necessary.

9.4 Bayesian Marginalization

This section describes a method to handle uncertainty in the set of estimated imaging parameters (the geometric and photometric registrations, and the point-spread function) in a principled manner by Bayesian marginalization. Such parameters are sometimes known as *nuisance parameters* because they are not directly part of the desired output of the algorithm, which in this case is the high-resolution image.

We also describe the alternative Bayesian approach of Tipping and Bishop [15], which marginalizes over the high-resolution image in order to make a *maximum marginal likelihood* point estimate of the imaging parameters. This gives an improvement in the accuracy of the recovered registration (measured against known truth on synthetic data) compared to the Maximum a Posteriori (MAP) approach, but has two important limitations: (i) it is restricted to a Gaussian image prior in order for the marginalization to remain tractable, whereas others have shown improved image super-resolution results are produced using distributions with heavier tails, and (ii) it is computationally expensive due to the very large matrices required by the algorithm, so the registration is only possible over very small image patches, which take a long time to register accurately. In contrast our registration-marginalizing approach allows for a more realistic image prior and operates with smaller matrix sizes.

Both the registration-marginalizing and image-marginalizing derivations proceed along similar lines mathematically, though the effect of choosing a different variable of integration makes a large difference to the way in which the super-resolution algorithms proceed.

9.4.1 Marginalizing over Registration Parameters

The goal of super-resolution is to obtain a high-resolution image, and so approach treats the other parameters – geometric and photometric registrations and the point-spread function – as "nuisance variables" that might be marginalized out of the problem.

If these parameters are collected together into a single vector, ϕ, these marginalizations can be expressed simply as

$$
\begin{aligned}
p\left(\boldsymbol{x}\,\middle|\,\left\{\boldsymbol{y}^{(k)}\right\}\right) &= \int p\left(\boldsymbol{x},\boldsymbol{\phi}\,\middle|\,\left\{\boldsymbol{y}^{(k)}\right\}\right)\mathrm{d}\boldsymbol{\phi} \\
&= \int \frac{p\left(\boldsymbol{x},\boldsymbol{\phi}\right)}{p\left(\left\{\boldsymbol{y}^{(k)}\right\}\right)}p\left(\left\{\boldsymbol{y}^{(k)}\right\}\,\middle|\,\boldsymbol{x},\boldsymbol{\phi}\right)\mathrm{d}\boldsymbol{\phi} \\
&= \frac{p\left(\boldsymbol{x}\right)}{p\left(\left\{\boldsymbol{y}^{(k)}\right\}\right)}\int p\left(\left\{\boldsymbol{y}^{(k)}\right\}\,\middle|\,\boldsymbol{x},\boldsymbol{\phi}\right)p\left(\boldsymbol{\phi}\right)\mathrm{d}\boldsymbol{\phi}. \quad (9.41)
\end{aligned}
$$

Notice that $p\left(\boldsymbol{x},\boldsymbol{\phi}\right) = p\left(\boldsymbol{x}\right)p\left(\boldsymbol{\phi}\right)$ (because the super-resolution image and registration parameters are independent) and that $p\left(\boldsymbol{x}\right)$ and $p\left(\left\{\boldsymbol{y}^{(k)}\right\}\right)$ can be taken outside the integral. This leaves one free to choose any suitable super-resolution image prior $p\left(\boldsymbol{x}\right)$, rather that being constrained to picking a Gaussian merely to make the integral tractable, as in the image-marginalizing case discussed later.

A prior distribution over $\boldsymbol{\phi}$, which appears within the integral, must be specified. We assume that a preliminary image registration (geometric and photometric) and an estimate of the PSF are available, and also that these registration values are related to the ground truth registration values by unknown zero-mean Gaussian-distributed additive noise. The registration estimate can be obtained using classical registration methods, either intensity-based [10] or estimation from image points [8]. There is a rich literature of *Blind Image Deconvolution* concerned with estimating an unknown blur on an image [11].

We introduce a vector $\boldsymbol{\delta}$ to represent the perturbations from ground truth in the initial parameter estimate. For the parameters of a single image, this gives

$$
\begin{bmatrix} \boldsymbol{\theta}^{(k)} \\ \lambda_1^{(k)} \\ \lambda_2^{(k)} \end{bmatrix} = \begin{bmatrix} \bar{\boldsymbol{\theta}}^{(k)} \\ \bar{\lambda}_1^{(k)} \\ \bar{\lambda}_2^{(k)} \end{bmatrix} + \boldsymbol{\delta}^{(k)} \tag{9.42}
$$

where $\bar{\boldsymbol{\theta}}^{(k)}$ and $\bar{\boldsymbol{\lambda}}^{(k)}$ are the estimated registration, and $\boldsymbol{\theta}^{(k)}$ and $\boldsymbol{\lambda}^{(k)}$ are the true registration. The stacked vector $\boldsymbol{\delta}$ is then composed as

$$
\boldsymbol{\delta}^T = \left[\boldsymbol{\delta}^{(1)T}, \boldsymbol{\delta}^{(2)T}, \ldots, \boldsymbol{\delta}^{(K)T}, \delta_\gamma \right] \tag{9.43}
$$

where the final entry is for the PSF parameter so that $\gamma = \bar{\gamma} + \delta_\gamma$, and $\bar{\gamma}$ is the initial estimate.

The vector $\boldsymbol{\delta}$ is assumed to be distributed according to a zero-mean Gaussian:

$$
\boldsymbol{\delta} \sim \mathcal{N}(\mathbf{0}, \boldsymbol{V}). \tag{9.44}
$$

The covariance matrix $\boldsymbol{V}$ is assumed to be diagonal, and its elements are chosen to reflect the confidence in each parameter estimate, for example by setting the entries corresponding to translation parameters to give a standard deviation of a tenth of a low-resolution pixel, which should be correct given a good registration as a starting point.

The final step before the integral of (9.41) can be evaluated is to bring out the dependence on $\boldsymbol{\phi}$ in $p\left(\{\boldsymbol{y}^{(k)}\} | \boldsymbol{x}, \boldsymbol{\phi}\right)$. Starting with

$$
p\left(\{\boldsymbol{y}^{(k)}\} | \boldsymbol{x}, \boldsymbol{\phi}\right) = \left(\frac{\beta}{2\pi}\right)^{-\frac{KM}{2}} \exp\left\{-\frac{\beta}{2} e(\boldsymbol{\delta})\right\} \tag{9.45}
$$

where

$$
e(\boldsymbol{\delta}) = \sum_{k=1}^{K} \left\| \boldsymbol{y}^{(k)} - \lambda_1^{(k)} \boldsymbol{W}\left(\boldsymbol{\theta}^{(k)}, \gamma\right) \boldsymbol{x} - \lambda_2^{(k)} \right\|_2^2 \tag{9.46}
$$

and where where $\boldsymbol{\theta}$, $\boldsymbol{\lambda}$ and γ are functions of $\boldsymbol{\delta}$ and the initial registration values, we can then expand the integral in (9.41) to an integral over $\boldsymbol{\delta}$,

$$
p\left(\boldsymbol{x} | \{\boldsymbol{y}^{(k)}\}\right) = \frac{p(\boldsymbol{x}) |\boldsymbol{V}^{-1}|^{1/2} \beta^{KM/2}}{p(\{\boldsymbol{y}^{(k)}\}) (2\pi)^{(KM+Kn+1)/2}}
$$
$$
\times \int \exp\left\{-\frac{\beta}{2} e(\boldsymbol{\delta}) - \frac{1}{2}\boldsymbol{\delta}^T \boldsymbol{V}^{-1} \boldsymbol{\delta}\right\} d\boldsymbol{\delta}. \tag{9.47}
$$

We can expand $e(\boldsymbol{\delta})$ as a second-order Taylor series about the parameter estimates $\left\{\bar{\boldsymbol{\theta}}^{(k)}, \bar{\boldsymbol{\lambda}}^{(k)}\right\}$ and $\bar{\gamma}$ in terms of the vector $\boldsymbol{\delta}$, so that

$$
e(\boldsymbol{\delta}) = f + \boldsymbol{g}^T \boldsymbol{\delta} + \frac{1}{2}\boldsymbol{\delta}^T \boldsymbol{H} \boldsymbol{\delta}. \tag{9.48}
$$

Values for f, g and H can be found numerically (for geometric registration parameters) and analytically (for the photometric parameters) from x and $\{y^{(k)}, \theta^{(k)}, \lambda^{(k)}\}$.

We are now in a position to evaluate the integral in (9.47) using the identity [2]

$$\int \exp\left\{-bx - \frac{1}{2}x^T A x\right\} dx \;=\; 2\pi^{\frac{d}{2}} |A|^{-\frac{1}{2}} \exp\left\{b^T A^{-1} b\right\}, \quad (9.49)$$

where d is the dimension of the vector b.

The exponent in the integral in (9.47), becomes

$$a \;=\; -\frac{\beta}{2}e(\delta) - \frac{1}{2}\delta^T V^{-1}\delta \qquad (9.50)$$

$$\;=\; -\frac{\beta}{2}f - \frac{\beta}{2}g^T\delta - \frac{1}{2}\delta^T\left[\frac{\beta}{2}H + V^{-1}\right]\delta. \qquad (9.51)$$

so that

$$\int \exp\{a\}\,d\delta \;=\; \exp\left\{-\frac{\beta}{2}f\right\}\int \exp\left\{-\frac{\beta}{2}g^T\delta - \frac{1}{2}\delta^T S\delta\right\}d\delta \qquad (9.52)$$

$$\;=\; \exp\left\{-\frac{\beta}{2}f\right\}(2\pi)^{\frac{nK+1}{2}}|S|^{-\frac{1}{2}}\exp\left\{\frac{\beta^2}{8}g^T S^{-1}g\right\} \qquad (9.53)$$

where $S = \left[\frac{\beta}{2}H + V^{-1}\right]$ and n is the number of registration parameters (geometric and photometric) per image. Using this integral result along with (9.47), the final expression for our conditional distribution of the super-resolution image is

$$p\left(x \,\middle|\, \{y^{(k)}\}\right) \;=\; \frac{p(x)}{p(\{y^{(k)}\})}\left(\frac{\beta^{KM}|V^{-1}|}{(2\pi)^{KM}|S|}\right)^{\frac{1}{2}}$$

$$\times \exp\left\{-\frac{\beta}{2}f + \frac{\beta^2}{8}g^T S^{-1}g\right\}. \qquad (9.54)$$

To arrive at an objective function that we can optimize using gradient descent methods, we take the negative log likelihood and neglect terms that are not functions of x. Using the Huber image prior from Section (9.1.3), this gives

$$\mathcal{L} \;=\; \frac{\nu}{2}\rho(Dx, \alpha) + \frac{\beta}{2}f + \frac{1}{2}\log|S| - \frac{\beta^2}{8}g^T S^{-1}g. \qquad (9.55)$$

This is the function we optimize with respect to x to compute the super-resolution image. The dependence of the various terms on x can be summarized

$$f(x) = \sum_{k=1}^{K} \left\| y^{(k)} - \lambda_1^{(k)}(\delta) W^{(k)}(\delta) x - \lambda_2^{(k)}(\delta) \right\|_2^2 \quad \text{(scalar)} \quad (9.56)$$

$$g(x) = \frac{\partial f(x)}{\partial \delta} \quad (p \times 1 \text{ gradient vector}) \quad (9.57)$$

$$H(x) = \frac{\partial g(x)}{\partial \delta} \quad (p \times p \text{ Hessian matrix}) \quad (9.58)$$

$$S(x) = \frac{\beta}{2} H(x) + V^{-1} \quad (p \times p \text{ matrix}), \quad (9.59)$$

where δ is the p-element vector of "nuisance variables" (e.g., registrations and PSF size), which is assumed to be Gaussian distributed with covariance V. A detailed derivation of the gradient of (9.55) can be found in [14].

9.4.2 Marginalizing over the High-Resolution Image

We will outline the marginalization method used in [15] here, since it is useful for comparison with our method, and also because the model used here extends theirs, by adding photometric parameters, which introduces extra terms to the equations.

The prior used in [15] takes the form of a zero-mean Gaussian over the pixels in x with covariance Z_x. A simplified version has already been discussed in Section 9.1.3 (equations (9.15) and (9.23)), but if we consider the exact form of the probability and its normalizing constant, it is

$$p(x) = (2\pi)^{-\frac{N}{2}} |Z_x|^{-\frac{1}{2}} \exp \left\{ -\frac{1}{2} x^T Z_x^{-1} x \right\}. \quad (9.60)$$

This is used to facilitate the marginalization over the super-resolution pixels in order to arrive at an expression for the marginal probability of the low-resolution image set conditioned only on the set of imaging parameters:

$$p\left(\left\{ y^{(k)} \right\} \middle| \left\{ \theta^{(k)}, \lambda^{(k)} \right\} \right) = \int p\left(\left\{ y^{(k)} \right\} \middle| x, \left\{ \theta^{(k)}, \lambda^{(k)} \right\} \right) p(x) \mathrm{d}x. \quad (9.61)$$

The data likelihood component of the integral is exactly as in equation (9.3) and the full derivation can be found in [14].

Taking y to be a stacked vector of all the input images $\left\{ y^{(k)} \right\}$, and λ_2 to be a stack of the $\lambda_2^{(k)}$ vectors, this marginal distribution over y conditioned on the registration parameters turns out to be

$$y \sim \mathcal{N}(\lambda_2, Z_y), \quad (9.62)$$

where

$$Z_y = \beta^{-1} I + \Lambda_1 W Z_x W^T \Lambda_1^T. \quad (9.63)$$

Here Λ_1 is a matrix whose diagonals are given by the $\lambda_1^{(k)}$ values of the corresponding low-resolution images, and W is the stack of individual $W^{(k)}$ matrices.

The objective function, which does not depend on x, is optimized with respect to $\{\theta^{(k)}, \lambda^{(k)}\}$ and γ, and is given by

$$
\mathcal{L} = \frac{1}{2} \left[\beta \sum_{k=1}^{K} \left\| y^{(k)} - \lambda_1^{(k)} W^{(k)} \mu - \lambda_2^{(k)} \right\|_2^2 + \mu^T Z_x^{-1} \mu - \log |\Sigma| \right] \quad (9.64)
$$

where

$$
\Sigma = \left[Z_x^{-1} + \beta \sum_{k=1}^{K} \lambda_1^{(k)2} W^{(k)T} W^{(k)} \right]^{-1} \quad (9.65)
$$

$$
\mu = \beta \Sigma \left(\sum_{k=1}^{K} \lambda_1^{(k)} W^{(k)T} \left(y^{(k)} - \lambda_2^{(k)} \right) \right). \quad (9.66)
$$

The expression for the posterior mean μ is the closed form of the overall MAP solution for the super-resolution image. However, in [15], the optimization over registration and blur parameters is carried out with low-resolution image patches of just 9×9 pixels, rather than the full low-resolution images, because of the computational cost involved in computing the terms in (9.64) — even for a tiny 50×50-pixel high-resolution image, the Z_x and Σ matrices are 2500×2500. The full-sized super-resolution image can then be computed by fixing the optimal registration and PSF values and finding μ using the full-sized low-resolution images, $y^{(k)}$, rather than the 9×9 patches. This is exactly equivalent to solving the usual MAP super-resolution approach of Section 9.1.2.3, with $p(x)$ defined as in (9.15), using the covariance of (9.23).

In comparison, the dimensionality of the matrices in the terms comprising the registration-marginalizing objective function (9.55) is in most cases much lower than those in (9.64). This means the terms arising from the marginalization are far less costly to compute, so our algorithm can be run on entire low-resolution images, rather than just patches.

9.4.3 Implementation Notes

The objective function (9.55) can be optimized using *Scaled Conjugate Gradients* (SCG) [12], noting that the gradient can be expressed as

$$
\frac{d\mathcal{L}}{dx} = \frac{\nu}{2} D^T \frac{d}{dx} \rho (Dx, \alpha) + \frac{\beta}{2} \frac{df}{dx} - \frac{\beta^2}{4} \xi^T \frac{dg}{dx}
$$
$$
+ \left[\frac{\beta}{4} \text{vec} \left(S^{-1} + \frac{\beta^2}{8} \xi \xi^T \right) \right]^T \frac{d\text{vec}(H)}{dx}, \quad (9.67)
$$

where

$$
\xi = S^{-1} g, \quad (9.68)
$$

and where *vec* is the matrix vectorization operator. Derivatives of f, g, and H with respect to x can be found analytically for photometric parameters, and numerically (using the analytic gradient of $e^{(k)} \left(\delta^{(k)} \right)$ with respect to x) with respect to the geometric parameters.

The upper part of H is block-diagonal $nK \times nK$ sparse matrix, and the final $(nK + 1)^{\text{th}}$ row and column are non-sparse, assuming that the blur parameter is shared between the images, as it might be in a short video sequence, for instance, and that the image registration errors for two different images are independent. Notice that the value f in (9.55) is simply the reprojection error of the current estimate of x at the mean registration parameter values, i.e., the value of (9.46) evaluated at $\bar{\theta}^{(k)}$, $\bar{\lambda}^{(k)}$, and $\bar{\gamma}$. Gradients of this expression with respect to the λ parameters, and with respect to x can both be found analytically. To find the gradient with respect to a geometric registration parameter $\theta_i^{(k)}$, and elements of the Hessian involving it, a central difference scheme involving only the kth image is used.

9.4.4 Experimental Evaluation

Results from two experiments show the marginalization methods working on synthetic and real data. The first example uses a synthetic dataset to allow a quantitative measure of performance to be taken with respect to known ground truth high-resolution images. The second example shows the full system working on real data, and compares the results to the standard Huber-MAP method, and to the approach of [15].

Synthetic Eyechart Sequence

Sixteen low-resolution "eyechart" inputs were created from the ground truth image of an eye chart, as used for Figure 9.12. Each image is generated at a zoom factor of 4, again using the model with 2 translational degrees of freedom and two photometric degrees of freedom. A Gaussian point-spread function with a standard deviation of 0.4 low-resolution pixels is used, and Gaussian noise (30dB; standard deviation equivalent to approximately 3.4 gray levels) is added to the intensity of each low-resolution pixel independently.

Geometric and photometric registration parameters were initialized to the identity, and the images were registered using an iterative intensity-based scheme [3]. The resulting parameter values were used to recover two sets of super-resolution images: one using the standard Huber-MAP algorithm, and the second using our extension integrating over the geometric and photometric registration uncertainties. The Huber parameter α was fixed at 0.01 for all runs, and ν was varied over a range of possible values representing ratios between ν and the image noise precision β.

The images giving lowest RMS error from each set are displayed Figure 9.17. Visually, the differences between the images are subtle, though the bottom row of letters is better defined in the output from our algorithm. Plotting

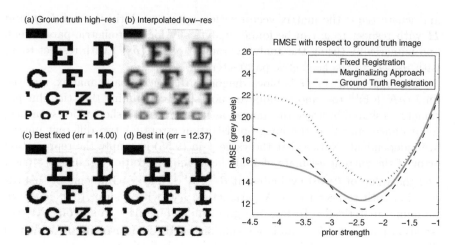

FIGURE 9.17: Super-resolving the synthetic eyechart dataset. (a) ground truth image; (b) interpolated low-resolution image; (c) best (minimum MSE) image from the regular Huber-MAP algorithm; and (d) best result using our approach of integrating over θ and λ. As well as having a lower RMSE, note the improvement in black-white edge detail on some of the letters on the bottom line. Right: variation of RMSE with prior strength for the standard Huber-prior MAP super-resolution method and our approach integrating over θ and λ.

the RMSE as a function of ν, it is clear that the registration-marginalizing approach achieves a lower error compared to the ground truth high-resolution image than the standard Huber-MAP algorithm for any choice of prior strength, $\log_{10}(\nu/\beta)$. Because ν and β are free parameters in the algorithm, it is an advantage that the marginalizing approach is less sensitive to variation in their values.

Real Data

The final example uses real data with a 2D translation motion model and a 2-parameter lighting model exactly as above; the low-resolution images appear in Figure 9.18. Homographies were provided with the data, but were not used. Instead, an iterative illumination-based registration was used on the subregion of the images chosen for super-resolution, and this agreed with the provided homographies to within a few hundredths of a pixel.

Super-resolution images were created for a number of image prior strengths, and equivalent values to those quoted in [4] were selected for the Huber-MAP recovery, following a subjective evaluation of other possible parameter settings. For the registration-marginalizing approach, a similar parameter error distribution as that used in the synthetic experiments was assumed. Finally, Tipping and Bishop's method, extended to cover the illumina-

FIGURE 9.18: Two of the ten input images in the real dataset.

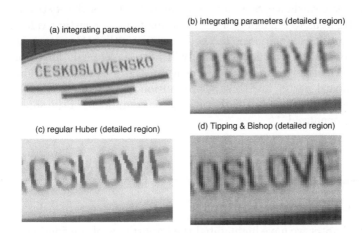

FIGURE 9.19: Super-resolving the "Československo" sequence. (a) Full output from our algorithm. (b) Detail of the central letters, again with our algorithm. (c) Detail with the regular Huber-MAP super-resolution image. (d) Detail with Tipping and Bishop's method of marginalization. The Gaussian form of their prior leads to a more blurred output, or one that overfits to the image noise on the input data if the prior's influence is decreased.

tion model, was used to register and super-resolve the dataset, using the same PSF standard deviation (0.4 low-resolution pixels) as the other methods.

The three sets of results on the real data sequence are shown in Figure 9.19. To facilitate a better comparison, a sub-region of each is expanded to make the letter details clearer. The Huber prior tends to make the edges unnaturally sharp, though it is very successful at regularizing the solution elsewhere. Between the Tipping and Bishop image and the registration-marginalizing approach, the text appears more clear in our method, and the regularization in the constant background regions is slightly more successful. Also note that the Gaussian prior on the image-marginalizing method is zero-mean (see Equation (9.23)), so in this case having a strong enough prior to suppress the background noise has also biased the output image towards the mid-gray zero value, making the white regions appear darker than they do in the other methods.

9.4.5   Discussion

It is possible to interpret the extra terms introduced into the objective function in the registration-marginalizing method as an extra regularizer term or image prior. Considering (9.55), the first two terms are identical to the standard MAP super-resolution problem using a Huber image prior. The two additional terms constitute an additional distribution over x in the cases where the parameter covariance S is not dominated by V; as the distribution over θ and λ tightens to a single point, the terms tend to constant values.

The intuition behind the method's success is that this extra prior resulting from the final two terms of (9.55) will favor image solutions which are not acutely sensitive to minor adjustments in the image registration. Since the checkerboard patter in ML super-resolution images die to ill-conditioning *is* very sensitive to the exact registration, this component of the super-resolution image is penalized.

9.5   Concluding Remarks

In this chapter we have highlighted the importance of considering latent quantities such as image registration or point-spread function size as part of the super-resolution problem instead of estimating and fixing them in advance. Within a probabilistic framework based on a generative model of the image formation process, two different algorithms were described: one which optimizes the latent variables at the same time as the super-resolution image, and one which marginalizes them out of the problem.

The registration-marginalizing approach to super-resolution shows several advantages over Tipping and Bishop's original image-integrating algorithm. These are a formal treatment of registration uncertainty, the use of a much more realistic image prior, and the computational speed and memory efficiency relating to the smaller dimension of the space over which it operates. Note that while the examples in the marginalization section concentrated on a translation-only motion model, there is no constraint in the mathematical derivation that prevents it from being applied to more complex parametric motion models such as affine or planar projective homographies.

The simultaneous super-resolution algorithm that was presented earlier is conceptually simpler to understand and implement than the marginalization approach, but likewise demonstrated a quantitative improvement in super-resolution image quality. While a combination of the two approaches may yield even more accurate results at a higher computational cost, the efficacy of the simultaneous approach makes it a reliable choice of super-resolution algorithm when a high degree of reconstruction accuracy is required.

Bibliography

[1] S. Baker and T. Kanade. Limits on super-resolution and how to break them. 24(9):1167–1183, 2002.

[2] C. Bishop. *Neural Networks for Pattern Recognition.* Oxford University Press, 1995.

[3] M. J. Black and P. Anandan. A framework for the robust estimation of optical flow. pages 231–236, 1993.

[4] D. P. Capel. *Image Mosaicing and Super-resolution (Distinguished Dissertations).* Springer, ISBN: 1852337710, 2004.

[5] S. Farsiu, M. Elad, and P. Milanfar. A practical approach to super-resolution. In *Proc. of the SPIE: Visual Communications and Image Processing,* San-Jose, 2006.

[6] S. Farsiu, M. D. Robinson, M. Elad, and P. Milanfar. Fast and robust multiframe super resolution. 13(10):1327–1344, October 2004.

[7] R. C. Hardie, K. J. Barnard, and E. E. Armstrong. Joint MAP registration and high-resolution image estimation using a sequence of undersampled images. 6(12):1621–1633, 1997.

[8] R. I. Hartley and A. Zisserman. *Multiple View Geometry in Computer Vision.* Cambridge University Press, ISBN: 0521540518, 2nd ed., 2004.

[9] P. J. Huber. *Robust Statistics.* John Wiley and Sons, 1981.

[10] M. Irani and S. Peleg. Improving resolution by image registration. *Graphical Models and Image Processing,* 53:231–239, 1991.

[11] D. Kundur and D. Hatzinakos. Blind image deconvolution. *IEEE Signal Processing Magazine,* 13(3):43–46, May 1996.

[12] I. Nabney. *Netlab Algorithms for Pattern Recognition.* Springer, 2002.

[13] N. Nguyen, P. Milanfar, and G. Golub. Efficient generalized cross-validation with applications to parametric image restoration and resolution enhancement. 10(9):1299–1308, September 2001.

[14] L. C. Pickup. *Machine Learning in Multi-frame Image Super-resolution.* PhD thesis, University of Oxford, February 2008.

[15] M. E. Tipping and C. M. Bishop. Bayesian image super-resolution. In S. Thrun, S. Becker, and K. Obermayer, editors, *Advances in Neural Information Processing Systems,* volume 15, pages 1279–1286, Cambridge, MA, 2003. MIT Press.

[16] W. Triggs, P. McLauchlan, R. Hartley, and A. Fitzgibbon. Bundle adjustment: A modern synthesis. In W. Triggs, A. Zisserman, and R. Szeliski, editors, *Vision Algorithms: Theory and Practice*, LNCS, pages 298–375. Springer Verlag, 2000.

10

Variational Bayesian Super-Resolution Reconstruction

S. Derin Babacan

Northwestern University

Rafael Molina

Universidad de Granada

Aggelos K. Katsaggelos

Northwestern University

CONTENTS

10.1 Introduction

In many imaging applications, acquiring an image of a scene with high spatial resolution is not possible due to a number of theoretical and practical limitations. These limitations include for instance the sensor resolution, the Rayleigh resolution limit, the increased cost, data transfer rate, and the amount of shot noise due to the size of the digital sensor, among others. In these cases, super-resolution (SR) methods can be utilized to process one or more low-resolution (LR) images of the scene together to obtain a high-resolution (HR) image. The basic principle of super-resolution is that changes in the LR images caused by the blur and the (camera or scene) motion provide additional data that can be utilized to reconstruct the HR image from the set of LR observations. Super-resolution methods are widely utilized in a number of imaging fields, such as surveillance, remote sensing, medical and nano-imaging.

Although the super-resolution literature is rich (see [26, 25, 34] for reviews) it is still an open and widely investigated topic. The challenges in formulating and solving the super-resolution problem are the accurate modeling of the generative process, the description and incorporation of prior knowledge about unknown variables into the restoration process (regularization), the adaptation of the solution to the characteristics of the unknowns (spatially adaptive or nonstationary solutions), the automatic determination of the parameters of the problem, the adaptation of the estimates of these parameters as the iterations progress, and the efficient implementation of the solution approaches.

super-resolution is a highly ill-posed problem, especially when the motion and blur parameters are estimated along with the HR image solely from the LR images. A number of methods perform the estimation of these parameters in a separate preprocessing stage [15, 53, 17, 49]. However, these parameters are generally very hard to estimate using only LR observations, which makes estimation errors unavoidable in many practical systems. The errors in estimating the blur and registration parameters cause significant drawbacks in super-resolution, leading to instabilities in the recovery of the HR image and significantly affecting the robustness of the restoration procedures.

Some methods utilize robust image estimation methods to alleviate the problems caused by the errors in the motion estimates. A robust backprojection method is proposed in [53] based on median estimators. Farsiu et al. [17] proposed to use an observation model based on l_1-norms and image priors based on bilateral total-variation (BTV) functions, whose combination makes the algorithm robust to motion outliers. Other methods employ regularization by modeling the registration errors as Gaussian noise [28, 21]. All methods in this category attempt to reduce the effect of estimation errors and noise by decreasing the weight of unreliable observations in the restoration process, but they do not attempt to correct the errors in the motion estimation process.

Another class of SR methods estimate both the HR image and the motion parameters simultaneously. The most common approach in this category is alternating minimization (AM), where at each iteration, the estimates of the HR image and the motion parameters are improved progressively in an alternating fashion [20, 35, 43, 42, 44, 52, 24, 48, 38]. Some methods in this category also employ explicit models of the errors in motion estimates. In [48] and [38], the errors in motion and blur parameters are assumed to follow Gaussian distributions. In [48], the HR image is marginalized out from the joint distribution and the motion and blur parameters are estimated from this marginal distribution. A major disadvantage of this method is that the marginalization of the HR image requires the utilization of a Gaussian image prior, which overpenalizes strong image edges and therefore reduces the quality of the estimated HR image. In [38], this problem is overcome by marginalizing the motion and blur parameters, and employing a Huber prior to model the HR image. Recently, a joint identification method is proposed in [22] where the optimization problem is solved simultaneously for both the HR image and motion parameters. Finally, methods that do not utilize explicit knowledge of the motion estimation parameters have been proposed in [46, 39].

A major drawback of most super-resolution methods is that they employ a number of parameters that need to be tuned. This tuning process can be cumbersome and time-consuming since the parameter values have to be chosen differently for each image and degradation condition. Moreover, the algorithmic performance depends significantly on the appropriate choice of parameters, such that generally a long supervised process is needed to obtain useful results.

The Bayesian framework provides a powerful means for addressing all the challenges described above. First, a systematic modeling of the observations, the sought after unknown variables, and the algorithm parameters can be achieved using a Bayesian formulation. Second, the Bayesian methodology offers a variety of inference methods for the estimation of the unknowns, which have a number of advantages over more traditional (deterministic) methods. These include accounting for the estimation errors that increases robustness; providing distribution estimates instead of point estimates, which provides a measure of uncertainty of the estimation process; and providing a general framework for quantitative selection and comparison of models and inference strategies based on specific applications.

In this chapter, we describe the Bayesian framework for systematically modeling the observed LR images, the unknown HR image, and the motion and blur parameters. Using this model we develop two SR algorithms that jointly estimate the HR image and all algorithmic parameters. Through the utilization of variational Bayesian analysis, the proposed framework provides uncertainties of the estimates during the restoration process, which helps to prevent error-propagation and improves robustness. All required algorithmic parameters are estimated along with the HR image, and therefore algorithms do not require user supervision. Moreover, the parameters are estimated optimally in a stochastic sense, which provides high-reconstruction performance. We demonstrate experimentally that the proposed methods provide HR images with high quality and compare favorably to existing SR methods. Finally, we discuss extensions of the model to incorporate the estimation of the motion and blur parameters.

10.2   Problem Formulation

In super-resolution, the general discrete model of the generation of LR images y_k, $k = 1, \ldots, L$ from the HR image x caused by warping, blurring, and downsampling is given in matrix-vector form by

$$y_k = AH_kC(s_k)x + n_k = B_k(s_k)x + n_k, \qquad (10.1)$$

where A is the $N \times PN$ downsampling matrix, H_k is the $PN \times PN$ blurring matrix, $C(s_k)$ is the $PN \times PN$ warping matrix generated by the motion vector s_k, and n_k is the $N \times 1$ acquisition noise. Note that the matrices H_k and $C(s_k)$ and the noise n_k can be different for each LR image y_k. The LR images y_k and the HR image x consist of N and PN pixels, respectively, where the integer $P > 1$ is the factor of increase in resolution.

The effects of downsampling, blurring, and warping can be combined into a single $N \times PN$ system matrix $B_k(s_k)$, such that each row in matrix $B_k(s_k)$ maps the pixels in the HR image x to one pixel in the LR image y_k. Given (10.1), the super-resolution problem is to find an estimate of the HR image x from the set of LR images $\{y_k\}$ using prior knowledge about $\{B(s_k)\}$, $\{n_k\}$, and x.

In super-resolution algorithms, the blurring matrices H_k and the motion vectors s_k are generally assumed to be known. However, this assumption does not hold in many practical systems since exact blurring and motion information are very hard to obtain. Therefore, a more practical solution is to incorporate their estimation along with the estimation of the HR image. The estimation of the blur and motion can be performed in two ways: First, they can be identified separately from x, and these estimates can later be utilized in an HR image estimation algorithm. Algorithms in a second category estimate the unknown image, blur and motion parameters simultaneously. The most common approach in this category utilizes *alternating minimization* methods, where the formulated optimization problem is solved for the HR image, motion and blur in an alternating fashion. Recently, a joint identification method is proposed in [22] where the optimization problem is solved simultaneously for both the HR image and motion parameters.

10.3   Bayesian Framework for Super-Resolution

The fundamental principle of Bayesian methods is to treat all parameters and observable variables as unknown stochastic quantities and assign probability distributions to them. Therefore, in super-resolution, the unknown image x, the motion parameters $\{s_k\}$, and the blurring matrices H_k (if they are assumed unknown) are all treated as samples of random fields, with corresponding *prior* probability density functions that incorporate our knowledge about the imaging process and the nature of the image into the estimation process.

These prior distributions depend on parameters Ω, which are termed *hyperparameters*. Most super-resolution algorithms assume that these parameters are known, or are estimated separately from the image, motion, and blur. On the other hand, in the Bayesian framework an additional modeling stage can be incorporated, where *hyperprior* distributions are assigned to model the hyperparameters. This additional modeling stage leads to a *hierarchical* Bayesian framework, which generally increases robustness to errors when there is uncertainty, and it is essential when the confidence in the observed data is low (for instance, due to high-acquisition noise, or very low-spatial resolution).

Specifically, the hierarchical Bayesian framework for super-resolution consists of two stages and is formulated as follows. In the the first stage the prior distributions $p(\boldsymbol{x}|\alpha_{\text{im}})$, $p(\boldsymbol{s}_k, \boldsymbol{H}_k|\Omega)$, and the conditional distribution $p(\boldsymbol{y}|\boldsymbol{x}, \{\boldsymbol{s}_k\}, \{\boldsymbol{H}_k\}, \{\beta_k\})$ are defined. The hyperparameters Ω are modeled by the hyperprior distribution $p(\Omega)$ in the second stage. This hierarchical modeling allows us to write the joint distribution as

$$p(\boldsymbol{x}, \{\boldsymbol{s}_k\}, \{\boldsymbol{H}_k\}, \Omega, \boldsymbol{y}) = p(\boldsymbol{y}|\boldsymbol{x}, \{\boldsymbol{s}_k\}, \{\boldsymbol{H}_k\}, \Omega)\, p(\boldsymbol{x}|\Omega)\, p(\{\boldsymbol{s}_k\}, \{\boldsymbol{H}_k\}|\Omega)\, p(\Omega)\,.$$

$$(10.2)$$

In the following subsections we provide the description of the individual distributions used to model the unknowns.

10.3.1 Observation Models

Due to the model in (10.1), the conditional distribution of the observed images $\boldsymbol{y}_k$ is related to the noise $\boldsymbol{n}_k$. A typical model for $\boldsymbol{n}_k$ is zero-mean independent white Gaussian noise, which results in the conditional distribution of the LR image $\boldsymbol{y}_k$ given by

$$p(\boldsymbol{y}_k|\boldsymbol{x}, \boldsymbol{s}_k, \boldsymbol{H}_k, \beta_k) \propto \beta_k^{N/2} \exp\left[-\frac{\beta_k}{2} \parallel \boldsymbol{y}_k - \boldsymbol{B}_k(\boldsymbol{s}_k)\boldsymbol{x} \parallel^2\right], \qquad (10.3)$$

with β_k the inverse variance (precision). Assuming statistical independence of the noise among the LR image acquisitions, the conditional probability of the set of LR images y given x can be expressed as

$$
\begin{aligned}
p(y|x, \{s_k\}, \{H_k\}, \{\beta_k\}) &= \prod_{k=1}^{L} p(y_k|x, s_k, H_k, \beta_k) \\
&= \left[\prod_{k=1}^{L} \beta_k^{N/2} \right] \exp\left[-\frac{1}{2} \sum_{k=1}^{L} \beta_k \parallel y_k - B_k(s_k)x \parallel^2 \right].
\end{aligned}
$$

(10.4)

The independent Gaussian model in (10.4) is used in most of the existing super-resolution methods [28, 48, 21, 38, 48, 11, 24]. Some methods utilized l_1-norm based observation models that take both acquisition and registration noise into account [53, 17]. Alternatively, an explicit modeling of the registration errors can be utilized so that (10.4) only models the acquisition noise [48, 38].

10.3.2   Image Models

The prior distribution $p(x|\Omega)$ reflects our prior knowledge about the nature of the HR image x and constrains the space of possible solutions to the most probable ones. Therefore, the quality of the estimated HR image, as well as the accuracy in the estimates of other unknowns depends on the incorporation of accurate image models. Typical descriptions of the natural images used in the super-resolution literature are smooth, piecewise-smooth, or textured, among others. These descriptions are embedded into the Bayesian framework using priors, which typically specify probabilistic relations between neighboring pixels or their derivatives.

An important consideration while choosing the image prior is the analytical tractability of the Bayesian inference. This generally limits the possible choices. The simplest possible image prior is the noninformative flat prior [45, 47, 23]

$$
p(x) \propto \text{constant.} \tag{10.5}
$$

The most common model is the class of Gaussian models, often termed as *Simultaneous Autoregression* (SAR) or *Conditional Autoregression* (CAR) models, expressed as

$$
p(x|\alpha_{\text{im}}) \propto \exp\left[-\frac{\alpha_{\text{im}}}{2} \parallel Cx \parallel^2 \right], \tag{10.6}
$$

where C is the discrete Laplacian operator. The SAR model is suitable for the image if it is assumed that the luminosity distribution is smooth on the image domain. However, it is also well-known that generally this prior leads to oversmooth image estimates, where the image edges are overpenalized and not well-preserved. Non-quadratic image priors are generally considered more appropriate for natural images, as they aim at preserving edges by not over-penalizing discontinuities, i.e., outliers in the image gradient distribution. One of the most popular non-quadratic image priors is the total variation (TV) image prior. The TV function is utilized successfully in a number of image recovery methods including denoising [41], blind deconvolution [5, 13], inpainting, and super-resolution [36, 12]. The TV prior is given by

$$p(\boldsymbol{x}|\alpha_{\text{im}}) = \frac{1}{Z(\alpha_{\text{im}})} \exp\left[-\frac{1}{2}\alpha_{\text{im}}\text{TV}(\boldsymbol{x})\right], \tag{10.7}$$

where $Z(\alpha_{\text{im}})$ is the partition function and

$$\text{TV}(\boldsymbol{x}) = \sum_{i=1}^{PN} \sqrt{(\Delta_i^h(\boldsymbol{x}))^2 + (\Delta_i^v(\boldsymbol{x}))^2}. \tag{10.8}$$

The operators $\Delta_i^h(\boldsymbol{x})$ and $\Delta_i^v(\boldsymbol{x})$ correspond, respectively, to the horizontal and vertical first order differences at pixel i. For a fully-Bayesian analysis, the explicit form of the partition function $Z(\alpha_{\text{im}})$ is needed but it cannot be calculated. It can, however, be approximated by a quadratic function [5]

$$p(\boldsymbol{x}|\alpha_{\text{im}}) \propto c\,\alpha_{\text{im}}^{PN/2} \exp\left[-\frac{1}{2}\alpha_{\text{im}}\text{TV}(\boldsymbol{x})\right], \tag{10.9}$$

where c is a constant.

Many other image priors have been proposed in the literature, including bilateral TV priors [17], Huber functions [38], anisotropic diffusion [27], and compound models [40]. Unfortunately, most of these complex priors cannot be directly used within Bayesian formulations, since they do not result in tractable inference procedures. In the case of the TV prior in (10.9) [2] and wavelet-based priors based on l_p norms ($0 < p \leq 1$) [3, 19], majorization-minimization approaches can be used to obtain variational approximations. A general framework for variational methods for nonquadratic image priors can be found in [37].

10.3.3 Blur Models

As mentioned above, the blur PSF is assumed to be known in most super-resolution algorithms. On the other hand, if they are assumed unknown, their estimation can also be incorporated into the Bayesian formulation using blur prior distributions. Depending on the nature of the blur the imaging device is expected to introduce, the blur prior can be chosen from one of the priors presented in Section 10.3.2. For instance, SAR priors are commonly used for smooth blur PSFs without discontinuities, including lens blur and blur due to atmospheric turbulence [33]. On the other hand, TV-based blur priors can be utilized for blur PSFs with discontinuities, such as box-shaped PSFs [13, 5]. Other choices include Dirichlet distributions [31], mixture distributions of exponentials [6, 18], among others (see [10] for a review of blur models).

10.3.4   Motion (Registration) Models

super-resolution is a highly ill-posed problem, especially when the motion parameters are estimated along with the HR image solely based on the LR images. In many cases, the motion is modeled using parametric models, such as translational, affine, projective motion, which significantly reduces the number of parameters to be estimated. These models are effective in capturing the real motion in the case of image super-resolution, whereas more general per-pixel dense motion fields are needed for video super-resolution.

Even when parametric motion models are utilized, the registration parameters are generally very hard to estimate using only LR observations, which makes estimation errors unavoidable in many practical systems. The errors in estimating the registration parameters cause significant drawbacks in super-resolution, leading to instabilities in the recovery of the HR image and significantly affecting the robustness of the restoration procedures.

The Bayesian methodology can be utilized to model the estimation errors in the registration parameters by treating these parameters as stochastic variables and modeling their uncertainty using probability distributions. Let us denote by $\bar{s}_k^p$ the estimate of s_k obtained from LR observations in a pre-processing step, using registration algorithms, such as the ones reported in [50, 29]. We can model the motion parameters as following Gaussian distributions with *a priori* means set equal to the preliminary motion parameters $\bar{s}_k^p$, that is,

$$p(s_k) = \mathcal{N}(s_k | \bar{s}_k^p, \Lambda_k^p), \qquad (10.10)$$

with Λ_k^p the prior covariance matrix. The parameters $\bar{s}_k^p$ and Λ_k^p incorporate prior knowledge about the motion parameters into the estimation procedure. If such knowledge is not available, $\bar{s}_k^p$ can be set equal to a zero vector, and $(\Lambda_k^p)^{-1}$ equal to a positive definite matrix with elements close to zero. This makes the observations solely responsible for the estimation process. Utilizing Gaussian distributions to model the uncertainty in the motion parameters have also been used in [28, 48, 38].

10.3.5   Hyperpriors on the Hyperparameters

The hyperparameters are crucial in determining the performance of the SR algorithm. Most super-resolution methods leave their estimation to the user, which requires a long parameter-tuning process and therefore limits the applicability of the super-resolution method. On the other hand, employing a fully Bayesian analysis allows their estimation as well. To obtain tractable Bayesian inference, generally *conjugate* hyperprior distributions are utilized, which lead to straightforward calculation or approximation of the posterior distribution $p(x, s_k, H_k, \Omega | y)$. Conjugate priors allow one to begin with a certain functional form for the conditional and prior distributions and end up with the posterior of the same functional form, but with parameters updated from the observed samples.

A large part of the Bayesian literature is devoted to finding conjugate hyperprior distributions (see [8]). The most commonly used hypeprior distribution is the uninformative prior model

$$p(\Omega) = \text{const.} \tag{10.11}$$

More flexible hyperprior distributions can also be utilized that allow for the incorporation of prior knowledge about the hyperparameters. For instance, for parameters ω corresponding to the precisions (inverse variances) of Gaussian distributions, the Gamma distribution is used, given by

$$p(\omega) = \Gamma(\omega|a_\omega^o, b_\omega^o) = \frac{(b_\omega^o)^{a_\omega^o}}{\Gamma(a_\omega^o)} \omega^{a_\omega^o - 1} \exp\left[-b_\omega^o \omega\right], \tag{10.12}$$

where $a_\omega^o > 0$ and $b_\omega^o > 0$ are the shape and scale parameters, respectively. By appropriately selecting these parameters one can make the estimation process rely on information provided by the observed data and prior knowledge about the hyperparameters.

10.4 Bayesian Inference

Let us denote the set of all unknowns by $\Theta = \{x, \{s_k\}, \{H_k\}, \Omega\}$ for clarity. As is widely known, the Bayesian inference is based on the posterior distribution

$$p(\Theta \mid y) = \frac{p(\Theta, y)}{p(y)}. \tag{10.13}$$

There is a number of methods that can be utilized to obtain estimates of the HR image x and possibly the motion parameters and blur using (10.13). Depending on the prior models used, analytical solutions may be hard to find, so often approximations are needed.

Most methods in the literature seek point estimates of the unknowns, which are generally obtained by finding values that maximize the posterior distribution

$$\hat{\Theta} = \underset{\Theta}{\text{argmax}}\, p(\Theta \mid y). \tag{10.14}$$

Maximum likelihood (ML) and *maximum a posteriori* (MAP) solutions correspond to (10.14), which reduce the problem to one of optimization. Regularization-based approaches frequently found in the literature fall into this category. Methods providing point estimates to the unknowns might suffer in some cases from a number of disadvantages. Common problems are overfitting in the presence of high noise, error propagation among the estimates of different unknowns, and lack of providing uncertainties of the estimates.

On the other hand, the Bayesian framework provides other methodologies for estimating the *distributions* of the unknowns, which deal better with uncertainty. Approximating or simulating the posterior distributions are two options. For instance, marginalization can be utilized to perform inference on a subset of unknowns. One option is to marginalize out the unknown HR image x, that is,

$$\hat{s}_k, \hat{H}_k, \hat{\Omega} = \underset{s_k, H_k, \Omega}{\operatorname{argmax}} \int_x p(x, \{s_k\}, \{H_k\}, \Omega, y) \, dx, \qquad (10.15)$$

and then to select as the HR image estimate

$$\hat{x} = \underset{x}{\operatorname{argmax}} \, p(x|\hat{\Omega}) p(y|x, \{\hat{s}_k\}, \{\hat{H}_k\}, \hat{\Omega}). \qquad (10.16)$$

This method is also called the *evidence*-based analysis and is utilized in [48]. A major disadvantage of this method is that the marginalization of the HR image requires for tractability the utilization of a Gaussian image prior, which overpenalizes strong image edges and therefore reduces the quality of the estimated HR image. In [38], this problem is overcome by adopting an *empirical*-based approach where first the motion and blur parameters are marginalized out, and the HR image is then estimated by employing a Huber image prior.

In *evidence*- and *empirical*-based approaches the marginalized variables are called *hidden variables*. The *expectation-maximization* (EM) algorithm, first described in [14], is a very popular method in signal processing for iteratively solving ML and MAP problems that include hidden variables. The basic principle of the EM algorithm is to first integrate out hidden variables from the joint distribution to obtain *marginal* distributions, and then maximize the posterior distribution obtained from the marginal distribution to provide estimates of the unknowns. The EM algorithm has the advantage of guaranteed convergence to local maxima of the posterior distribution, and it is particularly suited for inverse problems in image processing since the unknown image x is a natural choice for hidden data. However, in many applications, the calculation of the posterior distribution is intractable, which severely limits the application of the EM method.

To overcome this problem, *variational Bayesian* methods can be utilized. Variational Bayesian methods are generalizations of the EM algorithm where the intractable posterior distribution $p(\Theta|y)$ is approximated by a tractable distribution $q(\Theta)$. This approximating distribution is found by minimizing the Kullback-Leibler (KL) distance between $q(\Theta)$ and the posterior $p(\Theta \mid y)$, given by

$$
C_{KL}(q(\Theta) \parallel p(\Theta|\boldsymbol{y})) = \int q(\Theta) \log \left(\frac{q(\Theta)}{p(\Theta|\boldsymbol{y})} \right) d\Theta
$$

$$
= \int q(\Theta) \log \left(\frac{q(\Theta)}{p(\Theta, \boldsymbol{y})} \right) d\Theta + \text{const}, \quad (10.17)
$$

which is always nonnegative and equal to zero only when $q(\Theta) = p(\Theta \mid \boldsymbol{y})$, which corresponds to the EM result.

In order to reduce computational complexity and to obtain analytical solutions for the approximations of the parameter distributions, the distribution $q(\Theta)$ is assumed to have a factorized form using the mean field approximation [9, 7, 30]. For instance, in super-resolution, we may use the following factorization of $q(\Theta)$

$$
q(\Theta) = q(\boldsymbol{x}, \{\boldsymbol{s}_k\}, \Omega) = q(\boldsymbol{x})q(\Omega) \prod_{k=1}^{L} q(\boldsymbol{s}_k)q(\boldsymbol{H}_k). \quad (10.18)
$$

For a parameter $\theta \in \Theta$, let us denote by Θ_θ the subset of Θ with θ removed. An iterative procedure can be obtained by minimizing (10.17) using the factorization (10.18). At each iteration, the distribution $q(\theta)$ of the parameter θ can be estimated using the current estimates of other parameters Θ_θ as follows [9]

$$
q(\theta) = \underset{q(\theta)}{\arg\min}\, C_{KL}(q(\Theta_\theta)q(\theta) \parallel p(\Theta|\boldsymbol{y})) \quad (10.19)
$$

$$
= \text{const} \times \exp\left(\langle \log p(\Theta, \boldsymbol{y}) \rangle_{q(\Theta_\theta)} \right), \quad (10.20)
$$

where $\langle \cdot \rangle_{q(\Theta_\theta)}$ denotes expected value with respect to the distribution $q(\Theta_\theta)$. Alternating minimization strategies are generally employed to provide estimates for all unknown distributions.

Finally, sampling methods can be utilized to approximate the posterior distributions. Sampling methods represent the most general approaches to performing inference, and in theory they allow for the utilization of arbitrarily complex observation models and prior distributions. Sampling (or simulating) the posterior can provide solutions closer to the optimal ones than other methods, but they are computationally very intensive and their convergence is very hard to establish.

Variational Bayesian methods have certain advantages over other methods. First, they provide accurate approximations to the whole posterior mass. If the posterior distribution is sharply peaked about its maximum value this does not provide any advantage. However, in the case of high noise or heavy-tailed posterior using the maximum posterior values as estimates can be unreliable. For instance, as mentioned in [32], for a Gaussian in high dimensions most of the probability mass is concentrated away from the probability density peak, which corresponds to the solutions provided by the ML and MAP methods. In this case, variational Bayesian methods are expected to provide more accurate and robust solutions. The second advantage of the variational Bayesian methods is that they account for the uncertainties in the estimation processes through the use of distribution estimates. Finally, they are computationally much more efficient than sampling approaches.

10.5 Variational Bayesian Inference Using TV Image Priors

The general variational Bayesian analysis can be directly utilized for Gaussian observation models and Gaussian image priors (such as SAR). However, as mentioned above, Gaussian priors tend to result in oversmooth image estimates with reduced image quality. Unfortunately, non-quadratic (and non-convex) image priors cannot be directly utilized within variational Bayesian methods, as the calculation of the KL distance in (10.17) is intractable. This difficulty can be overcome by resorting to majorization-minimization (MM) [19] approaches (also referred to as variational approximations [37]). The main goal in these methods is to obtain bounds to the joint distribution that are easier to evaluate analytically than the original forms.

MM methods have been proposed for a number of non-Gaussian image priors. Examples include priors based on l_1-norms, l_p-norms with $0 < p < 1$ [2, 19] and TV image priors [3, 4]. In the following we will present an MM approach for TV image priors in combination with variational Bayesian inference for super-resolution reconstruction. We will first assume that the blurring matrices $\boldsymbol{H}_k$ and the registration parameters $\boldsymbol{s}_k$ are known, and later we will provide an outline of how this framework can be extended to include their estimation as well.

The main principle of the MM approach is to find a bound of the joint distribution in (10.2), which makes the minimization of (10.17) tractable. A lower bound of the distribution in (10.2) can be found as follows. Let us first define a functional $\mathrm{M}(\alpha_{\mathrm{im}}, \boldsymbol{x}, \boldsymbol{w})$ with a $PN-$dimensional vector $\boldsymbol{w} \in (R^+)^{PN}$, with components w_i, $i = 1, \ldots, PN$, as follows

$$\mathrm{M}(\alpha_{\mathrm{im}}, \boldsymbol{x}, \boldsymbol{w}) = c\,\alpha_{\mathrm{im}}^{PN/2} \exp\left[-\frac{\alpha_{\mathrm{im}}}{2} \sum_i \frac{(\Delta_i^h(\boldsymbol{x}))^2 + (\Delta_i^v(\boldsymbol{x}))^2 + w_i}{\sqrt{w_i}}\right],$$

(10.21)

where c is the same constant as in (10.9). As will become clear later, the auxiliary variable $\boldsymbol{w}$ is a quantity that needs to be computed and it has an intuitive interpretation related to the unknown HR image $\boldsymbol{x}$. Let us next consider the following inequality, derived from the geometric-arithmetic mean inequality, which states that for real numbers $a \geq 0$ and $b > 0$

$$\sqrt{ab} \leq \frac{a+b}{2} \Rightarrow \sqrt{a} \leq \frac{a+b}{2\sqrt{b}}. \tag{10.22}$$

Using $a = (\Delta_i^h(\boldsymbol{x}))^2 + (\Delta_i^v(\boldsymbol{x}))^2$ and $b = w_i$ in the inequality (10.22) it is easy to show that the functional $\mathrm{M}(\alpha_{\mathrm{im}}, \boldsymbol{x}, \boldsymbol{w})$ is a lower bound of the image prior $\mathrm{p}(\boldsymbol{x}|\alpha_{\mathrm{im}})$, that is,

$$\mathrm{p}(\boldsymbol{x}|\alpha_{\mathrm{im}}) \geq \mathrm{M}(\alpha_{\mathrm{im}}, \boldsymbol{x}, \boldsymbol{w}). \tag{10.23}$$

This lower bound can be used to find a lower bound for the joint distribution in (10.2), given by

$$\mathrm{p}(\boldsymbol{y}, \Theta) \geq \mathrm{p}(\boldsymbol{y}|\Theta)\mathrm{M}(\alpha_{\mathrm{im}}, \boldsymbol{x}, \boldsymbol{w})\mathrm{p}(\alpha_{\mathrm{im}}) \prod_{k=1}^{L} \mathrm{p}(\beta_k)$$

$$= \mathrm{F}(\Theta, \boldsymbol{w}, \boldsymbol{y}), \tag{10.24}$$

which results in an upper bound of the KL distance in (10.17) as

$$C_{KL}(\mathrm{q}(\Theta) \parallel \mathrm{p}(\Theta, \boldsymbol{y})) \leq C_{KL}(\mathrm{q}(\Theta) \parallel \mathrm{F}(\Theta, \boldsymbol{w}, \boldsymbol{y})). \tag{10.25}$$

The minimization of (10.17) can be replaced by the minimization of its upper bound in (10.25), since minimizing this bound with respect to the unknowns and the auxiliary variable $\boldsymbol{w}$ in an alternating fashion results in closer bounds at each iteration [5]. The bound in (10.25) is quadratic and therefore it is easy to evaluate analytically. Utilizing this bound, the standard solution of the variational Bayesian methods in (10.20) can be used to estimate the unknown distributions $\mathrm{q}(\theta)$ with $\theta \in \Theta$ as follows

$$\mathrm{q}(\xi) = \mathrm{const} \times \exp\left(\langle \log \mathrm{F}(\Theta, \boldsymbol{w}, \boldsymbol{y}) \rangle_{\mathrm{q}(\Theta_\xi)}\right). \tag{10.26}$$

In the following, the subscript of the expected value will be removed when it is clear from the context.

Let us now proceed with deriving the explicit forms of the solutions for each unknown using (10.26).

10.5.1 Estimation of the HR Image Distribution

From (10.26), the distribution $\mathrm{q}(\boldsymbol{x})$ of the HR image $\boldsymbol{x}$ can be found as

$$\mathrm{q}(\boldsymbol{x}) \propto \exp\left(-\frac{1}{2}<\alpha_{\mathrm{im}}> \sum_i \frac{(\Delta_i^h(\boldsymbol{x}))^2 + (\Delta_i^v(\boldsymbol{x}))^2}{\sqrt{w_i}}\right.$$

$$\left. -\frac{1}{2}\sum_k <\beta_k> \parallel \boldsymbol{y}_k - \boldsymbol{A}\boldsymbol{H}_k\boldsymbol{C}(\boldsymbol{s}_k)\boldsymbol{x} \parallel^2\right). \tag{10.27}$$

This distribution is a multivariate Gaussian distribution $q(\boldsymbol{x}) = \mathcal{N}(\boldsymbol{x}|\mu_{\boldsymbol{x}}, \Sigma_{\boldsymbol{x}})$ with parameters

$$\mu_{\boldsymbol{x}} = \Sigma_{\boldsymbol{x}} \left[\sum_k <\beta_k> \boldsymbol{B}_k(\boldsymbol{s}_k)^T \boldsymbol{y}_k \right], \tag{10.28}$$

$$\Sigma_{\boldsymbol{x}}^{-1} = \sum_k <\beta_k> \boldsymbol{B}_k(\boldsymbol{s}_k)^T \boldsymbol{B}_k(\bar{\boldsymbol{s}}_k) + <\alpha_{\text{im}}>(\Delta^h)^T \boldsymbol{W} \Delta^h + <\alpha_{\text{im}}>(\Delta^v)^T \boldsymbol{W} \Delta^v, \tag{10.29}$$

where

$$\boldsymbol{W} = \text{diag}\left(\frac{1}{\sqrt{w_i}}\right), \quad i = 1, \ldots, PN. \tag{10.30}$$

The elements w_i, $i = 1, \ldots, PN$, of the auxiliary vector $\boldsymbol{w}$ are calculated as

$$\begin{aligned} w_i &= \mathrm{E}_{\boldsymbol{x}}[(\Delta_i^h(\boldsymbol{x}))^2 + (\Delta_i^v(\boldsymbol{x}))^2] \\ &= (\Delta_i^h \mu_{\boldsymbol{x}})^2 + (\Delta_i^v \mu_{\boldsymbol{x}})^2 + \text{trace}\left[(\Delta_i^h)^T(\Delta_i^h)\Sigma_{\boldsymbol{x}}\right] + \text{trace}\left[(\Delta_i^v)^T(\Delta_i^v)\Sigma_{\boldsymbol{x}}\right]. \end{aligned} \tag{10.31}$$

It is clear from (10.31) that the vector $\boldsymbol{w}$ represents the local spatial activity in the HR image $\boldsymbol{x}$. Therefore, the matrix $\boldsymbol{W}$ introduces spatial adaptivity into the estimation process of the HR image in (10.28)–(10.29) by controlling the smoothing applied at different locations. Moreover, the uncertainty of the image estimate is also taken into account by the last two terms in (10.31) when calculating the spatial adaptivity vector $\boldsymbol{w}$ using the distribution $q(\boldsymbol{x})$.

10.5.2   Estimation of the Hyperparameter Distributions

In the last step of the algorithm, the distributions of the hyperparameters $q(\alpha_{\text{im}})$ and $q(\beta_k)$ are found from (10.26) as Gamma distributions, expressed as

$$q(\alpha_{\text{im}}) \propto \alpha_{\text{im}}^{PN/2-1+a_{\alpha_{\text{im}}}^o} \exp\left[-\alpha_{\text{im}}(b_{\alpha_{\text{im}}}^o + \sum_i \sqrt{w_i})\right], \tag{10.32}$$

and

$$q(\beta_k) \propto \beta_k^{N/2-1+a_\beta^o} \exp\left[-\beta_k(b_\beta^o + \frac{\mathrm{E}_{\boldsymbol{x}}\left[\| \boldsymbol{y}_k - \boldsymbol{B}(\boldsymbol{s}_k)\boldsymbol{x} \|^2\right]}{2})\right]. \tag{10.33}$$

The quantity $\mathrm{E}_{\boldsymbol{x}}\left[\| \boldsymbol{y}_k - \boldsymbol{B}(\boldsymbol{s}_k)\boldsymbol{x} \|^2\right]$ can be calculated as

$$\mathrm{E}_{\boldsymbol{x}}\left[\| \boldsymbol{y}_k - \boldsymbol{B}(\boldsymbol{s}_k)\boldsymbol{x} \|^2\right] = \| \boldsymbol{y}_k - \boldsymbol{B}(\boldsymbol{s}_k)\mu_{\boldsymbol{x}} \|^2 + \text{trace}\left[\boldsymbol{B}(\boldsymbol{s}_k)^T \boldsymbol{B}(\boldsymbol{s}_k)\Sigma_{\boldsymbol{x}}\right]. \tag{10.34}$$

The means of the distributions in (10.32) and (10.33), which are used as hyperparameter estimates, are given by

$$
<\alpha_{\text{im}}> = \frac{PN/2 + a^o_{\alpha_{\text{im}}}}{\sum_i \sqrt{w_i} + b^o_{\alpha_{\text{im}}}}, \tag{10.35}
$$

$$
<\beta_k> = \frac{N + 2a^o_\beta}{\mathrm{E}_{\boldsymbol{x}}\left[\| \, \boldsymbol{y}_k - \boldsymbol{B}(\boldsymbol{s}_k)\boldsymbol{x} \, \|^2\right] + 2b^o_\beta}. \tag{10.36}
$$

Note that the shape and scale parameters $a^o_{\alpha_{\text{im}}}$, a^o_β, $b^o_{\alpha_{\text{im}}}$, b^o_β can be used to incorporate prior knowledge about the variances of the HR image and observation noise, in case such knowledge is available. If they are set equal to $a^o_{\alpha_{\text{im}}} = a^o_\beta = 1$ and $b^o_{\alpha_{\text{im}}} = b^o_\beta = 0$, which corresponds to utilizing flat hyperprior distributions for the hyperparameters, the observed LR images are made solely responsible for the whole estimation process.

In summary, the algorithm iterates between estimating the HR image using (10.28) and (10.29), the spatial adaptivity vector $\boldsymbol{w}$ using (10.31), and finally the hyperparameters using (10.35) and (10.36). The algorithm is summarized below as Algorithm 1. A major computational difficulty with Algorithm 1 is the explicit construction of the matrix $\Sigma_{\boldsymbol{x}}$ in (10.29), which requires the inversion of a $PN \times PN$ matrix. To avoid this computation, we solve (10.28) efficiently using the conjugate gradient method, and in equations where the explicit form of $\Sigma_{\boldsymbol{x}}$ is needed, i.e., in (10.31) and (10.34), $\Sigma_{\boldsymbol{x}}$ is approximated by a diagonal matrix obtained by inverting the diagonal elements of (10.29). We have conducted extensive experiments with small images that permit the explicit inversion of (10.29) to verify the validity of this approximation. We have found out empirically that this approximation results in small errors, thus having a minor effect in the estimation process. Similar approximations have been utilized in other Bayesian recovery methods [48, 5, 24].

Algorithm 1 Variational Bayesian Super-Resolution

Calculate initial estimates of the HR image and hyperparameters
while convergence criterion is not met **do**
    1. Estimate the HR image distribution using (10.28) and (10.29).
    2. Compute spatial adaptivity vector $\boldsymbol{w}$ using (10.31).
    3. Estimate the distributions of the hyperparameters α_{im}, $\{\beta_k\}$ using (10.32) and (10.33).

It is worth emphasizing here that we did not assume *a priori* that q$(\boldsymbol{x})$ is a Gaussian distribution. This result is derived due to the minimization of the KL divergence with respect to all possible distributions according to the factorization q(α_{im})q$(\boldsymbol{x}) \prod_{k=1}^{L}$ q(β_k) [9]. We can, however, make an assumption that these distributions are *degenerate*, i.e., they take one value with probability one and the rest of the values with probability zero. Using this assumption, we obtain another algorithm very similar to the one presented above, with the only difference that the uncertainty terms arising from the covariance matrices

---

Algorithm 2 Variational Bayesian Super-Resolution with Degenerate Distributions

---

Calculate initial estimates of the initial HR image and hyperparameters
while convergence criterion is not met **do**

1. Calculate the HR image estimate $\hat{\boldsymbol{x}}$ using

$$\hat{\boldsymbol{x}} = \left[\sum_k \hat{\beta}_k \boldsymbol{B}_k(\boldsymbol{s}_k)^T \boldsymbol{B}_k(\boldsymbol{s}_k) + \hat{\alpha}_{\text{im}}(\Delta^h)^T \boldsymbol{W} \Delta^h + \hat{\alpha}_{\text{im}}(\Delta^v)^T \boldsymbol{W} \Delta^v \right]^{-1}$$

$$\times \left[\sum_k \hat{\beta}_k \boldsymbol{B}_k(\boldsymbol{s}_k)^T \boldsymbol{y}_k \right] \tag{10.37}$$

2. Compute spatial adaptivity vector $\boldsymbol{w}$ using

$$w_i = (\Delta_i^h(\hat{\boldsymbol{x}}))^2 + (\Delta_i^v(\hat{\boldsymbol{x}}))^2 \tag{10.38}$$

3. Compute hyperparameter estimates $\hat{\alpha}_{\text{im}}$, $\hat{\beta}_k$ using

$$\hat{\alpha}_{\text{im}} = \frac{PN/2 + a_{\alpha_{\text{im}}}^o}{\sum_i \sqrt{w_i} + b_{\alpha_{\text{im}}}^o}, \tag{10.39}$$

$$\hat{\beta}_k = \frac{N + 2a_\beta^o}{\| \boldsymbol{y}_k - \boldsymbol{B}(\boldsymbol{s}_k)\hat{\boldsymbol{x}} \|^2 + 2b_\beta^o}, \tag{10.40}$$

---

are removed. The derivation of this algorithm is very similar to the first one, and therefore we omit its details and provide the iterative procedure below in Algorithm 2.

It is clear that using a degenerate distribution for $\boldsymbol{x}$ in Algorithm 2 removes the uncertainty terms in the image and hyperparameters estimates. However, the incorporation of these uncertainties through the covariance of $\boldsymbol{x}$ improves the restoration performance, especially in cases when the observation noise is high. This is mainly due to the fact that poor estimations of one variable (due to noise or outliers) can influence the estimation of other unknowns, and as a result the overall performance can significantly be affected. By estimating the full posterior distribution of the unknowns instead of point estimates corresponding to the maximum probability (such as MAP estimates), the uncertainty of the estimates is incorporated into the estimation procedure to ameliorate the propagation of estimation errors among unknowns.

We conclude this section by commenting on the computational complexity of the algorithms. Algorithms 1 and 2 have similar complexities, with Algorithm 1 requiring more computations per iteration due to the incorporation of the covariance matrices. The majority of computations in both algorithms is performed for estimating the HR image, which is calculated efficiently using the conjugate gradient method. Therefore, the algorithms have computational

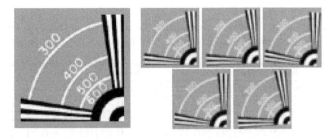

FIGURE 10.1: *(Left)* Original HR image, *(right)* Five synthetically generated LR images.

demands very similar to most existing SR algorithms in the literature (for instance, the AM methods [20, 35, 48, 38]).

10.6 Experiments

In this section, we will provide experimental results with both synthetic and real data to demonstrate the performance of the algorithms developed in the previous section. In synthetic experiments, the quality of the restored HR image is measured quantitatively by the peak signal-to-noise ratio (PSNR), which is defined as

$$\text{PSNR} = 10 \log_{10} \frac{NP}{\|\hat{\boldsymbol{x}} - \boldsymbol{x}\|^2} , \qquad (10.41)$$

where $\hat{\boldsymbol{x}}$ and $\boldsymbol{x}$ are the estimated and original HR images, respectively, and pixel values in both images are normalized to lie in the interval $[0, 1]$. In the following, Algorithm 1 will be abbreviated as *ALG1*, and Algorithm 2 as *ALG2*. We consider a motion model consisting of translational and rotational motion, so that $\boldsymbol{s}_k = (\theta_k, c_k, d_k)^T$, where θ_k is the rotation angle, and c_k and d_k are the horizontal and vertical translations of the kth HR image, respectively, with respect to the reference frame $\boldsymbol{x}$.

In all experiments reported below, the initial values of the algorithms *ALG1* and *ALG2* are chosen as follows: The registration parameters are estimated from the LR observations using the standard Lucas-Kanade method [29] (similar results were obtained with other registration algorithms such as [50]). The HR image estimate is then initialized using the *average image* [38], which is an oversmooth estimate of the HR image obtained using the LR

TABLE 10.1: Mean PSNR values and standard deviations of the estimated HRD images in 20 experiments provided by the SR algorithms at different SNR levels.

SNR	5dB	15dB	25dB	35dB	45dB
BCB	15.96 ± 0.08	17.02 ± 0.03	17.14 ± 0.01	17.16 ± 0.01	17.16 ± 0.01
ZMT	18.71 ± 0.08	20.48 ± 0.24	20.69 ± 0.38	20.55 ± 0.28	20.53 ± 0.01
RSR	19.07 ± 0.08	22.25 ± 0.05	26.40 ± 0.08	31.22 ± 0.06	33.67 ± 0.07
ALG1	20.34 ± 0.05	24.93 ± 0.13	28.66 ± 0.12	32.67 ± 0.11	36.85 ± 0.14
ALG2	17.48 ± 0.05	24.93 ± 0.12	28.48 ± 0.11	32.15 ± 0.09	36.05 ± 0.16

images as

$$x_a = s^{-1} \sum_{k=1}^{L} B_k(s_k)^T y_k, \tag{10.42}$$

where s is a diagonal matrix with the column sums of $B_k(s_k)$ as its elements. Note that this initial estimate is calculated very efficiently, and it generally increases the robustness of the algorithms to the noise. On the other hand, other initializations (such as bicubic interpolation) resulted in similar restorations.

The covariance matrices in *ALG1* are initially set equal to zero. The rest of the algorithm parameters are automatically calculated from the initial HR image estimate using the algorithmic steps provided in Algorithms 1 and 2. As the convergence criterion we used $\|x^n - x^{n-1}\|^2/\|x^{n-1}\|^2 < 10^{-5}$, where x^n and x^{n-1} are the image estimates at the nth and $(n-1)^{st}$ iterations, respectively.

For comparison, the following methods are used: (1) Bicubic interpolation (denoted by *BCB*), (2) the robust SR method in [53] (denoted by *ZMT*), which is based on backprojection with median filtering, and (3) the robust SR method in [17] (denoted by *RSR*), which is based on bilateral TV priors. We also experimented with other SR methods contained in the EPFL SR software [51], but they provided inferior results compared to *ZMT* and *RSR*, and therefore they are not reported here.

We generated 5 synthetic LR images from the HR image shown on the left in Figure (10.1) through warping, blurring, and downsampling by a factor of 2. The warping consists of both translation and rotation, where the translations are chosen as

$$\begin{pmatrix} 0.0 \\ 0.0 \end{pmatrix}, \begin{pmatrix} 0.0 \\ 0.5 \end{pmatrix}, \begin{pmatrix} 0.5 \\ 0.0 \end{pmatrix}, \begin{pmatrix} 1.0 \\ 0.0 \end{pmatrix}, \begin{pmatrix} 0.0 \\ 1.0 \end{pmatrix} \tag{10.43}$$

pixels, and the rotation angles are $(0°, 3°, -3°, 5°, -5°)$, respectively. For the blur we used a 3×3 uniform PSF. The LR images obtained after the warping, blurring and downsampling operations are further degraded by additive white Gaussian noise at SNR levels of 5dB, 15dB, 25dB, 35dB, and 45dB. Example LR images corresponding to the 25dB SNR case are shown in Figure (10.1).

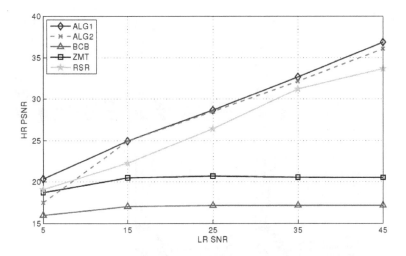

FIGURE 10.2: Mean PSNR values of SR algorithms for different input SNR levels (*ALG1* and *ALG2*: the variational Bayesian methods, *BCB*: bicubic interpolation, *ZMT*: method in [53], *RSR*: method in [17]).

Note that this resolution chart image is chosen for better illustration of the performance in resolution enhancement; similar results were obtained in experiments with other images.

We conducted simulations with 20 different noise realizations at each SNR level, and the average and variance of these experiments are reported. Since the algorithms *ZMT* and *RSR* contain algorithmic parameters, we exhaustively searched for the parameters resulting in the maximum PSNR value, so as to report their best performance. Moreover, we reported the maximum PSNR result obtained during their iterations rather than the PSNR result at convergence, and initialized the algorithms with both the bicubic interpolation result and the average image in (10.42), and chose the best resulting image among them. Note, however, that the parameters of *ALG1* and *ALG2* are estimated automatically, that is, there is no need for parameter tuning.

Mean PSNR values and standard deviations of the estimated HR images provided by the algorithms are shown in Table 10.1, and the mean PSNR values are plotted in Figure (10.2). As expected, all SR algorithms result in better reconstructions than bicubic interpolation. It is also clear that *ALG1* and *ALG2* provide the best performance among all methods across all noise levels. It should be emphasized that the PSNR values of the methods *ZMT* and *RSR* are obtained by exhaustively adjusting their parameters, which requires multiple runs, whereas the proposed methods provided their results in a fully-automated fashion in a single run. Therefore, even for the cases where the PSNR values are similar, algorithms *ALG1* and *ALG2* should be preferred as the methods of choice.

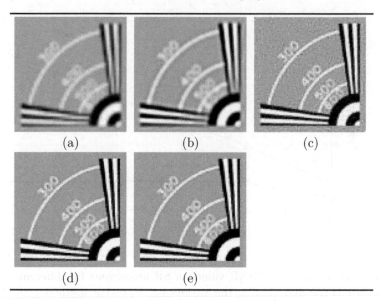

FIGURE 10.3: Example estimated HR images from different SR methods in the case when SNR=25dB. Results of (a) Bicubic interpolation -*BCB*- (PSNR = 17.14dB), (b) *ZMT* (PSNR = 20.55dB), (c) *RSR* (PSNR = 26.41dB), and the variational Bayesian methods (d) *ALG1* (PSNR = 28.75dB), and (e) *ALG2* (PSNR = 28.58dB).

In general, *ALG1* provides restored HR images with slightly higher quality than *ALG2*. This is especially evident in high-noise cases (e.g., SNR = 5dB), where the incorporation of the uncertainty prevents the algorithm from overfitting due to high noise.

Example HR restorations are shown in Figure (10.3) for the SNR = 25dB case, and in Figure (10.4) for the SNR = 45dB case. It is clear that *ALG1* and *ALG2* provide the most visually enhanced restorations with significantly reduced ringing artifacts and much sharper edges compared to the other methods. Restorations provided by *ALG1* and *ALG2* are very similar, with *ALG1* providing slightly sharper edges with less ringing artifacts.

Next we show example super-resolution results with a real-image dataset. 15 LR images were taken from the Adyoron dataset from [1]. The blur PSF is assumed to be a 5x5 Gaussian with variance 1. Algorithms *ZMT* and *RSR* are used again for comparing the performance of the algorithms, and we used the MDSP software [16] to obtain their results. We also provide results with the algorithm in [15], denoted by *EF*. The motion parameters are estimated from the LR images using the MDSP software.

The reconstructed HR images with a resolution enhancement by a factor equal to three obtained by bicubic interpolation and SR algorithms are shown in Figure (10.5). It is clear that methods *ALG1* and *ALG2* provide HR image

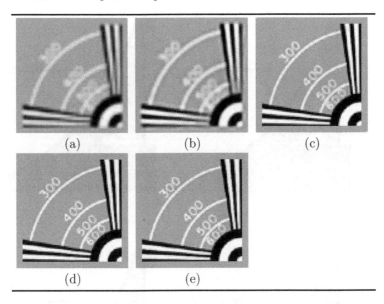

FIGURE 10.4: Example estimated HR images from different SR methods in the case when SNR=45dB. Results of (a) Bicubic interpolation -*BCB*- (PSNR = 17.16dB), (b) *ZMT* (PSNR = 20.53dB), (c) *RSR* (PSNR = 33.56dB), and the variational Bayesian methods (d) *ALG1* (PSNR = 36.81dB), and (e) *ALG2* (PSNR = 35.85dB).

estimates with sharper edges and fewer ringing artifacts than other methods. This is especially clear around the edges and around the letters in both images. Another observation is that *ALG1* and *ALG2* are very effective in preserving sharp image features while suppressing noise and motion artifacts. *ALG1* and *ALG2* provide very similar results, but *ALG1* results in slightly sharper images and the ringing artifacts around the edges are more suppressed than in the results of *ALG2*.

In summary, experimental results with both synthetic and real-image sets demonstrate that algorithms *ALG1* and *ALG2* are very effective in providing high quality super-resolution results, and they compare favorably to some of the state-of-the-art super-resolution methods.

10.7   Estimation of Motion and Blur

In Section 10.5, we developed two super-resolution algorithms that jointly estimate the HR image and the hyperparameters, where we assumed that the motion and blur information is known. However, both motion and blur can

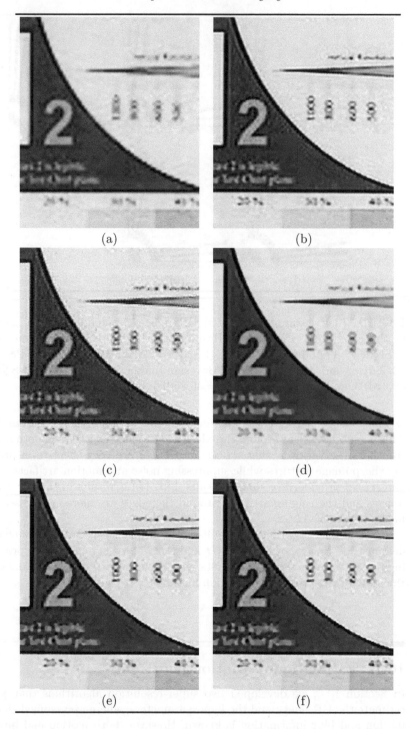

FIGURE 10.5: Super-resolution results (3x resolution increase) by (a) bicubic interpolation -*BCB*-, (b) *EF* [15], (c) *ZMT* [53], (d) *RSR* [17], and the variational Bayesian methods (e) *ALG1* and (f) *ALG2*.

also be modeled and estimated using the framework presented in Section 10.5. This incorporation is done by utilizing prior distributions on the unknown blur and motion parameters, as mentioned in Sections 10.3.3 and 10.3.4. In this section, we will provide an overview of how the motion parameters can be estimated simultaneously with the unknown image and the hyperparameters. The blur can be estimated using similar principles.

Let us again consider a motion model consisting of translational and rotational motion, so that $s_k = (\theta_k, c_k, d_k)^T$, where θ_k is the rotation angle, and c_k and d_k are the horizontal and vertical translations of the kth HR image, respectively, with respect to the reference frame x. As explained in Section 10.3.4, we first calculate the motion parameters from the LR observations. These parameters, denoted by $\bar{s}_k^p$, are considered inaccurate and assumed to follow Gaussian distributions given in (10.10).

Using the same prior distributions as before, the joint distribution can be formed as

$$\mathrm{p}(x, s_k, \Omega, y) = \mathrm{p}(y|x, \{s_k\}, \{\beta_k\}) \, \mathrm{p}(x|\alpha_{\mathrm{im}}) \, \mathrm{p}(s_k) \, \mathrm{p}(\alpha_{\mathrm{im}}) \prod_{k=1}^{L} \mathrm{p}(\beta_k). \quad (10.44)$$

We can then proceed in the same fashion as presented in Section 10.5. The only major exception is that since now the registration parameters s_k are also stochastic variables, the expectations should also be taken with respect to them. This might constitute some problems, since generally the terms containing $C(s_k)x$ are nonlinear with respect to s_k, and therefore the expectations with respect to s_k cannot be taken in a straightforward manner. However, this problem can be overcome by approximating $C(s_k)x$ with bilinear interpolation schemes, as was proposed in [22]. In this chapter, we will only provide an overview of the estimation procedure; detailed derivation and results can be found in [4].

Let us now provide the forms of the distribution approximations resulting from this formulation. The image distribution is found as

$$\mathrm{q}(x) \propto \exp\left(-\frac{1}{2}<\alpha_{\mathrm{im}}> \sum_i \frac{(\Delta_i^h(x))^2 + (\Delta_i^v(x))^2}{\sqrt{w_i}} \right.$$
$$\left. -\frac{1}{2} \sum_k <\beta_k> \mathrm{E}_{s_k}\left[\| \, y_k - AH_kC(s_k)x \, \|^2\right] \right). \quad (10.45)$$

The explicit form of this distribution depends on the expectation $\mathrm{E}_{s_k}\left[\| \, y_k - AH_kC(s_k)x \, \|^2\right]$. By utilizing a bilinear interpolation scheme and Taylor series expansions, it can be shown that [4] the image distribution is again a multivariate Gaussian distribution $\mathrm{q}(x) = \mathcal{N}(x|\mu_x, \Sigma_x)$ with parameters

$$\mu_x = \Sigma_x \left[\sum_k <\beta_k> B_k(<s_k>)^T y_k \right], \tag{10.46}$$

$$\Sigma_x^{-1} = \sum_k <\beta_k> B_k(<s_k>)^T B_k(<s_k>) + \sum_k <\beta_k> f(\lambda_k, <s_k>, x)$$
$$+ <\alpha_{\text{im}}> (\Delta^h)^T W \Delta^h + <\alpha_{\text{im}}> (\Delta^v)^T W \Delta^v, \tag{10.47}$$

where the function $f(\lambda_k, <s_k>, x)$ introduces the uncertainty of the registration parameters s_k, and λ_k are posterior precisions of the distributions of s_k. Note also that when estimating this distribution, the mean values $<s_k>$ of the registration parameters are used.

Next, we estimate the registration parameters as follows. The posterior distribution approximations of the parameters s_k are found as Gaussian distributions, that is,

$$q(s_k) = \mathcal{N}(s_k | <s_k>, \Lambda_k), \tag{10.48}$$

where when calculating the parameters $<s_k>$ and Λ_k, the uncertainty of the image estimate is also taken into account through the covariance Σ_x. We do not provide explicit forms of these parameters here for brevity. After the estimation of the HR image and the motion parameters, the hyperparameters λ_k can also be estimated using the same formulation. Details can be found in [4].

In summary, in addition to the HR image, the developed framework is rich enough to incorporate the modeling and estimation of the motion and blur parameters as well. Moreover, all required hyperparameters are also estimated using this framework. In addition to freeing the user from cumbersome parameter-tuning processes, the Bayesian framework allows for the estimation of these parameters optimally in a stochastic sense, i.e., without resorting to ad-hoc methods.

10.8   Conclusions

In this chapter, we presented the Bayesian framework for super-resolution reconstruction. We have shown that the Bayesian framework provides a powerful means to systematically model the low-resolution image acquisition process, the unknown high-resolution image, and the required algorithmic parameters. Moreover, if the motion and blur information is also unknown, these can also be modeled and estimated using the Bayesian formulation. A number of inference procedures are presented, and the advantages of utilizing Bayesian inference are discussed. We then presented a novel framework for utilizing total variation image priors within a Bayesian formulation, and developed two super-resolution algorithms with variational Bayesian analysis. Both algorithms estimate all unknowns and algorithmic parameters solely from the observed low-resolution images without prior knowledge or user intervention. The presented methods have a number of advantages: First, this framework allows for the estimation of distributions of unknowns, which prevent the propagation of estimation errors within the estimation procedure. This is especially useful when the acquisition noise is heavy. Second, all required parameters of the algorithms are calculated automatically so that they do not require user supervision unlike most existing super-resolution methods. Experimental results with both synthetic and real images are provided to demonstrate that despite the lack of manual parameter tuning, the methods provide super-resolution results superior to existing algorithms.

10.9 Acknowledgments

This work was supported in part by the Comisión Nacional de Ciencia y Tecnología under contract TIC2007-65533 and the Spanish research programme Consolider Ingenio 2010: MIPRCV (CSD2007-00018).

Bibliography

[1] MDSP super-resolution and demosaicing datasets. http://www.soe.ucsc.edu/ milanfar/software/sr-datasets.html, 2007.

[2] S. D. Babacan, R. Molina, and A. K. Katsaggelos. Generalized Gaussian Markov field image restoration using variational distribution approximation. In *IEEE International Conf. on Acoustics, Speech, and Signal Processing (ICASSP'08)*, Las Vegas, Nevada, February 2008.

[3] S. D. Babacan, R. Molina, and A. K. Katsaggelos. Total variation super resolution using a variational approach. In *IEEE International Conf. on Image Processing 2008*, October 2008.

[4] S. D. Babacan, R. Molina, and A.K. Katsaggelos. Variational Bayesian super resolution. Submitted to *IEEE Trans. on Image Processing*, 2009.

[5] S.D. Babacan, R. Molina, and A.K. Katsaggelos. Variational Bayesian blind deconvolution using a total variation prior. *IEEE Transactions on Image Processing*, 18:12–26, January 2009.

[6] S.D. Babacan, J. Wang, R. Molina, and A.K. Katsaggelos. Bayesian blind deconvolution from differently exposed image pairs. In *IEEE International Conference on Image Processing*, Cairo, Egypt, July 2009.

[7] M.J. Beal. *Variational Algorithms for Approximate Bayesian Inference.* PhD thesis, The Gatsby Computational Neuroscience Unit, University College London, 2003.

[8] J. O. Berger. *Statistical Decision Theory and Bayesian Analysis*, chapter 3 and 4. New York, Springer Verlag, 1985.

[9] C.M. Bishop. *Pattern Recognition and Machine Learning.* Springer, 2006.

[10] T. E. Bishop, S. D. Babacan, B. Amizic, A. K. Katsaggelos, T. Chan, and R. Molina. Blind image deconvolution: problem formulation and existing approaches. In P. Campisi and K. Egiazarian, editors, *Blind image deconvolution: theory and applications*, chapter 1. Boca Raton, FL: CRC Press, 2007.

[11] T. F. Chan, M. K. Ng, A. C. Yau, and A. M. Yip. Superresolution image reconstruction using fast inpainting algorithms. *Applied and Computational Harmonic Analysis*, 23(1):3–24, 2007. Special Issue on Mathematical Imaging.

[12] T. F. Chan, N. Ng, A. Yau, and A. Yip. Superresolution image reconstruction using fast inpainting algorithms. *Applied and Computational Harmonic Analysis*, 23(1):3–24, July 2007.

[13] T. F. Chan and C.K. Wong. Total variation blind deconvolution. *IEEE Transactions on Image Processing*, 7(3):370–375, Mar. 1998.

[14] A. D. Dempster, N. M. Laird, and D. B. Rubin. Maximum likelihood from incomplete data via the E-M algorithm. *Journal of the Royal Statistical Society, Series B*, 39:1–37, 1977.

[15] M. Elad and A. Feuer. Restoration of a single superresolution image from several blurred, noisy, and undersampled measured images. *IEEE Transactions on Image Processing*, 6(12):1646–1658, Dec 1997.

[16] S. Farsiu. *MDSP resolution enhancement software.* University of California at Santa Cruz, 2004.

[17] S. Farsiu, M. D. Robinson, M. Elad, and P. Milanfar. Fast and robust multiframe super resolution. *IEEE Transactions on Image Processing*, 13(10):1327–1344, Oct. 2004.

[18] R. Fergus, B. Singh, A. Hertzmann, S. T. Roweis, and W.T. Freeman. Removing camera shake from a single photograph. *ACM Transactions on Graphics, SIGGRAPH 2006 Conference Proceedings, Boston, MA*, 25:787–794, 2006.

[19] M.A.T. Figueiredo, J.M. Bioucas-Dias, and R.D. Nowak. Majorization-minimization algorithms for wavelet-based image restoration. *IEEE Transactions on Image Processing*, 16(12):2980–2991, Dec. 2007.

[20] R.C. Hardie, K.J. Barnard, and E.E. Armstrong. Joint MAP registration and high-resolution image estimation using a sequence of undersampled images. *IEEE Transactions on Image Processing*, 6(12):1621–1633, 1997.

[21] H. He and L. P. Kondi. An image super-resolution algorithm for different error levels per frame. *IEEE Transactions on Image Processing*, 15(3):592–603, March 2006.

[22] Y. He, K. H. Yap, L. Chen, and L. P. Chau. A nonlinear least square technique for simultaneous image registration and super-resolution. *IEEE Transactions on Image Processing*, (11):2830–2841, 2007.

[23] M. Irani and S. Peleg. Motion analysis for image enhancement: Resolution, occlusion, and transparency. *Journal of Visual Communication and Image Representation*, 4(4):324–335, 1993.

[24] A. Kanemura, S.-I. Maeda, and S. Ishii. Superresolution with compound Markov random fields via the variational EM algorithm. *Neural Networks*, 22(7):1025–1034, 2009.

[25] M.G. Kang and S. Chaudhuri (Eds.). Super-resolution image reconstruction. *IEEE Signal Processing Magazine*, 20(3), 2003.

[26] A. K. Katsaggelos, R. Molina, and J. Mateos. *Super Resolution of Images and Video*. Morgan and Claypool, 2007.

[27] H. Kim, J.-H. Jang, and K.-S. Hong. Edge-enhancing super-resolution using anisotropic diffusion. In *Proceedings of the IEEE Conference on Image Processing*, volume 3, pages 130–133, 2001.

[28] E. S. Lee and M. G. Kang. Regularized adaptive high-resolution image reconstruction considering inaccurate subpixel registration. *IEEE Transactions on Image Processing*, 12(7):826–837, July 2003.

[29] B.D. Lucas and T. Kanade. An iterative image registration technique with an application to stereo vision. In *Proceedings of Imaging Understanding Workshop*, pages 121–130, 1981.

[30] J. Miskin. *Ensemble Learning for Independent Component Analysis*. PhD thesis, Astrophysics Group, University of Cambridge, 2000.

[31] R. Molina, A. K. Katsaggelos, J. Abad, and J. Mateos. A Bayesian approach to blind deconvolution based on Dirichlet distributions. In *1997 International Conference on Acoustics, Speech and Signal Processing (ICASSP'97)*, volume 4, pages 2809–2812, Munich (Germany), 1997.

[32] R. Molina, A. K. Katsaggelos, and J. Mateos. Bayesian and regularization methods for hyperparameter estimation in image restoration. *IEEE Transactions on Image Processing*, 8(2):231–246, 1999.

[33] R. Molina, J. Mateos, and A.K. Katsaggelos. Blind deconvolution using a variational approach to parameter, image, and blur estimation. *IEEE Transactions on Image Processing*, November 2006.

[34] M. Ng, T. Chan, M.G. Kang, and P. Milanfar. Super-resolution imaging: Analysis, algorithms, and applications. *EURASIP Journal on Applied Signal Processing*, 2006:Article ID 90531, 2 pages, 2006.

[35] M. Ng, J. Koo, and N. Bose. Constrained total least-squares computations for high-resolution image reconstruction with multisensors. *International Journal of Imaging Systems and Technology*, 12(1):35–42, 2002.

[36] M. K. Ng, H. Shen, E. Y. Lam, and L. Zhang. A total variation regularization based super-resolution reconstruction algorithm for digital video. *EURASIP Journal on Advances in Signal Processing*, (74585), 2007.

[37] J. Palmer, D. Wipf, K. Kreutz-Delgado, and B. Rao. Variational EM algorithms for non-Gaussian latent variable models. In Y. Weiss, B. Schölkopf, and J. Platt, editors, *Advances in Neural Information Processing Systems 18*, pages 1059–1066. MIT Press, Cambridge, MA, 2006.

[38] L. C. Pickup, D. P. Capel, S. J. Roberts, and A. Zisserman. Bayesian methods for image super-resolution. *The Computer Journal*, 52:101–113, 2009.

[39] M. Protter, M. Elad, H. Takeda, and P. Milanfar. Generalizing the nonlocal-means to super-resolution reconstruction. *IEEE Transactions on Image Processing*, 18(1):36–51, Jan. 2009.

[40] D. Rajan and S. Chaudhuri. Generation of super-resolution images from blurred observations using an MRF model. *Journal of Mathematical Imaging and Vision*, 16:5–15, 2002.

[41] L. I. Rudin, S. Osher, and E. Fatemi. Nonlinear total variation based noise removal algorithms. *Physica D*, pages 259–268, 1992.

[42] C. A. Segall, R. Molina, A. K. Katsaggelos, and J. Mateos. Bayesian resolution enhancement of compressed video. *IEEE Transactions on Image Processing*, 13(7):898–911, 2004.

[43] C.A. Segall, R. Molina, and A.K. Katsaggelos. High-resolution images from low-resolution compressed video. *IEEE Signal Processing Magazine*, 20:37–48, 2003.

[44] F. Šroubek and J. Flusser. Multichannel blind deconvolution of spatially misaligned images. *IEEE Transactions on Image Processing*, 7:45–53, July 2005.

[45] H. Stark and P. Oskoui. High resolution image recovery from image-plane arrays, using convex projections. *Journal of the Optical Society of America A*, 6:1715–1726, 1989.

[46] H. Takeda, P. Milanfar, M. Protter, and M. Elad. Super-resolution without explicit subpixel motion estimation. *IEEE Transactions on Image Processing*, 18(9):1958–1975, Sept. 2009.

[47] A.M. Tekalp, M.K. Ozkan, and M.I. Sezan. High-resolution image reconstruction from lower-resolution image sequences and space varying image restoration. In *Proceedings of the IEEE International Conference on Acoustics, Speech and Signal Processing*, volume 3, pages 169–172, 1992.

[48] M. E. Tipping and C. M. Bishop. Bayesian image super-resolution. In *Advances in Neural Information Processing Systems 15 (NIPS)*. MIT Press, 2003.

[49] P. Vandewalle, L. Sbaiz, J. Vandewalle, and M. Vetterli. Super-resolution from unregistered and totally aliased signals using subspace methods. *IEEE Transactions on Signal Processing*, 55(7, Part 2):3687–3703, 2007.

[50] P. Vandewalle, S. Süsstrunk, and M. Vetterli. A frequency domain approach to registration of aliased images with application to super-resolution. *EURASIP Journal on Applied Signal Processing (special issue on Super-resolution)*, 2006:Article ID 71459, 14 pages, 2006.

[51] P. Vandewalle, P. Zbinden, and C. Perez. Superresolution v2.0. http://lcavwww.epfl.ch/software/superresolution/index.html, 2006.

[52] N.A. Woods, N.P. Galatsanos, and A.K. Katsaggelos. Stochastic methods for joint registration, restoration, and interpolation of multiple undersampled images. *IEEE Transactions on Image Processing*, 15:201–213, 2006.

[53] A. Zomet, A. Rav-Acha, and S. Peleg. Robust super-resolution. In *IEEE Computer Society Conference on Computer Vision and Pattern Recognition (CVPR 2001)*, pages 645–650, 2001.

11

Pattern Recognition Techniques for Image
Super-Resolution

Karl Ni

MIT Lincoln Laboratories

Truong Q. Nguyen

The University of California at San Diego

CONTENTS

Pattern recognition and machine learning define the act of taking data collected a priori, observing relationships inside the data, and generalizing the learned relationships. The output of such algorithms can either be discrete or continuous, categorized as *classification* or *regression* problems. Because

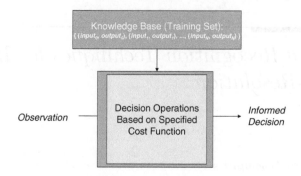

FIGURE 11.1: Statistical and machine-learning general framework.

super-resolution is an ill-posed problem, different combinations approaching from both types of paradigms exist with the same purpose: enhancing resolution. This chapter explores techniques and methodologies involved in formulating the problem through frameworks involving classification techniques, regression techniques, and combinations thereof.

11.1 Introduction

Human beings have survived for millions of years in large part by our ability to recognize patterns and draw inferences from them. In an attempt to comprehend what is involved in generalizing to real-world phenomena, computer scientists study statistical and machine learning principles. The underlying methodology behind statistically modeling data is to extract functional relationships from a controlled database and apply them to unseen test points. Based on what has been previously observed, an informed decision can be made. In pattern recognition terms, Figure 11.1 shows the block diagram that is universal to machine-learning algorithms.

The attribute that distinguishes pattern recognition techniques from other types of algorithms is a training set. Hence, the topics in this chapter refer to algorithms that use collected training data as a tool for image processing. For this reason, learning-based super-resolution algorithms with the framework shown in Figure 11.1 are often called *example-based* super-resolution algorithms. In this respect, relationships between low and high-resolution images in the algorithms discussed in this chapter are derived, for the most part, *empirically*.

Whereas computer vision fully embraces pattern recognition and statistical and machine-learning, the equivalent effort in image processing in the past has largely been left unexplored, and only recently have conferences been dedicated to the topic (see *EURASIP's Special Issue on Machine Learning in Image Processing*). Despite the limited scientific exposure, it seems only natural to utilize learning because humans use a similar process in assessing the quality of images. When the underlying solutions mimic how human learning would process the image, the results should reflect a more favorable human visual system (HVS) evaluation, i.e., effect a more visually pleasing image.

Because pixels can take on a wide range of values, the most direct approach to solving for high-resolution information is through regression. That is, we can replace the central block in Figure 11.1 with a function estimation block. Pixel values can be found directly from a group of features that relate to the high-resolution labels based on spatial proximity. The most common form of regression in image processing use linear filters, where high-resolution pixel values are derived from a linear combination of observed features. However, super-resolution is arguably a more complicated process than can be expressed by a single linear scheme. Therefore, a number of ideas brought forth in this chapter describe nonlinear regression algorithms whose parameters have been primarily determined through relationships learned from a training set.

On the other hand, not all the information in the training set may be relevant to the observation all the time, and it makes sense to pare down the information to only what is relevant for a given input. The choice of points within a training set define another problem: classification. Because learning algorithms were initially applied to problems that ask questions of a discriminating nature, nearest neighbor, classification, and clustering techniques are well-studied and easily applicable. Such techniques offer ways to partition the feature domain so that different types of image content can be treated differently. With respect to images, segmenting the input space is intimately connected to determining the "type" of image patch that the input feature represents. An action can then be taken based on the class that the input feature falls into. Moreover, the need for regression is not excluded (although classification may, in fact, sufficiently cover the nonlinear aspect of the problem).

Figure 11.2 depicts a common framework and puts Figure 11.1 into more mathematical terms. To relate to super-resolution, the feature space or input, $\mathbf{x}_{obs}$, represents low-resolution image content while the output denoted by $\mathbf{y}_{out}$ represents high-resolution image content. The learned relationship could result in pixel values, high-frequency frequency components, or any other useful information necessary to create high-resolution images, which are all derived from a training set Ω. The varied algorithms in this chapter are interesting because they decide what form Ω, may take, how it is obtained, and most importantly, where and the manner in which it is applied.

The mathematical terminology in pattern recognition allows freedom in assigning vectors $\mathbf{x}_{obs}$ and $\mathbf{y}_{out}$ to nearly any feature-space and co-domain. Specific to our case, the input observation vector $\mathbf{x}_{obs}$ in Figure 11.2 refers to any collection of low-resolution information taken from the test image. The

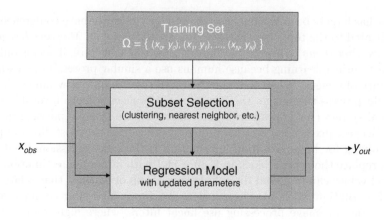

FIGURE 11.2: Common learning-based super-resolution framework

algorithms described in this chapter, with the exception of Markov-related optimizations, operate locally. That is to say, we super-resolve individual subsets $\{\mathbf{x}_i\}$ of pixels in the low-resolution image to determine pixel values in $\mathbf{y}_i$, which are, in turn, subsets of the high-resolution image. The simplest example definitions $\mathbf{x}$ and $\mathbf{y}$ are vectorized collections of contiguous pixels, or *image patches*.

11.2   Nearest Neighbor Super-Resolution

A few years ago, Freeman, Jones, and Pasztor [25] introduced a learning-based algorithm for the explicit purpose of super-resolving images. While applying learning to super-resolution was hardly a novel concept [11, 16], the article titled "Example-based Super-Resolution" became seminal to the community in that its discussion initiated a dialogue arguing that image content may be too rich to solve analytically.

The algorithm that Freeman et al. themselves propose utilizes the simplest learning technique: the nearest neighbor algorithm. That is,

Definition 1 *Given a training set* $\Omega = \{(\boldsymbol{x}_1, \boldsymbol{y}_1), (\boldsymbol{x}_2, \boldsymbol{y}_2), \cdots, (\boldsymbol{x}_N, \boldsymbol{y}_N)\}$, *the nearest neighbor pair to an observed test point* $\boldsymbol{x}_{obs}$ *is the pair that satisfies*

$$(\boldsymbol{x}_{nn}, \boldsymbol{y}_{nn}) = \operatorname*{argmin}_{(\boldsymbol{x}_i, \boldsymbol{y}_i)} ||\boldsymbol{x}_i - \boldsymbol{x}_{obs}||, \quad \forall \boldsymbol{x}_i \in \Omega \qquad (11.1)$$

The vectors $\mathbf{x}$ and $\mathbf{y}$ denote vectors corresponding to 5×5 low-resolution and 7×7 high-resolution image patches, respectively.

(a) Original

(b) 2× Interpolation   (c) High-Frequencies

FIGURE 11.3: Super-resolution results using a single neighbor.

Algorithms described in this chapter require some kind of prepro-cessing to be applied to **x** (and likewise, some postprocessing of **y**). One suggestion is to filter out the low-frequencies because it is plausible that the highest frequency components are the most important in predict-ing higher resolution information. Other preprocessing methods include subtracting out the center pixel, subtracting out DC values, and normal-izing the variance of the low-resolution image patch. There are a variety of ways to extract and represent the feature space, depending on the technique used.

Unfortunately, a single neighbor, even among several candidates, is insuf-ficiently descriptive of the relationship between low and high resolution, often leading to poor results. This is because a single choice causes increased sus-ceptibility to noise. Using a single neighbor is problematic also due, in part, to the ill-posed nature of the problem; often, a single $\mathbf{x}_{obs}$ could correspond to multiple $\mathbf{x}_i \in \Omega$.

Figure 11.3 shows the results using a single neighbor with only local and individual patch information with no spatial neighbor patches. On the bottom, the high-frequencies of local reconstruction efforts look "like oatmeal" as aptly put by Freeman et al. [25] Hence, the use of a Markov Random Field (to be discussed later) serves as a way to somewhat globalize the effort, improving the situation greatly.

11.2.1 k-Nearest Neighbor

In an effort to stem the effects of estimation noise that so obviously ails single neighbor super-resolution, we can use a decently sized *subset* of the training set Ω. That is, instead of a single training point, we draw information from k training points that are closely related to the input vector, $\mathbf{x}_{obs}$, in Figure 11.2. For image interpolation, once the relevant training samples are found, filters are specially designed to determine high-resolution values after identifying low-resolution content.

The k-nearest neighbor (k-NN) [21] rule is among the simplest statistical learning tools in density estimation, classification, and regression. Trivial to train and easy to code, the nonparametric algorithm is surprisingly competitive and fairly robust to errors given good cross-validation procedures.

Definition 2 *Given a training set* $\Omega = \{(\boldsymbol{x}_1, \boldsymbol{y}_1), (\boldsymbol{x}_2, \boldsymbol{y}_2), \cdots, (\boldsymbol{x}_N, \boldsymbol{y}_N)\}$*, the* k *nearest neighbors to an observation test point* $\boldsymbol{x}_{obs}$ *can be mathematically expressed as the set of points that satisfy:*

$$\boldsymbol{x}_j = \underset{\boldsymbol{x}_i \in \Omega_j}{\operatorname{argmin}} K_{\mathbb{F}}(\boldsymbol{x}_{obs}, \boldsymbol{x}_i) \tag{11.2}$$

where $1 \leq j \leq k$*,* $\boldsymbol{x}_j$ *is the* j^{th} *vector of* k *nearest neighbors, and* $\Omega_j = \{\Omega_{j-1} \backslash \boldsymbol{x}_{j-1}\} \subset \Omega$*.*

The Radial Basis Function

Regression, classification, or density estimation algorithms sometimes turn to *functionals* to represent a nonlinear relationship. In order to maintain generality, representations like that of (2) often do not specify a particular function or vector space for use. In practice, the most popular function in image processing used almost to exclusivity is the *radial basis function* (RBF).

The kernel function K used in many of the discussed algorithms is the RBF, stated in (11.3)

$$K_{\mathbb{F}}(\mathbf{x}_i, \mathbf{x}_j) = \frac{1}{2\pi \|\Sigma\|} \exp\left\{-d_{\mathbb{F}}(\mathbf{x}_i, \mathbf{x}_j)\right\} \leq 1 \tag{11.3}$$

where $d_{\mathbb{F}}(\mathbf{x}_i, \mathbf{x}_j)$ is the Mahalanobis distance or weighted Euclidean distance specified by $\frac{1}{2}(\mathbf{x}_i - \mathbf{x}_j)^T \Sigma^{-1}(\mathbf{x}_i - \mathbf{x}_j)$. Unfortunately, in the absence of prior knowledge, most k-NN algorithms determine proximity through unweighted Euclidean distances. With image data, a scaled version of the covariance matrix from the training set may be determined, though in most cases, the situation does not improve much with known Σ.

In Definition 2, K represents a kernel that integrates to one and is used as an appropriate similarity metric. Conceptually, the k-NN in (2) are simply the k closest training points to $\mathbf{x}_{obs}$ with respect to the kernel. Incidentally, K can also be used to estimate the density [21], where kernels are placed around every training point to approximate the probability density function (PDF).

11.2.2 k-Nearest Neighbor Regression

Statistical regression is defined as the estimation of a function that fits data to a model so that a continuous domain may be mapped onto a continuous range. Regression models estimate the relationship between observed values of $\mathbf{y}$ from select features of $\mathbf{x}$ by which the most probable value of $\mathbf{y}$ can be predicted for all values of $\mathbf{x}$. There are various techniques to obtain a regression model, and regression by k-NN is examined in this section.

Definition 3 *Let Ω be a training set of N input-output pairs. Then,*

$$\Omega = \{(\mathbf{x}_1, \mathbf{y}_1), (\mathbf{x}_2, \mathbf{y}_2), \cdots, (\mathbf{x}_N, \mathbf{y}_N)\} \tag{11.4}$$

The k-NN regression estimate $g(\mathbf{x}_{obs})$ of $\mathbf{y}$ at test point $\mathbf{x}_{obs}$ is given by [19, 18]:

$$\hat{\mathbf{y}} = g(\mathbf{x}_{obs}) = \frac{1}{k} \sum_{i=1}^{N} W_i(\mathbf{x}_{obs}, \Omega) \mathbf{y}_i \tag{11.5}$$

where $W_i \in \{0,1\}$ depending on whether or not $\mathbf{x}_i$ is among the k nearest neighbors of $\mathbf{x}_{obs}$.

For the problem specifically relating to image super-resolution, $\mathbf{x}_i$ is comprised of the ith low-resolution image patch in the training set. Likewise, $\mathbf{y}_i$ defines the ith high-resolution image patch. During runtime, where we denote runtime values with subscript *obs*, the adaptation of k-NN determines the high-resolution image patch $\mathbf{y}_{out}$ from a single low-resolution image patch $\mathbf{x}_{obs}$. For mathematical reasons, it is easier to represent image patches $\mathbf{x}_{obs}$ and $\mathbf{y}_{out}$ as vectors instead of square patches. Therefore, in subsequent derivations $\mathbf{x}$ and $\mathbf{y}$ are both vectors that have been rearranged from image blocks into a single column.

Naturally, the definition of (11.5) could be extended by not necessarily limiting W_i to 0 or 1, but rather the constraint $\sum_{i=1}^{N} W_i = k$. In fact, there are several common weighting schemes, ranging from posterior probability like expressions [21] to iteratively determined convex solutions [2], all functions of distances or weights that can used to minimize some criterion as in [12].

Large Sample Risk

For (x, y) jointly normal, under the squared-error loss case, the unconditional, large sample risk R as $N \to \infty$ of the k-NN estimate satisfies

$$R_N^{(k)} = \left(1 + 1/k + \frac{\sigma_1^2}{\sigma_2^2} E\left[x_{obs} - \frac{1}{k}\sum_{i=1}^{k} x_{i,N}\right]^2\right) R^* \qquad (11.6)$$

Here, R^* is the Bayes risk (minimum expected loss), $x_i \in \Omega$, and parameters σ_1 and σ_2 are standard deviation parameters in probability distribution functions (PDFs) $f(y)$ and $f(y|x)$. To minimize risk, (11.6) suggests a tradeoff between the $1/k$ term that limits erroneous reconstruction and the final term in (11.6), which simultaneously favors a larger k. This type of tradeoff is common in k-NN problems, and provides much need for cross-validation as will be seen even for adaptable k values [14].

One k-nearest neighbor algorithm [12], called Local Linear Embedding (LLE), involves manifold embedding. The training pairs in Ω refer to two spaces $\mathcal{X}$ and $\mathcal{Y}$ on which points $\mathbf{x}_i$ and $\mathbf{y}_i$ lie for all i. LLE depends on assumptions that data points sampled near each other from space $\mathcal{X}$ lie on or close to a manifold that is locally linear (or one that can be approximated as such), the geometry of which can be characterized through linear reconstruction weights. The manifold, which has high dimensionality, pulled from the sampled points are assumed to have an *intrinsic* dimensionality of much lower order, so that we might apply the same weights to manifolds on $\mathcal{Y}$.

Suppose $\mathbf{x}_{obs}$ is the observation vector to be super-resolved to $\mathbf{y}_{out}$, and let $\mathcal{N}_{x_{obs}}$ be the neighborhood in $\mathcal{X}$ that surrounds $\mathbf{x}_{obs}$. Then the optimal least squares weights w can be obtained by minimizing the L_2 reconstruction error for $\mathbf{x}_{obs}$:

$$\theta = \operatorname*{argmin}_{\boldsymbol{\theta}} ||\mathbf{x}_{obs} - \sum_{\mathbf{x}_i \in \mathcal{N}_{x_{obs}}} \theta_{\mathbf{x},i} \mathbf{x}_i|| \qquad (11.7)$$

Let X denote a matrix whose columns are the k nearest neighbor to $\mathbf{x}_{obs}$. The local Gram matrix $T_{\mathbf{x}}$ can be defined as

$$T_{\mathbf{x}} = (\mathbf{x}_{obs} \mathbf{1}^T - X)^T (\mathbf{x}_{obs} \mathbf{1}^T - X). \qquad (11.8)$$

Then, the solution is

$$\theta = \frac{T_{\mathbf{x}}^{-1} \mathbf{1}}{\mathbf{1}^T T_{\mathbf{x}}^{-1} \mathbf{1}}, \qquad (11.9)$$

and we can write:

$$\mathbf{y} = \sum_{\mathbf{x}_i \in \mathcal{N}_{x_{obs}}} \boldsymbol{\theta}^T \mathbf{y}_i \qquad (11.10)$$

Unfortunately, one of the paramount assumptions in neighbor embedding algorithms is isometry, which recent studies [54] have shown does not hold under the L_2-norm. One could, of course, use a different metric, but a more direct approach would be to embed the neighbors in a space that is not necessarily linear. Weinberger and Saul have promoted a large body of work devoted to kernel manifold learning (e.g., [60, 62, 61]).

Alternatively, another family of solutions of (11.5) known as *locally weighted regression* (LWR) offers more flexibility in application. This is done by replacing W by a particular model class $g(\mathbf{x}_{obs}, \boldsymbol{\vartheta})$, in which $\mathbf{y}_{out}$ is determined locally by a function g with parameter vector $\boldsymbol{\vartheta}$ based on how similar point $\mathbf{x}_i$ is to $\mathbf{x}_{obs}$ [53]. (Depending on the choice of function, $\boldsymbol{\vartheta}$ may differ in size and kind.) Then, the task of k-NN for regression becomes estimating select parameters for reconstruction in (11.11).

$$\boldsymbol{\vartheta}^* = \underset{\vartheta}{\arg\min} \sum_{\mathbf{x}_i \in \mathcal{N}(\mathbf{x}_{obs})} d_{\mathbb{R}}\left(g(\mathbf{x}_i, \boldsymbol{\vartheta}), \mathbf{y}_i\right) K_{\mathbb{F}}\left(\mathbf{x}_i, \mathbf{x}_{obs}\right) \qquad (11.11)$$

where $d_{\mathbb{R}}$ is a distance metric in the range (high resolution) and feature space (low resolution), respectively, and $\mathcal{N}(\mathbf{x}_{obs})$ is the neighborhood of $\mathbf{x}_{obs}$. It is this general regression form that garners the most attention in what follows in Section 11.2.3.

11.2.3 Adaptive k-NN for Super-Resolution

The crux is to achieve specificity with regard to image content without any loss of generalization of application. That is, how detailed can we make an image look while still maintaining a broad base of applicability?

The answer to this question is intimately related to the number of training samples used per reconstruction filter. In images, more training points per filter, i.e., k is large, implies better generalization, where estimation errors are diminished. Likewise, fewer training points per filter, i.e., k is small, implies better specificity, where image reconstruction is clearer and more detailed. Hence, it is reasonable to conclude that to accommodate a variety of possible test inputs, k must be variable.

In initializing an adaptive k-NN algorithm, modifications can be made to (11.11) for a viable alternative over LLE. With LWR, the only required assumption is that linear filtering yields an excellent approximation for local image construction as opposed to assuming some kind of duality between low-resolution and high-resolution manifolds in [12]. Hence, the $g(\mathbf{x}, \boldsymbol{\vartheta})$ in (11.11) becomes the linear filter in question, which can be reduced to an MMSE filter formulation, and we can approximate $\mathbf{y}_{out}$ by

$$\mathbf{y} = E\left[\mathbf{y}_{out} | \mathbf{x}_{obs}\right] \approx g(\mathbf{x}_{obs}, \boldsymbol{\vartheta}) = \hat{G}\mathbf{x}_{obs} \qquad (11.12)$$

where $\hat{G}$ is a $u \times d$ matrix, u being the upsizing factor, and is constructed by probability parameters $\boldsymbol{\vartheta}$ and neighboring low-resolution and high-resolution pairs.

As a matrix, G is a linear regression, which can be found through least squares solutions. Eventually, preprocessing steps such as mean-shifting or variance normalization should be implemented to determine the feature space $\mathbb{F}$ for both k-NN identification and regression, but for simplicity, let us say that these steps have already been incorporated. Let X and Y define matrices formed by collections of neighbors $\mathbf{x}_i$ and $\mathbf{y}_i$, respectively, setting up auto- and cross-correlation matrices defined by $R_{XX} = XX^T$ and $R_{XY} = XY^T$.

Because training data is seldom uniformly distributed, it is desirable to weight the estimated parameters by training points according to their relevance. This can be done by constructing a matrix P for a given neighborhood of $\mathbf{x}_{obs}$. Let $\mathbf{p}$ be a vector of similarity measures whose i^{th} entry is the value $K_{\mathbb{F}}(\mathbf{x}_i, \mathbf{x}_{obs})$. Then, a proper weighting of $\mathbf{x}_i \in \mathcal{N}(\mathbf{x}_{obs})$ can be written:

$$P = \mathbf{1}^T \mathbf{p}, \tag{11.13}$$

where k is the number of neighbors to use and $\mathbf{1}$ is a k dimensional vector of all ones. Hence, P has dimension $d \times k$. A least squares-like filter formulation roughly equivalent to the derivations from [3] can then be written:

$$G = \left((P \circ X) X^T \right)^{-1} \left((P \circ X) Y^T \right) \tag{11.14}$$

where the "$\circ$" operator denotes the Hadamard product, the element-wise product of two matrices.

The goal, then, is to find the right k for a desired tradeoff. Let k^* be the ideal number of neighbors. Then, $k < k^*$ causes G to overfit; the manifestation is a grainy and discontinuous image. Furthermore, if k were exceedingly small, $k \ll k^*$, G could become singular. This is because training points near $\mathbf{x}_{obs}$ could be very close together causing (11.14) to be underdetermined. Analytically speaking, vectors in X that are too similar can mean that R_{XX} is rank deficient and thus noninvertible. This is a dilemma because while k-NN should find the most relevant data, it is designed such that the collected vectors based on $\mathbf{x}_{obs}$ are similar to each other. Hence, though it is counterintuitive, it is important to choose a large enough neighborhood in $\mathbb{F}$ so that diversity in the $\mathcal{N}(\mathbf{x}_{obs})$ exists.

Definition 4 *Given a training set* $\Omega = \{(\mathbf{x}_1, \mathbf{y}_1), (\mathbf{x}_2, \mathbf{y}_2), \cdots, (\mathbf{x}_N, \mathbf{y}_N)\}$, *the optimal* k^* *is determined by*

$$k^*(\mathbf{x}, \Omega) = \underset{k}{\text{argmin}} \sum_{i=1}^{N} W_i(\mathbf{x}, \Omega, k) K(\mathbf{x}_i, \mathbf{x}) \geq \eta$$
$$\text{where} \qquad W_i(\mathbf{x}, \Omega, k) \in \{0, 1\} \tag{11.15}$$

where η *is a cross-validated value expressing the minimum number of points to be used.*

The expression in (11.15) obtains k^* by finding the minimum number of neighbors whose sum of similarity measures exceeds a threshold η, which is obtained through cross-validation. Moreover, η is a minimum bound of k since $K(\mathbf{x}_i, \mathbf{x}_{obs}) \leq 1$ for all $\mathbf{x}_i$.

Analyzing (11.15) for a given $\mathbf{x}_{obs}$, if there are only a few $\mathbf{x}_i$ with high probability of being related to it, that is $\sum_i K(\mathbf{x}_i, \mathbf{x}_{obs})$ is small, then the proposed algorithm will need to consider more points in hopes of generalizing well. Alternatively, if there are many $\mathbf{x}_i$ that are related to $\mathbf{x}_{obs}$, i.e., $\sum_i K(\mathbf{x}_i, \mathbf{x}_{obs})$ is large, it is unnecessary to use other points where the similarity is low because the specialized filter generated by the points within $\sum_i K(\mathbf{x}_i, \mathbf{x}_{obs}) \leq \eta$ is very likely to be accurate. Conceptually, we can visualize a ring that extends further and further depending on whether or not there are enough points inside the ring.

11.2.4 Heuristics for Insufficient Training in Adaptive k-NN Regression

Problems may arise in k-NN in cases where training is insufficient. For the adaptive k defined in (11.15), the further the ring of values under consideration extends, the smaller the similarity values and the less suited any additional training point is to complete the task of reaching η. In extreme cases, η may not even be reached before the entire training set is exhausted.

Consequently, for suboptimal training sets, it makes sense to restrict k^* by letting ζ be a maximum limit on k. A generic technique, i.e., bicubic interpolation, may then be used for those $\mathbf{x}_{obs}$ that Ω does not represent well. The complexity reduction from prematurely stopping the neighbor search through ζ should be obvious. However, error minimization also occurs because we effectively acknowledge that for $\mathbf{x}_{obs}$, the original intention of the proposed algorithm cannot be carried out due to a less than competent training set. Therefore, for any $\mathbf{x}_{obs}$ that k-NN is ill-equipped to manage (i.e., $k^* > \zeta$), the errors are bounded by whatever interpolation algorithm replaces k-NN.

The question now becomes finding what kind of interpolation algorithm should replace k-NN. Is there a particular type of image patch that the k-NN algorithm consistently disfavors? Moreover, based on this bias, are there certain properties of these patches that allow us to tailor a solution using this knowledge? The answer is yes on both accounts. After running several tests, we came across a peculiar reoccurring theme in generic training and testing images: texture patches never reached ζ and appeared at high quality, but edge patches often did and needed attention.

Using 2×10^5 training points and observing similarity measures in (11.13) (which are based on Euclidean distances), the texture matches usually retain similarity values of $K(\mathbf{x}_{obs}, \mathbf{x}_i) \approx 0.93$ (out of 1.00), whereas edge matches usually satisfy $K(\mathbf{x}_{obs}, \mathbf{x}_i) \leq 0.40$. Furthermore, in viewing a single image, only a small percentage of image patches are actually edges, so accumulating

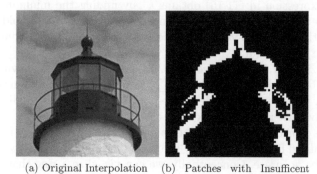

(a) Original Interpolation   (b) Patches with Insufficent Training

FIGURE 11.4: Areas of insufficient training for two hundred thousand points.

relevant image patches in (11.15) to surpass η is even more improbable. The situation is best described in Figure 11.4.

Though texture results in high peak signal to noise ratios (PSNR), unfortunately, the human visual system (HVS) focuses on edges [9, 50]. Fortunately, research into edge-oriented image filtering has been well-studied. In our framework, we can agglomerate a bank of edge-oriented filters that do "well-enough" when the "best" filter through k-NN is unavailable, effectively reducing the implementation to a specialized version of [3] with an added Markov Random Field (MRF) improvement (see the next section, Section 11.3).

With enough data points, however, replacing k-NN conditioned on $k^* > \zeta$ should occur relatively few times. That is, edges may and often are well-represented in the training set, which indicates the algorithm is operating closer to capacity.

11.3   Markov Random Fields and Approximations

It is widely acknowledged [49, 25, 38, 28] that local interpolation could benefit from global image information to predict high-resolution pixel values. Construction of high-resolution image detail from isolated, local low-resolution image patches (i.e., without information from adjacent patches and the image as a whole) using a single neighbor is shown in Figure 11.3, where overall results are described as looking "like oatmeal" from [25]. Without information from adjacent patches, high-frequency components of an image as a whole become patchy and discontinuous. The most common remedy is to globalize the effort by using surrounding window information, where many image pro-

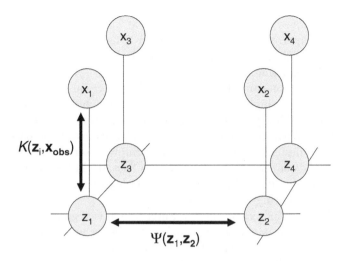

FIGURE 11.5: MRF model

cessing algorithms call upon Markov networks, specifically, in two-dimensions, the Markov Random Field (MRF).

The principle assertion of an MRF is that the distribution of any *image patch site* is conditional on the values of other image patches at neighboring image sites, likely chosen to be the ones closest in proximity. MRFs differ from other random fields in that a site distribution given all other site information depends solely on its neighbors. Which sites to call a neighbor, what relationships to enforce, and how they affect the output are intimately associated with its *local specifications*, which include model type and choice of *clique*.

The diagram in Figure 11.5 is very similar to the one in [25] (the difference being terminology). We can use $K(\mathbf{x}_{obs}, \mathbf{z}_i)$, where $\mathbf{x}_{obs}$ is the observation and $\mathbf{x}_i \in \Omega$ in Section 11.2 to determine possible states, $\{\mathbf{z}\}$. The Ψ assigns an inter-patch compatibility metric between output states. Functions K and Ψ often take on the same form, and a sensible model taken from [25] is given in (11.16).

$$P(\mathbf{z}|\mathbf{x}) = \frac{1}{Z} \prod_{ij} \Psi(\mathbf{z}_i, \mathbf{z}_j) \prod_i K(\mathbf{x}_{obs}, \mathbf{z}_i) \qquad (11.16)$$

where Z is a normalization factor, $\mathbf{z}$ is the output state, and $\mathbf{x}$ is the input patch. If Ψ and K make use of Gaussian subkernels, then (11.16) exemplifies the *Gibbs distribution*, a configuration that describes a global joint distribution of the graph that we assume the image fits into.

Parameters in MRFs can be determined by optimizing globally over all patches, a solution that must be iteratively determined. As one might guess, convergence to a single, meaningful solution result is likely intractable, so several techniques exist to approximate the MRF annealing process. Belief propagation approximates the solution in a few iterations and appears to perform very well [25]. The application MRF described in this section will be of the k-NN algorithm.

Super-resolution can seldom afford to nondeterministically iterate and anneal to a solution. Single-pass algorithms [25, 26] include extra arguments into the decision making process that increase propensity towards one neighbor over another. With single nearest neighbor algorithms, single-pass algorithms are simple to design and implement; one simply observes (11.16) without iterating. In the adaptive k-NN algorithm in Section 11.2 and in algorithms in sections to follow, designing a regression with globalization concerns can be done with a weighting matrix.

Recall (11.14) in Section 11.2.3. Filter design is augmented with a weighting matrix P. We could, in theory, use the belief propagation results or a single pass system to influence this P matrix to take advantage of an expression that is already designed to penalize or reward regression estimates. The logical course of action would be to reward those states that contain high values for $K(\mathbf{x}_i, \mathbf{z}_i)$ and $\Psi(\mathbf{z}_i, \mathbf{z}_j)$.

An Example Conditioning Scheme

Local relationships are governed by their proximity and likeness, and a simple weighting scheme takes advantage of both. A subtle weighting matrix for P in (11.14) that takes on the same form as (11.16) is shown here with some added flexibility. Keep in mind the scaling factor α should be rather small should we wish more attention on the observation patch $\mathbf{x}_{obs}$. The Gibbs distribution is

$$P_{(i, \cdot)} = \frac{1}{Z} \exp\left(||\mathbf{x}_{obs} - \mathbf{z}_i|| + \alpha \sum_{n \in \mathcal{N}} \sum_j ||\mathbf{z}_j^{(n)} - \mathbf{z}_i|| \right) \qquad (11.17)$$

Here, $\mathbf{z}_j^{(n)}$ refers to the jth candidate state (see Figure 11.5) of the nth low-resolution (LR) block in $\mathcal{N}$, the neighborhood of the input block.

There are efficient ways to implement Figure 11.6. For example, all neighboring low-resolution blocks are needed to calculate their high-resolution blocks in order to calculate P. Though the final output patch is calculated at every image patch site, the high-resolution regression values need only be calculated once. Hence, redundancy in a single-pass iterative algorithm is reduced.

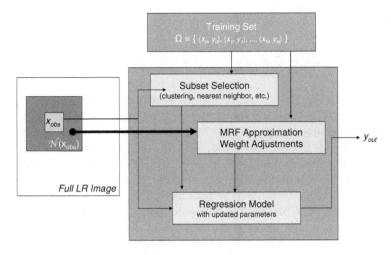

FIGURE 11.6: Applying Markov random field approximates.

11.4   Kernel Machines for Image Super-Resolution

Inferencing is a term often applied to machine learning and image processing. In super-resolution, we are inferring high-resolution pixel values from known low-resolution information. One branch of inferencing that has a privileged place in image processing are kernel methods because they are particularly effective in solving inverse problems like image restoration [63], deconvolution [40], and image super-resolution [45, 29, 34].

The most well-known kernel machine is the support vector machine (SVM) [15]. Initially, SVMs were extensively studied in classification problems [22, 46, 10], and later on, generalized to regression [42, 13, 58, 20] termed support vector regression (SVR). Because pixel values can take on a large number of values (depending on type of pixel representation), regression rather than classification makes the most sense in our problem space. Therefore, SVR is especially useful to image super-resolution because it can directly model and represent the highly nonlinear and complex relationship between low- and high-resolution image data.

It is the kernel function in SVR that offers algorithmic nonlinear capability, so naturally, a large body of work addresses kernel learning. But while kernel learning has been investigated for a number of applications, including classification [37] and dimensionality reduction [60], prior to Ni et al. [45], kernel learning for SVR has been largely left unstudied (save for some erroneous derivations in a technical report [48]). In fact, [45] specifically solves the image super-resolution problem via convex optimization framework, and this chapter reviews the work and its implications and conclusions.

Comparing Nearest Neighbor, k-NN and SVMs:

Suppose we have a training set Ω with N points. It is not always the case that N is large and for these problems, the large sample risk R_N of nearest neighbor (where $k = 1$) in (11.6) is usually smaller than the risk of k-NN [14], but overall, when N is large, k-NN is invariably the rule of choice over NN. Actually, when N is large and the dimensionality of $\mathbf{x}_i$ is small, k-NN is almost always preferable or at least competitive among other estimation techniques such as SVR. This is fairly intuitive because k-NN expects to blanket the entire domain with samples, possible only with N large, and easier if the number of dimensions were sufficiently small. However, SVR performs well when N is small and the dimensionality is large, and as will be shown, SVR is an excellent generalizing technique.

11.4.1   Support Vector Regression

The support vector machine (SVM), originally proposed in [57], is a learning algorithm with the ability to provide function estimation. By using a mapping, $\Phi : \mathcal{X} \rightarrow \mathcal{F}$, where $\mathcal{X}$ is the domain and $\mathcal{F}$ is usually a high-dimensional feature space, support vector regression (SVR) operates in feature space to approximate unknown functions in an output space $\mathcal{Y}$, thereby using nonlinear functions to linearly estimate an unknown regression. That is, to estimate a function, $f : \mathbf{x} \rightarrow y$, nonlinearly, $\mathbf{x}$ becomes embedded in a high-dimensional feature space $\mathcal{F}$ via mapping functions, $\phi(\mathbf{x}) : \mathcal{X} \rightarrow \mathcal{F}$.

As an ill-posed problem, image super-resolution is inherently a nonlinear operation. Yet, by taking training image patches, SVR models a way to generalize unseen inputs to what has been observed in the training set as the following optimization problem [51].

Definition 5 *Given a training set* $\Omega = \{(\boldsymbol{x}_1, \boldsymbol{y}_1), (\boldsymbol{x}_2, \boldsymbol{y}_2), \cdots, (\boldsymbol{x}_N, \boldsymbol{y}_N)\}$, *the support vector regression optimization problem pertains to finding the hyper-*

plane $\boldsymbol{w}$ and intersect b in the following problem:

$$\min_{w,b,\xi^{+/-}} \left(\frac{1}{2}||\boldsymbol{w}|| + C\sum_{i=1}^{N} (\xi_i^+ + \xi_i^-) \right)$$

$$s.t. \quad (\langle \boldsymbol{w}, \phi(\boldsymbol{x}_i) \rangle + b) - y_i \leq \varepsilon + \xi_i^+$$

$$y_i - (\langle \boldsymbol{w}, \phi(\boldsymbol{x}_i) \rangle + b) \leq \varepsilon + \xi_i^-$$

and

$$\xi_i^-, \xi_i^+ \geq 0, \tag{11.18}$$

for all $i \in [1, N]$ and where $\boldsymbol{w} \in \mathcal{F}$. The regression estimate is

$$y_i = \langle \boldsymbol{w}, \phi(\boldsymbol{x}) \rangle + b \tag{11.19}$$

Notice that for $y \in \mathcal{Y}, \forall\, i \in [1, N]$, there are two inequalities that bound the output training: one for the upper boundary and one for the lower boundary. Meanwhile, slack variable vectors ξ_i^+ and ξ_i^- correspond to the upper and lower parameters in which the function $g(\mathbf{x}) = \langle \mathbf{w}, \phi(\mathbf{x}) \rangle + b$ is allowed to deviate for a prespecified error and cost, $[\varepsilon, C] \geq \mathbf{0}^T$.

Theorem 1 *The dual optimization problem*

$$\max_{\alpha^+, \alpha^-} \quad -\frac{1}{2} \sum_{i,k} \left\{ (\alpha_i^+ - \alpha_i^-)(\alpha_k^+ - \alpha_k^-) K(x_i, x_j) \right\}$$

$$-\epsilon \sum_i (\alpha_i^+ + \alpha_i^-) + \sum_i y_i(\alpha_i^+ - \alpha_i^-)$$

$$s.t. \quad \sum_i (\alpha_i^+ - \alpha_i^-) = 0$$

$$0 \leq \alpha_i^{+/-} \leq C \tag{11.20}$$

with the regression estimate as

$$g(\boldsymbol{x}) = \sum_i (\alpha_i^+ - \alpha_i^-) K(\boldsymbol{x}, \boldsymbol{x}_i) + b \tag{11.21}$$

where a dot product in $\mathcal{F}$ is defined by $K(\boldsymbol{s}, \boldsymbol{t}) = \langle \phi(\boldsymbol{s}), \phi(\boldsymbol{t}) \rangle$, the kernel function.

Using a mapping $\Phi : \mathcal{X} \mapsto \mathcal{F}$, SVR is often better suited to represent complicated relationships that we otherwise could not realize linearly. Within $\mathcal{F}$, a kernel function written as a kernel matrix is defined to be a collection of dot products for an arbitrary Φ that may or may not be known. Using the kernel matrix K, computational complexity is reduced because the actual high-dimensional mapping in determining $d = \langle \phi(\mathbf{s}), \phi(\mathbf{t}) \rangle$, which is quite often intractable, is unnecessary when solving (11.20). This definition also allows Φ

to be unknown, in which case, K can be conceptually chosen to be a desired similarity metric depicting the "nearness" of two vectors. Thus, the selection of the kernel matrix K becomes important and should be sensitive to the training data.

Ordinarily, a single kernel matrix, usually a factor decided by human decision, is selected from a set of precalculated kernels to some degree of accuracy if sufficiently cross-validated. Rather than doing this, several works [37, 48, 45, 64, 35, 6] have explored the prospect of learning the kernel matrix. Of particular interest is [37] in which a linear combination of known kernels is optimized to produce a large kernel with good feature representation for the classification problem. The motivation behind this section is that the impossible task of cross-validating all possible combinations of precalculated kernels to determine an optimal one can be derived theoretically instead of analytically.

Within the ideas proposed in [37] is the possibility of incorporating multiple data sources to describe inherent vector space relationships by using multiple kernels. We can choose which features to use (i.e., dot products of $\mathbf{x}' = [x_1, x_3, \ldots]^T$ or $\mathbf{x}' = [x_2, x_5, \ldots]^T$, etc.) and how they will be used (i.e., RBF kernels, polynomial kernels, etc.), The extra degrees of freedom fit especially well with our design because the vector space for our particular problem is multidimensional, compounded by the seemingly desultory nature of local image content. As will become clearer later in this section, it is often the case in high-dimensional spaces that individual dimensions of the input vector $\mathbf{x}$ relate differently in the actual feature space. That is to say, individual features may be more relevant than others (feature selection or weighting) or contribute differently (Hilbert space selection).

Like the classification case in [37],[1] the analogous optimization for regression has been explored in [48], although errors lead the derivation to an incorrect outcome. The following sections present a reformulation of a semidefinite programming (SDP) and quadratically constrained quadratic programming (QCQP) problem to learn the kernel for regression in much the same way that it has been derived for classification.

11.4.2   Inductively Learning the Kernel Matrix for Regression

The nonlinear power of SVMs comes about through the kernel function or matrix. The diversity of image content implies a propensity towards including data from several sources: i.e., using not one but a convex function of multiple different kinds of kernel matrices. Hence, the capability to resolve high-resolution content relies on building an optimal feature space that is

---

[1]Lanckriet et al. [37] offers a method to both inductively and transductively learn a kernel matrix, but the domain in super-resolution is too large to predict input vectors beforehand and therefore induction is exclusively used.

both expressive and nonredundant. This section describes how to obtain this feature space by learning the kernel matrix in terms of a single convex optimization problem: a semidefinite programming problem.

Definition 6 *An SDP problem is a convex optimization problem of the form*

$$\min_{\boldsymbol{u}} \quad \boldsymbol{c}^T \boldsymbol{u}$$

$$s.t. \quad F^{(j)}(\boldsymbol{u}) = F_0^{(j)} + u_1 F_1^{(j)} + \ldots + u_m F_m^{(j)} \succeq 0$$

$$A\boldsymbol{u} = b, \qquad (11.22)$$

for a specified number of j and where $F_i^{(j)}$ are square matrices and $\boldsymbol{u} \in \mathbb{R}^m$.

As it turns out, learning the coefficients of a linear combination of K_i taken from a set $S = \{K_i\}$ of known kernels is a solvable SDP problem (11.35). In truth, any convex set of S would yield a convex optimization problem, though the problem may be exceedingly complex. This avoids a trivial solution, and thus, we now optimize with respect to the coefficients μ_i, of the linear combination in

$$K = \sum_i \mu_i K_i(\,\cdot\,,\,\cdot\,) \,. \qquad (11.23)$$

Theorem 2 *Given a labeled training set*

$$\Omega = \{(\boldsymbol{x}_1, \boldsymbol{y}_1), (\boldsymbol{x}_2, \boldsymbol{y}_2), \cdots, (\boldsymbol{x}_N, \boldsymbol{y}_N)\}\,, \qquad (11.24)$$

the SDP problem that optimizes for the kernel matrix in the SVR problem (11.18) is stated as

$$\min_{\mu, t, \lambda, q_u^+, q_l^-, q_u^-} \quad t$$

$$s.t. \quad \begin{pmatrix} 2\sum_i \mu_i K_i & \gamma \\ \gamma^T & t - 2C\boldsymbol{1}^T(\boldsymbol{q}_u^+ + \boldsymbol{q}_u^-) \end{pmatrix} \succeq 0$$

$$\boldsymbol{q}_u^+, \boldsymbol{q}_u^-, \boldsymbol{q}_l^- \succeq 0$$

$$\epsilon\boldsymbol{1} + \boldsymbol{q}_u^+ + \boldsymbol{q}_u^- - \boldsymbol{q}_l^- \succeq 0$$

$$\sum_i \mu_i K_i \succeq 0$$

$$trace(K) = c. \qquad (11.25)$$

Proof   Starting from the dual optimization problem in (11.20), for simplification, let $\boldsymbol{1}$ be a vector of ones and

$$\boldsymbol{\alpha}^+ + \boldsymbol{\alpha}^- = \boldsymbol{\beta}^+$$

$$\boldsymbol{\alpha}^+ - \boldsymbol{\alpha}^- = \boldsymbol{\beta}^- \qquad (11.26)$$

Recall the dual optimization problem in (11.20). Substituting for α^+ and α^- in (11.26), the optimization problem is written as

$$\max_{\beta^+,\beta^-} \quad -\frac{1}{2}\beta^{-T}K\beta^- + \mathbf{y}^T\beta^- - \epsilon \mathbf{1}^T\beta^+$$

$$\text{s.t.} \quad \mathbf{1}^T\beta^- = 0$$

$$0 \preceq \beta^+ + \beta^- \preceq 2C$$

$$0 \preceq \beta^+ - \beta^- \preceq 2C. \tag{11.27}$$

The optimal objective value of (11.27) is the point-wise supremum of affine functions in K, so it is a convex function of K [8]. Thus, it can be optimized with respect to K to yield an optimal kernel matrix. To ensure good generalization, the trace of K is constrained [37], and the corresponding optimization problem can be expressed in terms of its Lagrangian function as

$$\min_{K} \max_{\beta^{+/-}} \min_{\lambda, \mathbf{q}_{l,u}^{+/-}} \mathcal{L}(K, \beta^{+/-}, \lambda, \mathbf{q}_l^{+/-}, \mathbf{q}_u^{+/-}) \tag{11.28}$$

where we minimize with respect to K. Here, we have introduced the Lagrangian variables λ, $\mathbf{q}_{u/l}^{+/-} \succeq 0$, and constrained our solution with $trace(K) = c$. The Lagrangian $\mathcal{L}$ in (11.28) is written as

$$\mathcal{L}(K, \beta^{+/-}, \lambda, \mathbf{q}_l^{+/-}, \mathbf{q}_u^{+/-}) =$$
$$-\frac{1}{2}\beta^{-T}K\beta^- + \mathbf{y}^T\beta^- - \epsilon \mathbf{1}^T\beta^+ + \lambda e^T\beta^-$$
$$+ (\beta^+ + \beta^-)^T\mathbf{q}_l^+ + (\beta^+ - \beta^-)^T\mathbf{q}_l^-$$
$$- (\beta^+ + \beta^- - 2C\mathbf{1})^T\mathbf{q}_u^+$$
$$- (\beta^+ - \beta^- - 2C\mathbf{1})^T\mathbf{q}_u^-. \tag{11.29}$$

(11.27) is a convex optimization problem and the constraints are strictly feasible. Therefore, from Slater's conditions [56], strong duality holds and we can exchange the order of the maximum and minimum. In terms of β^+ and β^-, we have an unconstrained quadratic optimization problem, which can hence can be analytically solved. The global minimum is found by setting the derivative with respect to β^+ and β^- to zero. That is,

$$\left(\frac{\partial \mathcal{L}}{\partial \beta^+} \triangleq 0\right) \Rightarrow \quad -\epsilon \mathbf{1} + \mathbf{q}_l^+ + \mathbf{q}_l^- - \mathbf{q}_u^+ - \mathbf{q}_u^- = 0$$

$$\left(\frac{\partial \mathcal{L}}{\partial \beta^-} \triangleq 0\right) \Rightarrow \beta_{opt}^- = K^{-1}(\mathbf{y} + \lambda \mathbf{1} + \mathbf{q}_l^+ - \mathbf{q}_l^- - \mathbf{q}_u^+ + \mathbf{q}_u^-) \tag{11.30}$$

Aside from both β's, we can eliminate an additional variable, so we substitute for $\mathbf{q}_l^+$ such that

$$\mathbf{q}_l^+ = \epsilon \mathbf{1} + \mathbf{q}_u^+ + \mathbf{q}_u^- - \mathbf{q}_l^-$$

$$\Downarrow$$

$$\beta^{-*} = K^{-1}(\mathbf{y} + \lambda \mathbf{1} + \epsilon \mathbf{1} + 2\mathbf{q}_u^- - 2\mathbf{q}_l^-) \tag{11.31}$$

Let

$$\gamma = (\mathbf{y} + \lambda \mathbf{1} + \epsilon \mathbf{1} + 2\mathbf{q}_u^- - 2\mathbf{q}_l^-). \tag{11.32}$$

Then, we rewrite the objective function using (11.30) to obtain the expression:

$$\frac{1}{2}\gamma^T K^{-1} \gamma + 2C\mathbf{1}^T(\mathbf{q}_u^+ + \mathbf{q}_u^-) \tag{11.33}$$

The dual optimization problem after some relaxation is then a minimization of (11.33) with some added constraints. The variable K^{-1} in the first constraint brings up an important technique in the formulation of many SDP problems: the Schur complement lemma. The Schur complement lemma is useful in that it allows constraints to be expressed in linear matrix inequality (LMI) form. In terms of its usage with the problem at hand, $K \succeq 0$ implies that

$$t \geq \tfrac{1}{2}\gamma^T K^{-1} \gamma + 2C\mathbf{1}^T(\mathbf{q}_u^+ + \mathbf{q}_u^-)$$
$$\Updownarrow$$
$$\begin{pmatrix} 2K & \gamma \\ \gamma^T & t - 2C\mathbf{1}^T(\mathbf{q}_u^+ + \mathbf{q}_u^-) \end{pmatrix} \succeq 0 \quad, \tag{11.34}$$

where in (11.34), the positive semidefiniteness of the encompassing matrix in the bottom expression has been rewritten as an LMI by considering the Schur complement in the top expression. The final optimization problem takes the following form:

$$\min_{K,t,\lambda,\mathbf{q}_u^+,\mathbf{q}_l^-,\mathbf{q}_u^-} \quad t$$
$$\text{s.t.} \quad \begin{pmatrix} 2K & \gamma \\ \gamma^T & t - 2C\mathbf{1}^T(\mathbf{q}_u^+ + \mathbf{q}_u^-) \end{pmatrix} \succeq 0$$
$$\mathbf{q}_u^+, \mathbf{q}_u^-, \mathbf{q}_l^- \succeq 0$$
$$\epsilon e + \mathbf{q}_u^+ + \mathbf{q}_u^- - \mathbf{q}_l^- \succeq 0$$
$$trace(K) = c$$
$$K \succeq 0 \tag{11.35}$$

For the full derivation, see Ni and Nguyen's work [45] on the application of SVM's to image super-resolution.

11.4.3 The Quadratically Constrained Quadratic Programming Problem

There is an extra check in (11.25) as $K = \sum_{i=1}^{m} \mu_i K_i$ may not yield a positive definite kernel if some $\mu_i < 0$. Additionally, learning kernels from S under an SDP problem formulation with interior point methods and primal/dual optimization toolboxes are polynomial time, but have worst-case complexity of roughly $O((m + N)^2 N^{2.5})$, nearly intractable even with today's computational power. Introducing a constraint $\mu_i \geq 0$ will both ensure a positive definite kernel $K = \sum_{i=1}^{m} \mu_i K_i$ and reduce complexity to $O(mN^3)$ under interior point methods. The constraint will also lead to a quadratically constrained quadratic programming (QCQP) problem.

Definition 7 *A QCQP problem is defined to be a convex optimization problem of the form*

$$\min_{u} \quad f_0(u)$$
$$s.t. \quad f_j(u) \geq 0, \qquad j = 1, \ldots n \qquad (11.36)$$

for a specified number of j and where f_j are quadratic functions of the form $f_j(u) = (A_j u + b)^T (A_j u + b)$.

From (11.25), the QCQP for learning K arises from an added constraint, $\mu_i \geq 0$, which causes some loss of generality, though it does ensure positive definiteness when inductively applying the learned kernel. The intuition behind this is simple; a linear combination of kernels where the coefficients of the combination are guaranteed to be positive will always yield a positive-definite matrix and hence a valid kernel. Mathematically, this is

$$\mu_i \geq 0 \Rightarrow \left\{ \sum_i \mu_i K_i \succeq 0 \Leftrightarrow K \succeq 0 \right\} . \qquad (11.37)$$

On the other hand, the complexity of the kernel is never simplified because the positive eigenvalues of each $(\mu_i K_i)$ will never reduce kernel rank.

Theorem 3 *Let K_i be the ith example positive definite kernel function and $\mu_i \geq 0$. The QCQP problem is given as:*

$$\max_{\beta^+, \beta^-, p} \quad 2y^T \beta^- - 2\epsilon 1^T \beta^+ - cp$$
$$s.t. \quad p \geq \beta^{-T} K_i \beta^-$$
$$1^T \beta^- = 0$$
$$0 \preceq \beta^+ + \beta^- \preceq 2C$$
$$0 \preceq \beta^+ - \beta^- \preceq 2C \qquad (11.38)$$

A single optimization variable p may seem to suggest that only one dual variable μ_i is necessary, meaning that $\beta^{-T} K_i \beta^- = p$ is likely satisfied for one i. In low dimensional spaces in which there are fewer nonredundant K_i, this may be the case. In higher dimensional, more complicated spaces (including the vector space defined by our super-resolution approach), there may be several μ_i's that simultaneously satisfy equality in the constraint $p \geq \beta^- K_i \beta^-$.

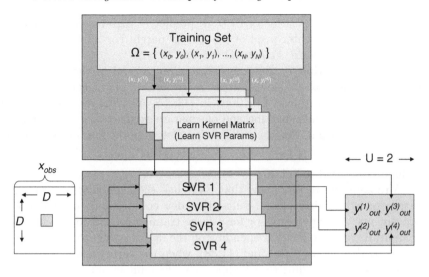

FIGURE 11.7: Kernel learning 2× image super-resolution.

This has important ramifications that justify both QCQP and SDP problems over single kernel cross-validation because it implies that there are several nonzero μ_i's. Consequently, several K_i matrices are required to fully describe a sufficiently descriptive Hilbert space, which through (11.38), can be theoretically obtained. The high probability that $\|\boldsymbol{\mu}\|_0$ (the L_0 norm of the weighting vector of μ_i scalars) is strictly greater than unity further validates the theoretical approach over the impossible task of cross-validating over every linear combination of K_i in $S = \{K_i\}$.

11.4.4 Applications to Super-Resolution

The feature space for nonlinear regression in image super-resolution is varied as there are many domains to which we can apply SVR. For example, we could apply SVR onto bicubic interpolations, effectively determining only "high" resolution content. Or, we could apply SVR to the higher order DCT coefficients of an 8×8 block. In this section, we examine the most straightforward administration of SVR and its optimal kernel to image super-resolution.

Given low and high-resolution image patches I_{LR} and I_{HR} with sizes $D \times D$ and $U \times U$ respectively, to super-resolve the single *center pixel* of I_{LR} by a factor of U, we define vectors

$$
\begin{aligned}
\mathbf{x} &= \text{vectorize}(I_{LR}) - \text{center pixel}(I_{LR}) \in \mathbb{R}^{D^2 \times 1} \\
\mathbf{y} &= \text{vectorize}(I_{HR}) - \text{center pixel}(I_{LR}) \in \mathbb{R}^{U^2 \times 1} \quad (11.39)
\end{aligned}
$$

in a given training set Ω of $\mathbf{x}_i$ feature and $\mathbf{y}_i$ label pairs. The task at hand is super-resolution by a factor of U to predict U high-resolution pixels corresponding to the center pixel of the $D \times D$ patch. For $2X$ super-resolution, this is shown in Figure 11.7 for $D = 5$ and $U = 2$.

This is a multiple output regression problem and in the literature it is often solved as separate single output regressions. There has been some work on learning vector valued functions in [43, 55, 59], but in this chapter, we only discuss the traditional method of separate single output regression problems for each output dimension. Therefore, learning the four outputs becomes $\left\{ y_{out}^{(j)} = g^{(j)}(\mathbf{x}) \right\} \subset \mathbb{R}$ for $j = 1, \ldots, 4$, given the input $\mathbf{x} \in \mathbb{R}^{D^2}$, and $g^{(j)}$ is estimated by SVR in (11.20).

Although the results show that SVR has the capability to provide this regression with fairly clear results, the idea could stand to gain from improvements. A single regressor for a large training set introduces substantial computational complexity. Depending on the dataset, the problem quickly become intractable in (11.35) and (11.38), when the kernel matrix size for each $K_i(\cdot, \cdot)$ scales according to N^2 where N is the number of training points. For $K(\cdot, \cdot)$ to be a sum of m small kernels, the required order of memory exceeds $m \cdot N^2$ without even considering other inequality constraints. Also, without further enhancements, this idea relies on the heavy machinery of SVR to recognize all types of image content, which affects the quality of the prediction due to the problem complexity and the large variety of $\mathbf{x}$ in $\mathcal{X}$.

A Word of Caution:

While SVR is ideally able to fit an arbitrarily complex system by properly cross-validating, without intimate knowledge of all the high-dimensional manifolds and subspaces of image patches (which involves a complex mixture of edges, gradients, texture, etc.), the perfect RKHS using known kernel functions for a single SVR may still not be solvable. Consequently, the sheer number of parameters for an SVR with an imperfect kernel space may render the optimization problem unmanageable. If such a problem is encountered, one should try boosting or using multiple SVRs trained on smaller datasets, a concept to be discussed in the next section.

11.5   Multiple Learners and Multiple Regressions

Diversity and complexity in image patches suggest that super-resolution is inherently a *nonlinear* operation. So far, the discussion has centered around regression techniques whose base function estimation is linear, and nonlinearity has been provided through some kind of embedding, kernel or otherwise. The quality of the nonlinear function estimation is only accurate to the extent of the capabilities of the high-dimensional mapping defined by the embedding. Additionally, if the underlying assumptions on the input space are incorrect, then the number of parameters could potentially become unmanageable even in a scalable estimation technique such as SVR.

Instead of embedding the image structure directly, the complicated structure defining image complexity can be modeled nonlinearly by using *many* function estimates instead of a single regression. Depending on how the learners are organized, there are many approaches under different names that work towards their goals through this concept: (ada)-boost [27], mixtures of experts [5, 36, 41], neural networks [1, 47], vector quantized regression [16], etc. This section examines the application of ensemble learning to the super-resolution problem, and goes through the most popular approaches.

Let $f(\mathbf{x})$ be the true function that maps low-resolution image patches to high-resolution image patches. Then, we substitute an all-encompassing function $g(\mathbf{x})$ that approximates $f(\mathbf{x})$ by aggregating several smaller functions $g_i(\mathbf{x})$. The implementation varies, but most often, the weak learners are content enhancers that sharpen, smooth, etc., depending on when it is necessary to do so. Additionally, it is necessary to weight individual regression results by a certain quantity that may be chosen in a sensible manner.

11.5.1 Neural Networks and Super-Resolution

No survey of pattern classification/recognition and machine learning algorithms is complete without including descriptions of neural networks [21, 23, 31]. Neural networks are rooted in artificial intelligence, and its unsupervised training simultaneously learns nonlinear relationships alongside the actual discriminating function.

Neural networks work by using a collection of *perceptron functions* or *neurons*. The neurons are grouped in a number of layers, where the only visible layers are input and output layers. The rest of the neurons are embedded in *hidden layers* as their activations cannot be directly seen. The construct can be viewed in Figure 11.8.

Neural networks can be described as feed-forward or as more complicated networks with feedback paths. Most image super-resolution algorithms [44, 16], to keep from overfitting and excessive training times, utilize three-layered feed-forward neural networks.

Again, while neural networks implement linear functions, they are frequently applied in nonlinear space. For example, each item in Figure 11.8 could potentially be an RBF, a common theme in early learning-based super-resolution works [1]. Additionally, the variety of nonlinearity is unlimited, and each neuron could potentially express a different kind of relationship. Like the kernel machines described in the previous section, to obtain the scal-

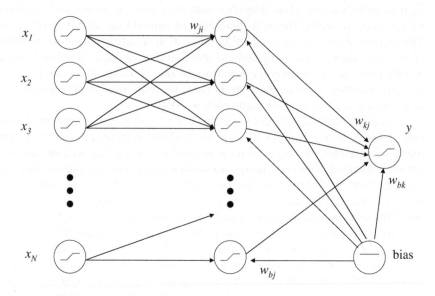

FIGURE 11.8: Three-layer feed-forward neural network framework.

ing weights of the nonlinear functions, multi-layer neural networks are often trained through iterative descent techniques like *back propagation* [32]. Unlike kernel machines, determination of the weights through training must be explicitly solved in nonlinear space.

Recently, we have seen the decline of neural networks. While the expressive power of neural networks is great, because one can simply add hidden units to represent complicated relationships, one of the more difficult problems to solve in using neural networks lies in regularizing the scale of the overall network. This is especially the case in training networks that allow for more circuitous feedback paths. Although the visible layers are fixed, bounds to the number of hidden neurons, weights, and parameters depend on a number of factors, which to date, can at best be solved heuristically. Hence, training can be painfully slow. For complicated relationships, "weeks" is quite common; a good neural network trains for "years." Nevertheless, with enough training time and data, neural networks tend to work extremely well.

11.5.2 Unsupervised Clustering

This section describes unsupervised methods to break up image content into classes. The intuition for multiple classes derives from the fact that different attributes may define the representations that an image patch can take on and therefore require different treatment. In other words, the *type* of image patch dictates how we super-resolve it. For example, take a 5×5 image patch $\mathbf{x}$ with low interpixel variance. The image patch is probably an area of an image that may be a portion of a wall, floor, sky, or some other area where no interesting image activity is occurring. We could label $\mathbf{x}$ as a surface patch. On the other hand, if the left half of $\mathbf{x}$ is dark and the right half is light, then $\mathbf{x}$ could be labeled a left-to-right edge "type."

Because there may be a considerable number of classes or "types," if not infinite, hand labeling training sets may not be entirely feasible. Instead, unsupervised clustering serves as an organizing tool to learn classes without having the user explicitly label them. Unsupervised learning in pattern classification is not unlike the problem of density estimation. In fact, one of the most successful super-resolution algorithms [4] assumes mixture distributions on $\mathbf{x}$ and that the individual components that make up the mixture describe a single class. A popular image patch distribution takes on the form of a Gaussian mixture, and can hence be modeled as a *multivariate Gaussian mixture model* (GMM).

Definition 8 *The multivariate random variable $\boldsymbol{x}$ is distributed as a Gaussian mixture of order "m" if it can be represented in the general form of*

$$\boldsymbol{x} = \sum_{j=1}^{m} \pi_j \mathcal{G}\left(\boldsymbol{x}, \mu_j, \Sigma_j\right) \tag{11.40}$$

where the jth Gaussian $\mathcal{G}$ is weighted by π_j and is of the form

$$\mathcal{G}(\boldsymbol{x}, \mu_j, \Sigma_j) = \frac{1}{(2\pi|\Sigma_j|)^{d/2}} exp\left(-\frac{1}{2}(\boldsymbol{x} - \mu_j)^T \Sigma_j^{-1}(\boldsymbol{x} - \mu_j)\right) \tag{11.41}$$

with mean μ_j and covariance matrix Σ_j.

The most common method to determine GMM parameters is via the expectation maximization (EM) algorithm [17]. The basic idea in EM extends maximum-likelihood techniques to learn a governing distribution from an incomplete training set with missing features. Solving for GMM parameters is a special case in which the "missing data" are vectors denoting the class information.

GMMs are especially attractive because they offer confidence measures representing how sure we are that $\mathbf{x}$ belongs to the mixture component j. The measures come directly out of the expectation step in EM, which essentially calculates Bayes's rule through the posterior probability:

$$
\begin{aligned}
h_{j|\mathbf{x}} &= Pr(j|\mathbf{x}) \\
&= \frac{Pr(j)Pr(\mathbf{x}|j)}{Pr(\mathbf{x})} \\
&= \frac{\pi_j \mathcal{G}(\mathbf{x}, \mu_j, \Sigma_j)}{\sum_j \pi_j \mathcal{G}(\mathbf{x}, \mu_j, \Sigma_j)},
\end{aligned}
\tag{11.42}
$$

where $P(j) = \pi_j$, the prior probability that an image content "type" arises.

To go from $\mathbf{x} \to \mathbf{y}$, i.e., super-resolve, the output $\mathbf{y}$ is represented as a *mixture of experts* in an expected value-like expression in (11.43).

$$
\begin{aligned}
g(\mathbf{x}) &= E[\mathbf{y}|\mathbf{x}] \\
&= \sum_j E[\mathbf{y}|\mathbf{x}, j] Pr(j|\mathbf{x}) \\
&= \sum_j h_{j|\mathbf{x}} g_j(\mathbf{x}, h_{j|\mathbf{x}}),
\end{aligned} \tag{11.43}
$$

with posterior probability $h_{j|\mathbf{x}}$ and individually trained regression, which we have equated with $E[y|\mathbf{x}, j]$. Training and applying function g to $\mathbf{x}$ to estimate $\mathbf{y}$ will be discussed in Section 11.5.4.

Unsupervised training to find multiple class descriptions can also be approached *discriminantly* rather than *generatively*. Under discriminant paradigms, there are no assumptions on underlying distributions (such as GMM), and the input space is partitioned in a relatively straightforward manner. Vector quantization (VQ) [33] and tree-based super-resolution [4] are examples with considerable success in image super-resolution.

Super-resolution with vector quantization [33] creates a number of proto-type vectors $\{\mathbf{v}_j\}$ to compare to the vectorized image patch, $\mathbf{x}_{obs}$. Typically, thousands of vectors are used to achieve specificity. The estimation uses the best choice of regression g_{j*} from a bank of pretrained regressors $\{g_j(\mathbf{x})\}$ corresponding to each $\mathbf{v}_j$. The output estimate $\mathbf{y}$ is given by:

$$
\mathbf{y} = g_{j*}(\mathbf{x}) \tag{11.44}
$$

where

$$
j^* = \operatorname*{argmin}_j \|\mathbf{v}_j - \mathbf{x}_j\| \tag{11.45}
$$

A more advanced classification scheme selects individual types of regressors using classification trees instead of vector quantization. The end application is still framed as (11.44), but the difference lies in how j^* is chosen and how the training classes are obtained. Tree-based resolution enhancement [4] iteratively partitions the training set into subsets according to the primary eigenvector of the subset's covariance matrix. The tree is grown and each subset is divided until the resultant division gives a regression error that exceeds the error prior to division. The tree is then pruned on a large cross-validation set, eliminating classes that yield excessive regression error. Runtime computational complexity is dependent on the tree depth, but usually, finding j^* is trivial.

11.5.3 Supervised Clustering

Numerical results for clustering (or unsupervised classification) algorithms yield very accurate results, but the class selection may not always entirely make visual sense as shown in Figure 11.10. As mentioned, we would like to avoid parsing the training set by hand, but human design of classes may be done quickly provided enough knowledge of the problem domain is present. We can design a group of functions to correlate with input vectors based on what properties exist in the image patch.

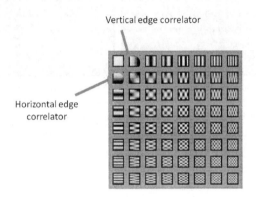

FIGURE 11.9: Example discriminant functions for vertical and horizontal soft edges selected from DCT coefficients.

There are several choices in how to analytically design classification schemes. Figure 11.9 is an example using Discrete Cosine Transform (DCT) coefficients. The devised algorithm correlates the input patch **x** with the primary and most descriptive DCT coefficients, and based on the class, a regression can be provided. The results are spotlighted in Figure 11.10.

In Figure 11.10a, Bouman's unsupervised EM software produces a four-class unsupervised classifier based on a GMM. There is apparently no *visually* discernable pattern in Figure 11.10. Other than the edge class on the upper-right corner,Figure 11.10a [3] lacks interesting visual patterns. Meanwhile, DCT classes in Figure 11.10b mark off vertical edges, horizontal edges, smooth areas, and texture, clockwise from the top-left quite well.

11.5.4 Integrating Regression

With a number of classes, the training set can be clustered or classified into a number of subsets, each of which requires a function estimator (or regressors) for final super-resolution.

Definition 9 *Suppose we have a training set with N points:*

$$\Omega = \{(\boldsymbol{x}_1, \boldsymbol{y}_1), (\boldsymbol{x}_2, \boldsymbol{y}_2), \ldots (\boldsymbol{x}_N, \boldsymbol{y}_N)\}, \tag{11.46}$$

where $\boldsymbol{x}_i$ and $\boldsymbol{y}_i$ denote low and high-resolution information, respectively. The

| (a) GMM Classification [3] | (b) Supervised Classification |

FIGURE 11.10: Class Comparisons in (a) unsupervised clustering using EM on a GMM assumption versus (b) supervised clustering based on DCT basis functions.

"m" subsets Ω_i classification schemes can divide the set Ω into m subsets:

$$\Omega = \left\{ \{(x_i^{(1)}, y_i^{(1)})\}, \{(x_i^{(2)}, y_i^{(2)})\}, \ldots (x_m^{(2)}, y_m^{(2)})\} \right\}$$

$$= \{\Omega_1, \Omega_2, \ldots, \Omega_m\} \tag{11.47}$$

then the jth regression is the regression that most accurately maps domain to range in Ω_j:

$$g_j : x^{(j)} \rightarrow y^{(j)} \tag{11.48}$$

Section 11.5.2 has introduced several clustering methods, most of which fall into one of four common categories [30]: exclusive, overlapping, hierarchical, and probabilistic clustering. Of the four types, only probabilistic clustering offers uncertainty measures of random patterns belonging to specified clusters. The attribute fits especially well with GMM clustering through EM described in Section 11.5.2 where posterior probabilities can produce a weighting system in producing optimal regression estimates.

Rather than independently (and blindly) training function estimators for each class, recall in Section 11.2 and Section 11.3 the introduction of the P matrix (or vector). During training, the same idea can be applied with scalar elements P_j of P corresponding to individual uncertainty measures for class j. One such choice of P in consideration of the training point $\mathbf{x}_i$ is a matrix of the posterior probabilities $h_{j|\mathbf{x}_i} = P(J = j|\mathbf{x}_i)$. This section explores choices of g_j, which come in the form of techniques described throughout Section 11.2 to Section 11.4 and the estimation of their parameters given that we know j and Ω_j.

Section 11.5.2 has introduced a mixture of experts framework in (11.43) used by Resolution Synthesis techniques [3]. The modified weighted least squares regression is applied in (11.49):

$$g(\mathbf{x}) = \sum_j h_{j|\mathbf{x}} G_j \mathbf{x}, \tag{11.49}$$

where $G_j\mathbf{x}$ substitutes for the conditional expectation, $E[\mathbf{y}|\mathbf{x}, j]$.

There are numerous methods to apply regression given the class. For specific linear regression techniques, we can modify frameworks to apply P to a wide variety of methodologies (RANSAC [24], MARS [39], etc.) The expression for a modified weighted least squares linear regression using P is exactly the solution in (11.14). In terms of nonlinear regression, we examine the SVRs introduced in Section 11.4 (though other nonlinear regression techniques [52, 7] have potential as well). Weighting points in the SVR framework is analogous to the effect of choosing C in the original primal problem (11.18) for *every* training point, (11.50), on the solution hyperplane.

$$\min_{\mathbf{w},b} \quad \frac{1}{2}\|\mathbf{w}\|^2 + \sum_i C_i(\xi_i^+ + \xi_i^-)$$

s.t.

$$y_i - (\langle \mathbf{w}, \phi(\mathbf{x}_i)\rangle + b) - \epsilon \le C_i\xi_i^+$$
$$(\langle \mathbf{w}, \phi(\mathbf{x}_i)\rangle + b) - y_i - \epsilon \le C_i\xi_i^-$$
$$C_i, \xi_i^-, \xi_i^+ \ge 0 \tag{11.50}$$

The larger C_i is, the more penalty is incurred for nonflat regression solutions, in effect restricting the freedom to closely fit the training data in the constraints. The optimization arises from individual weighting of slack variables $\xi_i^{+/-}$ for every point in the training set. Then, for the jth SVR, multiplying all ξ_i^- with the corresponding posterior probability of a particular class, $h_{j|\mathbf{x}_i}$ produces a desired effect.

Consequently, to obtain an expression with weighted importance on training points, let $\mathbf{P}_j$ be a vector of probabilities of all the points belonging to class j. Then, the final equations can be rewritten from the optimization problems in Section 11.4.

Theorem 4 *The weighted SDP problem for the cth support vector regression in a mixture of SVM's for image super-resolution is defined as*

$$\min_{\mu,t,\lambda,q_u^+,q_l^-,q_u^-} \quad t$$

s.t.

$$\begin{pmatrix} 2\sum_i \mu_i K_i & \gamma \\ \gamma^T & t - 2C\boldsymbol{P}_c^T(\boldsymbol{q}_u^+ + \boldsymbol{q}_u^-) \end{pmatrix} \succeq 0$$
$$trace(K) = c$$
$$\sum_i \mu_i K_i \succeq 0 \tag{11.51}$$
$$\boldsymbol{q}_u^+, \boldsymbol{q}_u^-, \boldsymbol{q}_l^- \succeq 0$$
$$\epsilon e + \boldsymbol{q}_u^+ + \boldsymbol{q}_u^- - \boldsymbol{q}_l^- \succeq 0$$

Theorem 5 *If $\mu_i > 0$, $\forall i$ in (11.51), then the equivalent SDP problem can be written as a QCQP problem, which is written as*

$$\max_{\beta^+,\beta^-,p} \quad 2\boldsymbol{y}^T\beta^- - 2\epsilon\boldsymbol{1}^T\beta^+ - cp$$

$$s.t. \quad \begin{aligned} & p \geq \beta^- K_i \beta^- \\ & \boldsymbol{1}^T\beta^- = 0 \\ & 0 \preceq \beta^+ + \beta^- \preceq 2\boldsymbol{P}_cC \\ & 0 \preceq \beta^+ - \beta^- \preceq 2\boldsymbol{P}_cC \end{aligned} \tag{11.52}$$

11.6   Design Considerations and Examples

A number of learning algorithms have been introduced throughout this chapter, all of which have involved training. Hence, in using data-driven techniques, the proper setup and design considerations must be implemented, of which cross-validation is an integral concept. In this section, we go over a cross-validation example and then draw up some comparisons that describe the merits and advantages of the nonlinearity modeling in image super-resolution through our various classification and regression frameworks. A short discussion follows with observations conveying the power of pattern recognition techniques in the super-resolution domain.

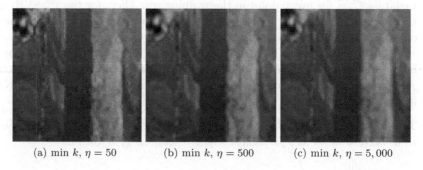

(a) min k, $\eta = 50$ (b) min k, $\eta = 500$ (c) min k, $\eta = 5,000$

FIGURE 11.11: Cross-validation: effect of varying η in the adaptive k-NN super-resolution scheme described in Section 11.2.3.

Cross-validation, an important component of all learning algorithms, assesses the fit of a model prior to its usage. Empirical methods for image and video processing, like the ones discussed in this chapter, must not only be tuned numerically but *visually* as well. Common forms of cross-validation are leave-one-out and k-fold cross-validation. Cross-validation guards against *overfitting* and *overgeneralizing*. For example, Fig. 11.11 aids in the design of the adaptive k-NN regression techniques described by Section 11.2.3, where the parameter η must be cross-validated. Too small an η and the image becomes visually noisy whereas too large an η and the image is blurred and washed out.

(a) Multiple Linear Regression   (b) Multiple Nonlinear SVR

FIGURE 11.12: Multiple learners in a mixture of experts method through a GMM as described in Section 11.5.4: linear vs. nonlinear regression.

On the whole, modified least squares regression estimates have been effective through the various communicated frameworks. More computationally-tasking, nonlinear SVR estimates, of course, perform better in the same frameworks given their expressive power. The examples in Figure 11.12 and Figure 11.13 implements a mixture of experts framework [5, 47, 27] yielding somewhat similar results (with a presharpening preprocessing step in the linear case), though the nonlinear case produces a somewhat crisper image. One must be careful, however, as poorly chosen models and training sets degrades both model complexity and reconstruction quality.

Each algorithm elucidates certain image attributes that are attractive depending on what the viewer notices the most. As expected, k-NN performs well in areas of texture (observe the leaves and bushes in Figure 11.13.) We have collected enough training data to avoid the heuristics of Section 11.2.4, but the concepts introduced in that section do appear to come into play around the edges. It is only noticeable upon closer-inspection, and quality-wise, adaptive k-NN approaches are competitive. Yet, their complexity at $O(N)$ *runtime* leaves much to be desired. A number of approximate neighbor methods address this issue. As an alternative, multiple regressors in Figure 11.13c and Figure 11.13d take advantage of the HVS model's sensitivity to edges, though the SVR produces better texture results.

(a) Single NN + MRF [25]   (b) Adaptive k-NN + MRF

(c) Multiple Linear Regressors [3]   (d) Multiple SVR Regressors [45]

FIGURE 11.13: Comparisons of various learning algorithms described throughout the chapter.

11.7 Remarks

We have a reviewed a number of learning-based image super-resolution algorithms. Although we have, by no means, addressed the application of all pattern recognition techniques to image super-resolution, we have covered the most successful and popular methods. Any of the techniques could easily be expanded into an entire book (see the references for resources), but the ideas and techniques should be adequate to start the reader on their way. As the field of machine learning continues to advance, we expect the powerful potential for applications to image super-resolution to do the same.

It is crucial to remember that learning algorithms, although powerful in generalization, do not exempt the user from detailed and intimate knowledge of the problem domain. Learning algorithms provide a gateway to infer relationships that may otherwise be too complicated to describe. Like any other *tool*, one needs to know how to use it for a given scenario. While throwing a learning algorithm at a problem as an initial step may work somewhat, one may never realize the full potential without careful study of the domain and co-domain.

11.8 Glossary

Clique: A grouping of image sites whose individual sites are not neighbors of themselves. Additionally, a clique's elements that are neighbors of other image sites have, in turn, those image sites as neighbors for themselves.

Cross-Validation: A technique to determine how well a statistical analysis generalizes using independent sets of data.

Expressive Power: The capability to implement different types of functions.

Functional: A functional is a real-valued function on a vector space. The vector space can and is often a space of continuous and differentiable functions.

Image Patch: An organized collection of pixel values that are taken from a single, spatially contiguous image location.

Image Patch Site: The (x, y)-position of an image patch.

Mixture of Experts: From an input vector, several functions, or "experts" pool their decisions to vote on whose outcome is most likely correct. The output is a result that is a weighted consideration of the experts' decision.

Overfitting: The event that a statistical model begins to describe the noise of a training set instead of its underlying relationships.

Perceptron Functions: A simple, single unit in a network of functions (i.e., neural network).

Bibliography

[1] F. Ahmed, S. Gustafson, and M. Karim. High-fidelity image interpolation using radial basis function neural networks. *Proceedings of IEEE National Aerospace and Electronics Conference*, 2:588–592, 1995.

[2] A. Atiya. Estimating the posterior probabilities using the k-nearest neighbor rule. *Neural Comp.*, 17(3):731–740, 2005.

[3] C. Atkins and C. Bouman. *Classification based methods in optimal image interpolation.* PhD thesis, Purdue University, 1998.

[4] C. Atkins, C. Bouman, and J. Allebach. Tree-based resolution synthesis. In *Proceedings of the Image Processing, Image Quality, Image Capture Systems Conference*, pages 405–410, 1999.

[5] R. Avnimelech and N. Intrator. Boosted mixture of experts: An ensemble learning scheme. *Neural Computation*, 11:483–497, 1999.

[6] F. R. Bach and M. I. Jordan. Predictive low-rank decomposition for kernel methods. In *ICML '05: Proceedings of the 22nd international conference on Machine learning*, pages 33–40, 2005.

[7] D. M. Bates and D. G. Watts. *Nonlinear Regression Analysis and Its Applications.* John Wiley and Sons, Inc., New York, 1988.

[8] S. Boyd and L. Vandenberghe. *Convex Optimization.* Cambridge University Press, New York, NY, 2004.

[9] D.C. Burr, M.C. Morrone, and D. Spinelli. Evidence for edge and bar detectors in human vision. *Vision Research*, 4:419–431, 1989.

[10] E. Byvatov and G. Schneider. Support vector machine applications in bioinformatics. *Applied Bioinformatics*, 2(2):67–77, 2003.

[11] F. M. Candocia and J. C. Principe. Super-resolution of images based on local correlations. *IEEE Transactions on Neural Networks*, 10(2):372, March 1999.

[12] H. Chang, D. Yeung, and Y. Xiong. Super-resolution through neighbor embedding. *IEEE Conference on Computer Vision and Pattern Recognition*, 1:275–282, 2004.

[13] S. Chen, K. Jeong, and W. Hardle. Recurrent support vector regression for a nonlinear ARMA model with applications to forecasting financial returns. SFB 649 Discussion Papers SFB649DP2008-051, Sonderforschungsbereich 649, Humboldt University, Berlin, Germany, July 2008.

[14] T. Cover. Estimation by the nearest neighbor rule. *IEEE Transactions on Information Theory*, IT-14(1), January 1968.

[15] N. Cristianini and J Shawe-Taylor. *An Introduction to Support Vector Machines and Other Kernel-Based Learning Methods.* Cambridge University Press, 1st edition, March 2000.

[16] C. A. Dávila and B. R. Hunt. Training of a neural network for image super-resolution based on a nonlinear interpolative vector quantizer. *Applied Optics*, 39:3473–3485, July 2000.

[17] A. P. Dempster, N. M. Laird, and D. B. Rubin. Maximum likelihood from incomplete data via the EM algorithm. *J. Royal Statistics Society*, 39:1–38, 1977.

[18] L. Devroye, L. Gyorfi, A. Krzyzak, and G. Lugosi. On the strong universal consistency of nearest neighbor regression function estimates. *The Annals of Statistics*, 22(3):1371–1385, 1994.

[19] L. Devroye and T. Wagner. Distribution-free consistency results in non-parametric discrimination and regression function estimation. *The Annals of Statistics*, 8(2):231–239, 1980.

[20] H. Drucker, C. Burges, L. Kaufman, A. J. Smola, and V. Vapnik. Support vector regression machines. In *NIPS*, pages 155–161, 1996.

[21] R. Duda, H. Hart, and D. Stork. *Pattern Classification*. Wiley Interscience, 2nd edition, 2000.

[22] N. Emelyanov and R. Gilbert. Astrometric results of observations of mutual occultations and eclipses of the Galilean satellites of Jupiter in 2003. *Astronomy and Astrophysics*, 453:1141–1149, 2006.

[23] L. Fausett, editor. *Fundamentals of neural networks: Architectures, algorithms, and applications*. Prentice-Hall, Inc., Upper Saddle River, NJ, USA, 1994.

[24] M. A. Fischler and R. Bolles. Random sample consensus: A paradigm for model fitting with applications to image analysis and automated cartography. *Communications of the ACM*, 24(6):381–395, 1981.

[25] W. Freeman, T. Jones, and E. Pasztor. Example-based super-resolution. *IEEE Computer Graphics and Applications*, 22(2):56–65, 2002.

[26] W.T. Freeman, E.C. Pasztor, and O.T. Carmichael. Learning low-level vision. *International Journal of Computer Vision*, 40(1):25–47, 2000.

[27] Y. Freund and R. Schapire. A Short Introduction to Boosting. *Journal of Japanese Society for Artificial Intelligence*, 14(5):771–780, 1999.

[28] S. Geman and D. Geman. Stochastic relaxation, Gibbs distribution, and the bayesian restoration of images. *IEEE Trans. Pattern Analysis and Machine Intelligence*, 6(4):721–741, November 1984.

[29] P. M. Goebel and A. N. Belbachir. Single image superresolution interpolation by wavelet support vector regression. *Wavelets and Applications Semester and Conference*, 2006.

[30] J. Hartigan. *Clustering Algorithms*. John Wiley and Sons, Inc., 1975.

[31] S. Haykin. *Neural Networks: A Comprehensive Foundation*. Macmillan, New York, 1994.

[32] R. Hecht-Nielsen. Theory of the backpropagation neural network. *Neural networks for perception (Vol. 2): Computation, learning, architectures*, pages 65–93, 1992.

[33] S. Hong, R. Park, S. Yang, and J. Kim. Image interpolation using interpolative classified vector quantization. *Image Vision Comput.*, 26(2):228–239, 2008.

[34] K. I. Kim, M. Franz, and B. Scholkopf. Kernel hebbian algorithm for single-frame super-resolution. *Statistical Learning in Computer Vision*, pages 135–149, 2004.

[35] B. Kullis, M. Sustik, and I. Dhillon. Learning low-rank kernel matrices. *International Conference on Machine Learning*, 2006.

[36] J. Kwok. Support vector mixture for classification and regression problems. In *Proceedings of the Fourteenth International Conference on Pattern Recognition*, volume 1, pages 255–258, Brisbane, Qld., Australia, 1998.

[37] G. Lanckriet, N. Cristianini, P. Bartlett, L. El Ghaoui, and M. Jordan. Learning the kernel matrix with semidefinite programming. *Journal of Machine Learning Research*, 5:27–72, 2004.

[38] A. B. Lee, K. Pedersen, and D. Mumford. The nonlinear statistics of high-contrast patches in natural images. *Int. J. Comput. Vision*, 54(1-3):83–103, 2003.

[39] P. Lewis and J. Stevens. Nonlinear modeling of time series using multivariate adaptive regression splines (Mars). *Journal of the American Statistical Association*, 86(416):864–877, 1991.

[40] D. Li. Support vector regression based image denoising. *Image Vision Comput.*, 27(6):623–627, 2009.

[41] Clodoaldo A.M. Lima, Andre L.V. Coelho, and Fernando J. Von Zuben. Hybridizing mixtures of experts with support vector machines: Investigation into nonlinear dynamic systems identification. *Information Sciences*, 177:2049–2074, 2007.

[42] H. Mansouri, R. Gilbert, T. Trafalis, L. Leslie, and M. Richman. Ocean surface wind vector forecasting using support vector regression. In C. H. Dagli, A. L. Buczak, D. L. Enke, M. J. Embrechts, and O. Ersoy, editors, *Smart Systems Engineering: Computational Intelligence in Architecturing Complex Engineering Systems*, volume 17 of *Intelligent Engineering*

Systems through Artificial Neural Networks, pages 333–338, New York, NY, USA, 2007.

[43] C. A. Micchelli and M. Pontil. On learning vector-valued functions. *Neural Computation*, 17, 2005.

[44] C. Miravet and F. Rodriguez. A hybrid mlp-pnn architecture for fast image superresolution. *Lecture Notes Computer Science*, 2714:401, 2003.

[45] K. Ni and T. Nguyen. Image superresolution using support vector regression. *IEEE Transactions on Image Processing*, 16(6):1596–1610, 2007.

[46] E. Osuna, R. Freund, and F. Girosi. Training support vector machines: An application to face detection. pages 130–136, 1997.

[47] N. Plaziac. Image interpolation using neural networks. *IEEE Transactions on Image Processing*, 8:1647–1651, November 1999.

[48] S. Qiu and T. Lane. Multiple kernel learning for support vector regression. Technical report, University of New Mexico, 2005.

[49] D. Rajan and S. Chaudhuri. An MRF-based approach to generation of super-resolution images from blurred observations. *J. Math. Imaging Vis.*, 16(1):5–15, 2002.

[50] A. Roorda. Human visual system - image formation. *The Encyclopedia of Imaging Science and Technology*, 1:539–557, 2002.

[51] B. Scholkopf. Statistical learning and kernel methods. Technical report, Microsoft Research Limited, 2000.

[52] G. Seber and C. Wild. *Nonlinear Regression*. John Wiley and Sons, Inc., New York, 2003.

[53] G. Shakhnarovich, T. Darrell, and P. Indyk. *Nearest-Neighbor Methods in Learning and Vision: Theory and Practice*. The MIT Press, 2005.

[54] K. Su, Q. Tian, Q. Que, N. Sebe, and J. Ma. Neighborhood issue in single-frame image superresolution. pages 1122–1125, Amsterdam, The Netherlands, 2005. IEEE Computer Society.

[55] S. Szedmak, J. Shawe-Taylor, and E. Parado-Hernandez. Learning via linear operators: Maximum margin regression. Technical report, PASCAL, Southampton, UK, 2005.

[56] L. Tuncel. On the slater condition for the SDP relaxations of nonconvex sets. *Operations Research Letters*, 29:181–186, 2001.

[57] V. Vapnik. *The Nature of Statistical Learning Theory*. Springer-Verlag, New York, 1995.

[58] V. Vapnik, S. Golowich, and A. Smola. Support vector method for function approximation, regression estimation and signal processing. In *NIPS*, pages 281–287, 1996.

[59] E. Vazquez and E. Walter. Multi-output support vector regression. 13th *IFAC Symposium on System Identification*, pages 1820–1825, 2003.

[60] K. Weinberger and L. Saul. Unsupervised learning of image manifolds by semidefinite programming. *Int. J. Comput. Vision*, 70(1):77–90, 2006.

[61] K. Weinberger and L. Saul. Fast solvers and efficient implementations for distance metric learning. In *ICML*, pages 1160–1167, 2008.

[62] K. Weinberger, F. Sha, and L. Saul. Learning a kernel matrix for nonlinear dimensionality reduction. In *ICML '04: Proceedings of the Twenty-First International Conference on Machine Learning*. ACM, 2004.

[63] C. Yao, P. Yu, and R. Hung. Extractive support vector algorithm on support vector machines for image restoration. *Fundam. Inf.*, 90(1-2):171–190, 2009.

[64] Z. Zhang, D. Yeung, and J. Kwok. Bayesian inference for transductive learning of kernel matrix using the Tanner-Wong data augmentation algorithm. *Proceedings of the 21st International Conference on Machine Learning*, 2004.

12

Super-Resolution Reconstruction of Multichannel Images

Osman G. Sezer

Georgia Institute of Technology

Yucel Altunbasak

Georgia Institute of Technology

CONTENTS

This chapter describes how super-resolution techniques can be applied to improve both spatial and spectral resolution of the multichannel images. The digital cameras that become a part of the daily life captures color images that have three channels (i.e., red, green, and blue), however the number of frequency channels can go up to hundreds for some multichannel data such as hyperspectral images. The multichannel data in remote sensing applications addresses the need to resolve not only the shape of a particular object but also its compositional features (e.g., concrete, asphalt, brick, etc.) With the advent of analysis tools, the multichannel imagery makes it possible to identify, recognize, detect, or classify objects or regions of interest for commercial, civilian and military reconnaissance purposes.

There are several agents that reduce the quality of multichannel imagery. These are sensor noise, solar illumination and atmospheric effects, changes in viewing angle, and secondary illumination coming from nearby objects. The resolution enhancement methods offer invaluable solutions to these limiting factors on the quality of final imagery without the need of expensive equipment. An important design factor that has to be considered for the enhancement of the multichannel data is their dimensionality. The analysis of high-dimensional data such as hyperspectral imagery becomes cumbersome due to the computational burden and the storage issues. On such cases, applying a dimension reduction technique with an appropriate subspace representation, which utilizes the inherent low-dimensionality of the data, has apparent computational and storage benefits for data analysis. Besides, the reduced dimensional subspace representation has been shown to provide some level of robustness against noise for the resolution enhancement applications as well [7], [20]. Therefore, improving the resolution of the multichannel data with subspace based super-resolution method has a high return value. This chapter is dedicated to formulate the super-resolution problem in a reduced dimensional spectral subspace mainly for hyperspectral data. Nevertheless, the introduced super-resolution method can be applied to improve the resolution of any multichannel imagery. First a generative image model is defined to approximate sensor acquisition process at spectral and spatial domains. Next, a blind source separation method is introduced to decompose the spectral information at each pixel into approximated constituent endmembers (pure materials). By only using a limited number of the approximated endmembers to represent spectral data at each pixel, the computational load and the noise inherent in the data will be reduced. Finally, the acquisition model with the reduced dimensional subspace representation is integrated into the super-resolution formulations to improve the spectral and spatial resolution of the given multichannel data.

12.1   Introduction

Before getting into the details of the super-resolution reconstruction of multichannel images, it would be useful to introduce the basics of the multichannel imaging and their applications. This summary does not intend to explain various interacting physical phenomena that affects the final imagery, hence the interested readers are encouraged to refer to [10], [21] for a better treatment.

Any imaging application relies on the fact that object reflects, absorbs, and emits electromagnetic radiation. The interaction between the composition of the constituent matter that forms the object and the shape of it determines the amount of electromagnetic energy reflected at each wavelength. During the evolution, the human visual system is adapted to its environment by having photoreceptor cells that are tuned to absorb light at different wavelengths. Similarly, a conventional camera records intensity images at red, green and blue spectral radiations scattered from the objects. Thus, the multichannel remote sensing with color is already a significant part of daily life.

In the engineering community, the name *remote sensing* is often associated with the field that does the analysis of the data obtained via the satellites, high/low altitude reconnaissance planes, etc. [12]. In the early stages of this field, only the gray-scale images of the Earth surface were available to researchers. Now, however, with the advanced sensing systems used in space technology, it is possible to capture various forms of the multichannel imagery such as multispectral, hyper-spectral, or color images, which supply vast amounts of information to the researchers about the characteristics of the objects and materials for further analysis.

For more accurate analysis of the nature of the observed objects, the scientists working in the remote sensing field proposed to measure the electromagnetic radiation reflected from objects or materials at a higher number of wavelength bands. In this effort, the hyperspectral imaging is introduced as an effective way to extract a contiguous spectral signature of the materials at a distance. Since each material has a different spectral reflectance property, even the detection of subtle objects become possible with hyperspectral imaging. The difference of hyperspectral sensors compared to conventional imaging devices comes from its ability to sample the spectrum of the reflected radiation at a higher number of wavelengths (orders of hundreds) at each pixel location. Therefore, a distinct spectral feature for the objects or materials of interest can be extracted. This enables superior analysis of ore fields, detailed mineral exploration, accurate measurement of the crop yield, identification of military vehicles even under camouflage, and so on.

For multichannel image data, there are two main agents that determine the final spectral characteristics at each pixel site. These are the reflectance properties of the constituent matters (endmembers) of the object or material of interest and the solar radiation at the spectral band that sensors are working on. There are various other factors that might effect the quality of the final imagery, such as the atmospheric effects that are caused by absorption and scatter of the reflected radiance, illumination from nearby objects and changes in viewing angle. Nevertheless, these effects are either considered negligible or compensated ahead of the resolution enhancement process. Thus, without going into complicated details of these effects, the image acquisition model is kept very straightforward; yet it captures the main characteristics of the imaging process in the spatial and spectral domains.

There are a wide variety of approaches that have been used to increase the resolution of multichannel data, especially after the recent interest on hyperspectral images. A very common approach for spectral resolution enhancement in the literature is to use the spatial information of the high-resolution imagery of a different sensor (such as panchromatic or multispectral) to improve the spectral characteristics of the hyperspectral data. For this, the component substitution methods replace the low-pass components with the high-resolution imagery [17], whereas the high-pass techniques improves the spectral bands of the low-resolution data by adding the high spatial frequency content of the high-resolution images [2]. Inherently these methods are similar to the sharpening methods employed to improve human interpretation of the multispectral images in the early literature. A different approach to enhance the resolution of the hyperspectral images is utilizing the spectral mixture analysis [18], [6]. Basically these methods employ the linear mixing model and find the fractions of the high-resolution endmembers at each pixel by a constrained nonlinear optimization. The presented super-resolution algorithm also uses very basic unmixing methods for spectral representation as a part of its reconstruction method. However, rather than fusing information from different types of sensors, the motion difference between the shots of the same scene is utilized to achieve high-resolution imagery.

The super-resolution algorithm of this chapter tries to improve the resolution of the imagery at spatial and spectral domains simultaneously via convex set projections similar to [1]. These convex sets put constraints on the statistics of the residual data, which is defined as the difference between the actual observation and the estimated one produced by the imaging model. Next, for each constraint set a corresponding projection operator is defined. With the successive application of the projection operation onto these convex constraint sets, the residual data approaches to the noise of the image acquisition model; hence, the estimated observations get close to actual ones. The projection operators alter the input of the imaging model, which is the desired high-resolution image. Therefore, if the outputs of the imaging system consistently match with the actual observations for a particular scene, the reconstruction with convex set projection will achieve the desired high-resolution image.

12.2   Notation

The multichannel images are formed by stacking up two-dimensional snap shots of the scene at different wavelengths, hence they are three-dimensional data (refer to Figure 12.1). It is important to introduce the notation that is adopted in this chapter before moving on to details of the super-resolution re-

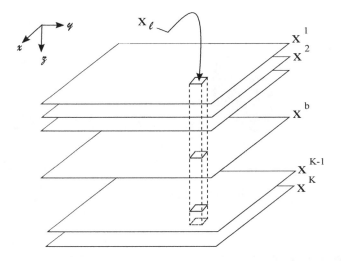

FIGURE 12.1: Hyperspectral data.

construction. Apparently a new convention for data representation is needed. Here, the 3D discrete data of size $M \times N \times K$ is represented by $\mathsf{X}[x, y, z]$ matrix notation. As show in the Figure 12.1, x and y are the spatial coordinates of the data X, while z is the spectral one. For the sake of simplicity, this 3D matrix is mapped to a 2D one by lexicographically arranging the spatial coordinates at each wavelength. This mapping of 3D coordinate system, $[x, y, z]$, into 2D system, $[\ell, z]$, can be expressed by the spatial coordinate mapping $\ell = f(x, y) = x + (y - 1) \times M$. The new vector of spatial data at band b is defined as

$$x^b = \left[\ \mathsf{X}[1, b], \mathsf{X}[2, b], \cdots, \mathsf{X}[M.N, b] \ \right]^T \tag{12.1}$$

where the spatial resolution is $M \times N$ pixels. x^b vector is the lexicographically ordered 2D image at band b. Next, for each co-registered pixel location, a spectral vector of size $K \times 1$ is defined as

$$x_\ell = \left[\ \mathsf{X}[\ell, 1] \quad \mathsf{X}[\ell, 2] \quad \cdots \quad \mathsf{X}[\ell, K] \ \right]^T. \tag{12.2}$$

In the text, this vector is often referred as the *spectral signature* at pixel ℓ. To better visualize the relation between these vectors their matrix representation is given below

$$\mathsf{X} = \begin{pmatrix} \underline{\quad\quad} & x^{1^T} & \underline{\quad\quad} \\ \underline{\quad\quad} & x^{2^T} & \underline{\quad\quad} \\ & \vdots & \\ \underline{\quad\quad} & x^{K^T} & \underline{\quad\quad} \end{pmatrix} = \begin{pmatrix} \vert & \vert & & \vert \\ x_1 & x_2 & \cdots & x_{M.N} \\ \vert & \vert & & \vert \end{pmatrix}. \tag{12.3}$$

With this matrix form, it is easy to see these equalities $\mathsf{X}[\ell, b] = x^b[\ell] = x_\ell[b]$.

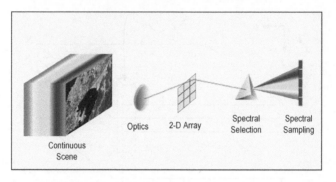

Continuous Scene

Optics   2-D Array   Spectral Selection   Spectral Sampling

FIGURE 12.2: Image acquisition.

12.3 Image Acquisition Model

The image acquisition process given in Figure 12.2 simply highlights some basic components that will be incorporated into an observation or imaging model. First, the continuous scene data passes through the optics, which is followed by some form of spatial sampling due to the spatial arrangements of the spectral sensors in two-dimensional array. These spectral sensors acts like a simple prism that disperse the incoming light into a spectrum of colors. Basically, in the remote sensing applications, the tools like prisms are exploited for the spectral selection. Especially for hyperspectral imaging, to produce a contiguous spectrum with high number of wavelength bands various spectral selection techniques other than prisms such as gratings, filter wheels, and interferometer are employed [3]. Finally, the dispersed energy of the incoming light is recorded at different wavelengths to form the multichannel images.

Assuming that the necessary atmospheric corrections related with the solar radiation is done, the observation model given in Figure 12.3 roughly approximates the rest of the imaging processes. Note that the order of the functional blocks of the observation model does not have one-to-one matching with the actual image acquisition picture at Figure 12.2. The main reason is that it is easy to work with discrete data. Therefore, the model starts with continuous to discrete sampling without aliasing,

$$\mathsf{X}[x, y, z] = \mathsf{X}(x.C_x, y.C_y, z.C_z) \tag{12.4}$$

where $\mathsf{X}(.)$ represent continuous data, the scalar C_z is the spectral sampling period, and the variables C_x and C_y control the spatial sampling density.

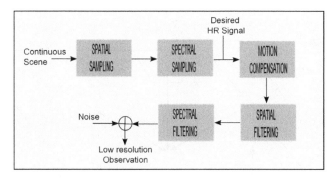

FIGURE 12.3: Observation model.

12.3.1 Motion Compensation

In this discrete signal domain, several methods are available to find the relative motion between the sensors and the observed scene. By finding the relative motion, the different observations of the same scene can be registered to a common two-dimensional grid (often at a higher resolution level). This registration process can be done by the standard motion estimation algorithms like blockmatching. However, since the scene is relatively static during the imaging process, finding the global motion parameters that describe the relative motion between the reference grid and the observations with a projective transform can provide a more accurate registration. As shown in Figure 12.3, the motion compensation functional block acts on high-resolution data. Unfortunately, the high-resolution data is not available; hence, the motion parameters can only be calculated with the low-resolution observations. The advantage of using the global projective mapping for the motion compensation of low-resolution observations is because it enables a very accurate estimation of individual pixel locations at the corresponding high-resolution grids.

The registration process is provided in Figure 12.4. Here, the ground plane denotes the region of interest for the super-resolution algorithm. The purpose is to improve the surface details of that region. The smaller planes with names R, O_1, and O_2 are the reference, the first and the second observation planes, respectively. These planes indicates the imaging planes of three multichannel sensors facing the ground. Thus, each sensor produces images of the same scene at different views. This sensor configuration can be the result of three different passes of the same sensor at that region or three completely different sensors collecting data simultaneously.

Considering that the pictures of Earth are taken at very high altitudes, assuming a planar region of interest for super-resolution reconstruction is in general reasonable. Therefore, it is possible to find a mapping between the pixels of the reference plane R and observation planes O_1 and O_2 simply by finding the 2D planar homography between the observations. Essentially, 2D

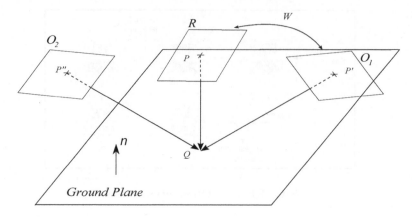

FIGURE 12.4: Registration of sensor imaging planes via planar homography.

planar homography is a projective transform that can locate the corresponding points in two planes. Let $(x_1, y_1, 1)$ and $(x_2, y_2, 1)$ be the homogeneous coordinates of reference and observation planes, then with 2D planar homography, these planes can be registered with

$$\begin{pmatrix} x_1 \\ y_1 \\ 1 \end{pmatrix} = T \begin{pmatrix} x_2 \\ y_2 \\ 1 \end{pmatrix} \tag{12.5}$$

The details of finding the planar homography matrix T that registers the observations into one reference grid can be found in [8]. Note that due to the mapping, the coordinate $(x_1, y_1, 1)$ may end up in noninteger location. For that case, a linear interpolation algorithm will be employed to find the intensity values at the pixel positions. The Figure 12.4 shows a point on the ground plane, Q, and its corresponding observations at different sensors (P, P', and P''), which can be located by the projective mapping. Similarly, every pixel at any imaging plane that faces the region of interest can be registered to a reference two-dimensional grid at plane R with this mapping. As it was mentioned before, the T matrix can be formed by using any other motion estimation algorithm, too.

12.3.2 Spatial Filtering

In the design of an imaging model for the super-resolution reconstruction, since the desired resolution level is higher than the resolution of the observed images, the blurring and the downsampling operations are usually incorporated into the model to explain the relation between the targeted and observed resolution. These operations helps to simulate the aggregate effect of many different physical processes on the final imagery. The optical blur, the point-spread function (PSF) of the imaging sensor, the motion blur and the distance between the scene and the sensor are a few of these physical effects. The spatial filtering functional block in Figure 12.3 encapsulates the blurring and the downsampling operations in the spatial domain. The corresponding mathematical operations for the image at band b can be given as

$$y^b[n_1, n_2] = \sum x^b[m_1, m_2]h^b[n_1, n_2; m_1, m_2] \qquad (12.6)$$

where the spatial filtering operator h^b acts on the band b and composes three functional building blocks. For a simpler analysis, the matrix notation will be used. Next, let the H^b be the convolution matrix of the operator h^b then,

$$H^b = DB^bW \qquad (12.7)$$

where B^b is a square matrix that blurs the pixels of the motion compensated x^b with a Gaussian filter of support $n \times n$. The actual blur is time and space varying and estimating that would require considerable effort. However, under certain conditions the spatial blurring filter in the model can be approximated with a linear space-time invariant Gaussian kernel, which also simplifies the reconstruction process (the justification is provided at [1]). Downsampling operation is denoted by the matrix D, reducing the sampling rate of the targeted imagery to the observed signal's resolution. Hence, to achieve a spatial resolution of 1080x1920 pixels while the observed resolution is 540x960 pixels, a downsampling factor of 2 will be used for both of the coordinate axes in the observation model. Finally, the matrix W does the motion compensation operation using the projective mapping matrix T given in Equation 12.5 and a linear interpolation method to find the intensity values at integer locations (i.e., at each pixel). In the equations that follow, as a convention, the character y will be reserved to denote degraded imagery (either in spatial or spectral domain) while letter x will be used for the high-resolution data.

12.3.3   Spectral Filtering

The observation model described up to this point is very general and can be found in the standard formulations of the super-resolution reconstruction. However, the relation between the desired and the observed spectral resolution has to be constructed for the multichannel data. In this text, the spectral resolution is defined as the number of contiguous bands within a given spectral window. As an example, The Airborne Visible/Infrared Imaging Spectrometer (AVIRIS) database of hyperspectral images has spectral radiance of 224 contiguous spectral bands between 400 to 2500 nanometer wavelengths [4]. Thus, the spectral resolution of the hyperspectral imagery at the AVIRIS database is 224.

The spectral vector at a pixel location is also referred to as the spectral signature of the material at that particular pixel location (see Equation 12.2). The level of absorption of electromagnetic radiation at each wavelength can provide distinct features of a particular material for data analysis such as material identification. However, some materials have resembling spectral signature that might hinder identification if appropriate spectral resolution is not available. In general, having a higher spectral resolution can significantly improve the accuracy of the data analysis tasks. Therefore, a spectral filtering functional block is introduced into the model so that the resolution of the observed scene at the spectral domain can be increased.

In Figure 12.5 the spectral signatures of two pure minerals are given. One of the distinguishing features of Kaolinite (a phylosilicate) from Microcline (a silicate) is the double-trough due to electromagnetic absorption between 2-2.5 micron(μm) wavelengths. On a spectral interval, if a sensor does not have enough spectral resolution to capture this feature, Kaolinite can be confused with other silicates. This demonstrates that for better analysis of the multichannel data, higher spectral resolution is needed. These subtle features become even more essential for making an accurate analysis of the spectra of the mixed (nonpure) materials.

As a part of the effort to increase the spectral resolution, the following model is used in the derivations. Here, the low-resolution spectral signature at ℓ'th pixel position, y_ℓ, is expressed in terms of its high-resolution counterpart, x_ℓ, as

$$y_\ell[k] = \sum x_\ell[l] h_\ell[k; l] \qquad (12.8)$$

where the one-dimensional operator h_ℓ first blurs and then subsamples the data x_ℓ. Using the convenient matrix notation, this relation can be explained

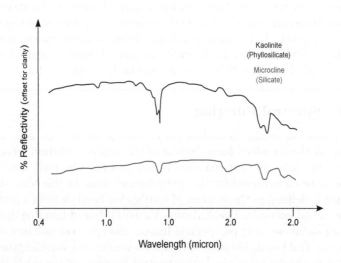

FIGURE 12.5: Pixel spectra of two pure materials (Microcline and Kaolinite) from the ASTER spectral library [11].

as,

$$y_\ell = H_\ell x_\ell = D_\ell B_\ell x_\ell. \tag{12.9}$$

where the convolution matrix B_ℓ blurs the high-resolution spectral signature vector, which is followed by the downsampling operator, D_ℓ. It is important to note that the performance of spectral resolution enhancement with the described model is tied to the performance of the spatial resolution enhancement. This is mainly because the observed multichannel images are assumed to be captured by the sensors that have same wavelength sampling patterns. Therefore, there is no extra information about the missing frequencies that can be fused at a higher resolution grid; hence only the increased accuracy of the existing frequencies provide the resolution improvement at the spectral domain. Nevertheless, the super-resolution algorithm described in this chapter can also be used to achieve higher spectral resolution if the multichannel data with different wavelength sampling patterns is available.

12.3.4   Multichannel Observation Model

The multichannel observation (or imaging) model tries to mimic the physical process that the high-resolution signal goes through in the imaging pipeline of the sensor. If the synthetic output of the observation model, $\hat{y}$, produces reliable estimates of the observed signal, y, then the model will be in agreement with the observations. Having this generative imaging model, the desired high-resolution data can be reconstructed by the tools designed for solving the inverse problems.

To finalize the multichannel observation model, the functional blocks that are described above are put together. Here the input signal of the observation model is the targeted high-resolution data. First, this signal is warped according to the relative motion between the observed image and the reference high-resolution grid. Then, the spatial filtering is applied to the warped input signal at all frequency bands. Next, with spectral filtering the observed spectral signature at pixel location ℓ is generated.

$$y_\ell = H_\ell \begin{pmatrix} H^1[\ell,.] & 0 & \cdots & 0 \\ 0 & H^2[\ell,.] & & \vdots \\ \vdots & & \ddots & 0 \\ 0 & \cdots & 0 & H^K[\ell,.] \end{pmatrix} \begin{pmatrix} x^1 \\ x^2 \\ \vdots \\ x^K \end{pmatrix} + \eta_\ell. \tag{12.10}$$

here $H^1[\ell,.]$ is the ℓ'th row of the spatial filter matrix applied to 1st spectral band. Also the term, η_ℓ, is introduced to account for the combination of the spatial and the spectral observation noise. Since the noise term sums up the effects of all the possible sources of distortion, a zero mean additive white Gaussian noise is assumed. For all the spectral signature vectors of the same scene, the motion compensation parameters will be the same due to the global projection. For simplicity, the observation model can be written as,

$$y_\ell = H_\ell M_\ell x + \eta_\ell. \tag{12.11}$$

where

$$M_\ell = \begin{pmatrix} H^1[\ell,.] & 0 & \cdots & 0 \\ 0 & H^2[\ell,.] & & \vdots \\ \vdots & & \ddots & 0 \\ 0 & \cdots & 0 & H^K[\ell,.] \end{pmatrix}, x = \begin{pmatrix} x^1 \\ x^2 \\ \vdots \\ x^K \end{pmatrix} \tag{12.12}$$

Note that the lexicographically ordered image of the scene at the b'th spectral band, x^b, is a large vector. However, since the support of the Gaussian blur is localized around the ℓ'th pixel, the most of the term of $H^b[\ell,.]$ vector is zero. Thus, the multiplication of the row vector $H^b[\ell,.]$ with x^b can be significantly simplified. Another simplification can be done by decomposing the spectral bands into a fewer number of basis planes by subspace representation. In the next section, the details of such a subspace representation based on blind source separation is given.

12.4   Subspace Representation

Employing a subspace representation method for the super-resolution reconstruction problem has two reasons. First, the multichannel data with a large number of spectral bands complicates the reconstruction problem. The fact that spectral bands have significant correlation, the inherent dimensionality of the data is expected to be less than the actual number of bands. With certain accuracy, it is possible to approximate the spectral bands by a linear combination of a small number of basis vectors, hence the complexity of the enhancement algorithm can be reduced considerably. Using a subspace approach to decrease the dimensionality of the spectral signatures might seem to contradict with the general goal of the super-resolution algorithm. Nevertheless, since the error of approximation is almost zero even with a few (often less than ten) basis vectors, the super-resolution reconstruction is not affected. The second reason for having a subspace representation is to gain robustness against the noise in the multichannel data. For the subspace representation methods such as Karhunen-Loeve Transform (KLT), the eigenvectors of the covariance matrix of the data with high eigenvalues are often retained in the reduced dimensional linear representation. In fact, the process of eliminating the eigenvectors with low eigenvalues from the representation acts like a denoising operation. This is because the eigenvectors with low eigenvalues are considered to capture the noise in the data. Therefore, removing them from the representation essentially denoises the data. Therefore, performing the super-resolution reconstruction in this reduced dimensional linear subspace adds robustness again the noise. In the Experiments and Discussions section, you can find results that are validating this observation.

Karhunen-Loeve Transform is the optimal method to reduce the dimensions of the data, in the minimum mean square error sense. However, it is argued at [13] that the metric induced by blind source separation/independent component analysis (BSS/ICA) is superior to KLT by providing a representation that is more robust to the effects of the noise. Also, as it was noted earlier, each pixel site in a hyperspectral data can possess more than one constituent material (endmember spectra). With the assumption that any observed spectrum at a co-registered pixel location is formed by linear combination of the spectrums of the endmembers, the utilization of blind source separation as a linear unmixing method is reasonable. Nevertheless, the spectral signatures of endmembers are not statistically independent; hence, one cannot expect the unmixed endmembers (or sources) of BSS to match exactly with the actual endmembers. A detailed treatment of application of BSS for unmixing hyperspectral data can be found in [14].

12.4.1 Blind Source Separation

Recently, blind source separation with independent component analysis attracted many researches in various fields from medical signal processing to speech processing as a data analysis tool. The name "blind" comes from the fact that both the source signal (in this case spectra of endmembers) and how they are mixed (abundance fraction) is unknown. The only assumption that will lead to the separation of the source signals is, that they are non-Gaussian and statistically independent. Therefore, the BSS with ICA reduces higher order dependencies compared to KLT that just makes signals uncorrelated. The basic linear mixture model of ICA for spectral analysis can be written as,

$$x_\ell = As_\ell + e_\ell \qquad (12.13)$$

where x_ℓ is the $K \times 1$ spectral signature vector that has the reflectance values at K different wavelengths, s_ℓ is the $k \times 1$ spectral independent component (IC) vector where $K \gg k$. The A matrix of size $K \times k$, on the other hand, has the source signature vectors (approximated endmember spectra) on its columns. The reconstruction error term, e_ℓ, is a result of the dimension reduction step applied prior to the core ICA algorithm. Essentially in this step, KLT is used to reduce the signal dimensionality.

There are various algorithms available to solve the BSS/ICA problem. In the experiments, a symmetric fixed-point algorithm with $f(x) = \tanh(x)$ nonlinearity is used for a fast and simple solution [9]. The algorithm starts with a random orthogonal matrix Q and at each iteration its rows, (q), are updated by

$$q := \mathbf{E}[\hat{x}f(q^T\hat{x})] - \mathbf{E}[f'(q^T\hat{x})]q \qquad (12.14)$$

where $\hat{x}$ the input data to core BSS/ICA algorithm. The expectations are calculated over a fairly large data set of spectral signatures. With this operation, the algorithm tries to find the direction that maximizes the negative entropy of the data. Next, after each iteration the Q matrix is orthonormalized by the following formula

$$Q = (QQ^T)^{-1/2}Q. \tag{12.15}$$

Finally, after the convergence of the iterations is achieved, the A matrix is constructed by

$$A = E\Lambda^{1/2}Q^T \tag{12.16}$$

where Λ denotes the diagonal matrix containing the k largest eigenvalues of the covariance matrix of the spectral signatures and the columns of E matrix have the eigenvectors of the k largest eigenvalues.

12.4.2   Observation Model with BSS

Having introduced the subspace model with the blind source separation, the next step is to integrate it into the observation model. Especially for hyperspectral images that have hundreds of spectral bands, BSS provides a linear representation with less then ten endmembers. As a result, the computational complexity of the super-resolution reconstruction algorithm will be reduced substantially. In order to decompose a given spectral signature vector, x_ℓ, into its independent component vector, s_ℓ, first an unmixing matrix, B, is calculated by taking the inverse of matrix A,

$$B = Q^{-T}\Lambda^{-1/2}E^T \tag{12.17}$$

then the s_ℓ vector will be found by

$$s_\ell = Bx_\ell \tag{12.18}$$

Since the number of constituent endmembers are expected to be significantly less the number of spectral bands, the dimension of the s_ℓ vector will be far smaller than the dimension of the spectral signature vector x_ℓ. The BSS will be applied to find A and B matrices of each scene. This way the compositional features of each scene can be captured to have a better subspace representation. The decomposition of spectral signatures with the BSS/ICA transform will be repeated for all pixels. By doing so, a new multichannel data will be formed, which will have k bands ($k \ll K$), where k is the number of endmembers assumed to exist in the scene and K is the spectral resolution. If the number of endmembers in the scene is unknown, one can find k value by keeping 99.9% of the energy of the representation at the dimension reduction stage of KLT. In the experiment, for hyperspectral images with $K = 244$, 99.9% energy preservation is achieved with k values around 7 to 9. Then using the notation introduced before, the observation model can be updated,

$$y_\ell = H_\ell AM_\ell(s + e) + \eta_\ell \tag{12.19}$$

where the s_ℓ's and e_ℓ's are rearranged to form the spectral basis image s^b and the reconstruction error at that band e^b. The corresponding notations used in the Eq. 12.19 are

$$M_\ell = \begin{pmatrix} H^1[\ell,.] & 0 & \cdots & 0 \\ 0 & H^2[\ell,.] & & \vdots \\ \vdots & & \ddots & 0 \\ 0 & \cdots & 0 & H^k[\ell,.] \end{pmatrix}, s = \begin{pmatrix} s^1 \\ s^2 \\ \vdots \\ s^k \end{pmatrix}, e = \begin{pmatrix} e^1 \\ e^2 \\ \vdots \\ e^k \end{pmatrix}$$

(12.20)

Note that although the diagonal terms of the M_ℓ does not change the number of rows reduces to k. In the next section, a detailed review for super-resolution reconstruction algorithm is given.

12.5   Reconstruction Algorithm

The super-resolution reconstruction is a signal restoration problem where the original signal is assumed to have a higher resolution compared to the observed signal. The observations are degraded and noisy versions of the original signal. The signal restoration is a common problem for various fields in signal processing including system identification, image and speech processing, etc. There are various techniques in the literature to solve the restoration problem under different degradation models. The introduced model for multichannel images is a generative imaging model in which the distortion mechanism is separated into the degradation operators and the noise process. Different than the standard super-resolution formulations, the imaging model describes the individual spectrum at each pixel location (i.e., the spectral signature at that pixel) as follows,

$$y_\ell^{(i)} = H_\ell M_\ell^{(i)} x + \eta_\ell^{(i)}.$$

(12.21)

where superscript (i) denotes the observation number. The spatial degradation matrix $M_\ell^{(i)}$ integrates the effects of motion, blur, and distance from the camera (by decimation or downsampling) for the (i)'th observation of ℓ'th spectral signature vector. The H_ℓ matrix, on the other hand, is the degradation operator at spectral domain that incorporates the quality of spectral data acquisition device into the model. The vector $\eta_\ell^{(i)}$, on the other hand, is the observation noise that combines the inaccuracy in the measurements due to the sensor properties and some secondary effects like atmospheric conditions into a single term. Here, the sensors are assumed to have the same physical characteristics, and only the motion compensation operation changes from one observation to the other, hence, the $M^{(i)}$ notation is used.

The restoration of the high-resolution data, f, by using the linear set of equations in Eq. 12.21 is an ill-posed problem. The solutions like the inverse filtering will just amplify the noise in the observations. Therefore, the regularization methods are needed to integrate a priori information about the distortion mechanism into the solution. In this chapter, the projection onto convex sets (POCS) method is employed to recover high-resolution multichannel imagery. POCS is an iterative method that employs a priori information about the degradation operator, the noise statistics, and the actual high-resolution image distribution. With this a priori information an estimate to the low-resolution observation is generated. The difference between the actual and the estimated low-resolution observations is called the residual of the imaging or observation model. In this case, the residual can be written as,

$$r^{(i)} = y_\ell^{(i)} - H_\ell M_\ell^{(i)} x \qquad (12.22)$$

where $y_\ell^{(i)}$ is the i'th low-resolution observation. To have a consistent estimate of the observations, the residual should have the same characteristics as the observation noise. However, the noise process cannot be known exactly. Generally, the only available information about the noise is its mean and variance. With this knowledge, a confidence interval can be defined for the residual data such that if an outlier is observed, the estimated high-resolution data, f, will be corrected accordingly. This correction operation is called the projection onto convex sets. The convex sets in this definition will be the constraint sets that are bounding the statistics of the residual signal (such as the outliers). Here, the assumption is that the solution (in this case, the high-resolution image) will be the the member of intersections of these constraint sets. The intersection set essentially satisfies all the constraints that can be imposed on the true solution. This intersection set is also called as the feasible region of the inverse problem. The method of POCS converges to a point in this feasible region called as the feasible solution by successive projection of an initial estimate of the solution onto the convex constraint sets. Depending on the initial estimate of the solution, and the order of projection, the solution produced by the POCS algorithm can change. Therefore, the solution provided by the method of POCS is not unique unless the feasible region has a single point. The fundamental mathematical concepts for POCS can be found in [19], [22].

12.5.1 The Subspace Observation Model

Before moving on to the description of the the super-resolution reconstruction algorithm with POCS, the residual signal has to be redefined for the subspace observation model that is given in Equation 12.19. Just to recap, the subspace observation model for the (i)'th observation is expressed as

$$y_\ell^{(i)} = H_\ell A M_\ell^{(i)}(s + e) + \eta_\ell^{(i)}. \qquad (12.23)$$

To be able to use the POCS-based reconstruction method, the noise and the degradation processes should be identified. The difference between the noise processes of two observation models at Eq. 12.21 and 12.23 is the reconstruction error term, $e^{(i)}$, due to reduced dimensional subspace representation. The noise term for the subspace observation model will be

$$\nu_\ell^{(i)} = H_\ell AM_\ell^{(i)} e + \eta_\ell^{(i)}. \qquad (12.24)$$

Since it is common to assume a Gaussian distribution for the reconstruction error, and all the operators acting on reconstruction error is linear, the final noise term ν_ℓ will be Gaussian as well. Next, the residual signal of the subspace observation model associated with an estimate $\hat{s}$ is given as,

$$r^{(i)} = y_\ell^{(i)} - H_\ell AM_\ell^{(i)} s. \qquad (12.25)$$

The POCS algorithm will try to make the statistical characteristics of this residual term approximately equal to those of the noise term, ν_ℓ, of the subspace observation model. Next, the details of two different POCS algorithms are provided. There is no a priori method to choose which one of these POCS algorithms will perform better for a given dataset. Therefore, the best way to choose the reconstruction algorithm is to test both and pick the better performing one.

12.5.2 POCS with Outliers of Residual

The POCS method relies on constraining the solution term when the residual signal deviates from the statistics of the noise process. Given that, to find the statistical characteristics of the noise process, an experimental set-up can be used to synthetically generate the noise term in Equation 12.24. Once the noise process is generated by a large number of synthetic observations, the noise statistics such as the variance and the mean can be calculated easily. The super-resolution algorithms that utilize the method of POCS often define constraints on the solution by limiting the extremum values of the residual signal (or the outliers) [16], [15], [5].

For the POCS with outliers of the residual, the constraints on the solution are performed by sequentially projecting the initial high-resolution estimate onto convex sets based on outliers when the residual deviates an unlikely amount from the mean. The convex set for the ℓ'th pixel location at the b'th band is defined as,

$$C_o^{[\ell,b]} = \left\{ s \Big| \left| y_\ell^{(i)}[b] - \left[H_\ell AM_\ell^{(i)} s \right]_b \right| \le \delta_o \right\} \qquad (12.26)$$

where $[v]_b$ is the b'th term of the vector v and δ_o is the statistical bound on the residual value. With the assumption that ν_ℓ has a Gaussian distribution, the confidence limit δ_o can be found by

$$\delta_o = 3\sigma_\nu \qquad (12.27)$$

where σ_ν is the mean standard deviation of the components of the noise vector ν_ℓ, and the bound δ_o reflects 99% confidence.

The feasible solution to the problem, $\mathbf{s}$, lies at the intersection of convex sets, $C_o^{[\ell,b]}$'s, for all $[\ell, b]$. Moving on the projection operator, first the residual signal with an estimate $\hat{\mathbf{s}}$ is given as,

$$\mathbf{r}^{(i)} = \mathbf{y}_\ell^{(i)} - \mathbf{V}_\ell^{(i)}\hat{\mathbf{s}}. \qquad (12.28)$$

where $\mathbf{V}_\ell = \mathbf{H}_\ell\mathbf{A}\mathbf{M}_\ell$. The corresponding projection operation for the set $C_o^{[\ell,b]}$ is

$$P_o^{[\ell,b]}\mathbf{s} = \begin{cases} \mathbf{s} + \frac{r^{(i)}[b]-\delta_o}{\|v_b\|^2}v_b & , \quad r^{(i)}[b] > \delta_o \\ \mathbf{s} & , \quad -\delta_o \leq r^{(i)}[b] \leq \delta_o \\ \mathbf{s} + \frac{r^{(i)}[b]+\delta_o}{\|v_b\|^2}v_b & , \quad r^{(i)}[b] < -\delta_o \end{cases} \qquad (12.29)$$

where the vector v_b has the entries of b'th row of matrix $\mathbf{V}_\ell$.

12.5.3 POCS with Variance of Residual

Another way to incorporate a priori information about the distortion mechanism of the imaging process into the POCS restoration is to use the variance of the residual signal. This time, the constraint requires that the variance of the residual be approximately equal to the noise variance. The constraint on the solution is defined as follows,

$$C_v^{[\ell]} = \left\{ \mathbf{s} \Big| \left\| \mathbf{y}_\ell^{(i)} - \mathbf{H}_\ell\mathbf{A}\mathbf{M}_\ell^{(i)}\mathbf{s} \right\|^2 \leq \delta_v \right\} \qquad (12.30)$$

where δ_v is the statistical confidence bound on the estimated variance of the residual, which is derived from noise statistics. The set $C_v^{[\ell]}$ denotes the variance constraint on the solution imposed by the spectral signature observation at the ℓ'th pixel location. The feasible solutions lies at the intersection of these constraint sets defined for all pixel locations, $\bigcap_\ell C_v^{[\ell]}$.

As mentioned before, noise is assumed to be uncorrelated Gaussian noise. Therefore the sample variance of residual will have a chi-square distribution. The confidence bound δ_v can be calculated by a Gaussian approximation of chi-square distribution. The formulation provided by [23] is as follows,

$$\delta_v = \sigma^2[\pm\lim_{0.95} + \sqrt{2(K-1)}]^2/2K \qquad (12.31)$$

where σ^2 is the mean sample variance of the residual signal and K is the dimensionality of the spectral signature of the observed data, $\mathbf{y}_\ell$. The notation $\lim_{0.95}$ is used to denote the 95% confidence limit for standard Gaussian distribution. The corresponding projection operator for the set $C_v^{[\ell]}$ is given as

$$P_v^{[\ell]}\mathbf{s} = \begin{cases} \mathbf{s} + (\mathbf{V}_\ell^T\mathbf{V}_\ell + \frac{1}{\lambda}\mathbf{I})^{-1}\mathbf{V}_\ell^T\mathbf{r}^{(i)} & , \quad \|\mathbf{r}^{(i)}\|^2 \leq \delta_v \\ \mathbf{s} & , \quad otherwise \end{cases} \qquad (12.32)$$

where λ is the Lagrange multiplier coming from the optimization formulation. For a detailed treatment of the derivation of the projection operators refer to [23]. Note that although less than or equal to sign is used in Equation 12.30 for defining the constraint on the solution, since the projection of any estimate of the solution that is outside the constraint set will be onto the boundary of the set, the approximate equality of the variance of the noise to the variance of residual signal can be achieved.

12.6 Experiments and Discussions

In order not to confuse the reader with too many results, the method of POCS with the variance of residual is chosen as the main super-resolution reconstruction approach, as it performed better than the method of POCS with the outliers of residual for the hyperspectral data used in the experiments. To test the performance of the super-resolution reconstruction algorithm two sets of experiments are designed. In the first setup, the subspace-based super-resolution algorithm is tested against the affects of the observation noise. For this, the different realizations of white Gaussian noise at various power levels are added to spectral bands. The robustness of the subspace-based and the pixel-based (standard) super-resolution algorithms are examined with quantitative measures. In the second experiment, the resolution improvement at spatial and spectral domain is compared with a 3D interpolation technique designed for multichannel data.

The experiments are conducted on the AVIRIS hyperspectral dataset [4] which provides a workbench for the targeted multichannel resolution enhancement methods. Two specific hyperspectral images are used to test the performance of the reconstruction algorithm at rural and urban settings. For the rural case, a portion of the AVIRIS database taken over Northwestern Indiana's Indian Pine Test Site is used. The Indian Pine hyperspectral data only has grass, corn, and soybean fields with very limited structural information. For urban settings, a portion of Moffett Field of AVIRIS database is extracted. Compared to Indian Pine data, the urban images of Moffett Field contain more structures with fine details such as buildings, roads, ponds, etc. The number of spectral bands, K, for the Moffett Field data is 224 while Indian Pine has 220 bands.

It is important to note that the hyperspectral image of the AVIRIS database is static and does not have any motion information. To implement the multichannel super-resolution reconstruction algorithm, the low-resolution images are needed to be generated synthetically. In this case, the original data will be the ground truth for the quantitative comparisons of the reconstruction performances. The super-resolution methods generate a high-resolution estimate of the scene by fusing the subpixel level information of the low-resolution observations. Therefore, the low-resolution versions of the original data have to possess relative subpixel motion with respect to the reference image. Otherwise, the super-resolution reconstruction cannot exceed the performance of standard interpolators. In the experiments, the observations are assumed to be captured by a hyperspectral sensor at different passes along a simple directional trajectory. Since the scene is expected to be stationary at different shots, the assumption of global motion becomes quite reasonable. After introducing various translational motion patterns, the original data will be blurred and subsampled in spatial and spectral domains to generate the low-resolution observations.

The most common measure to quantitatively understand the performance of the reconstruction is the peak-signal-to-noise-ratio. It can be defined as

$$PSNR = 20 \log_{10} \left(\frac{S_{peak}}{\sqrt{MSE}} \right) \tag{12.33}$$

where MSE is the mean square error between the ground truth and the estimated high-resolution signal, S_{peak} is the peak signal value. The peak signal value for each band can significantly change, which makes this measure biased towards bands with higher energy. To compensate this the definition of standard PSNR is changed in [1] as flows,

$$PSNR = 20 \log_{10} \left(\frac{\sum_{b=1}^{K} S_{peak,b}}{\sqrt{MSE}} \right) \tag{12.34}$$

where $S_{peak,b}$ is the peak signal value at b'th band. The PSNR measure used in the experiments are calculated by this new formulation.

12.6.1 Spectral Subspace

The important part of the experiments is to form the spectral subspace for the reconstruction algorithm. For each scene, first a reference observation is selected so that the rest of the observations will be registered with respected to the reference. Each spectral band of the reference observation is interpolated by a bicubic interpolator to the target high-resolution dimensions. The interpolated reference is used to calculate the mean value of each spectral band. The mean spectral signature is found by

$$\bar{x} = \frac{1}{M.N} \sum_{\ell=1}^{M.N} x_\ell. \tag{12.35}$$

Next, the covariance matrix of the spectral signature is calculated as

$$C = \frac{1}{M.N} \sum_{\ell=1}^{M.N} (x_\ell - \bar{x})(x_\ell - \bar{x})^T. \tag{12.36}$$

Finally, with the eigen-decomposition of the covariance matrix

$$C = E\Lambda E^T \tag{12.37}$$

the E matrix that has the eigenvectors in its columns and the eigenvalue matrix Λ can be found. Using only the eigenvector with k largest eigenvalues, one can obtain the mixing and unmixing matrices for blind source separation as given in Eq. 12.16 and Eq. 12.17. Note that in order to work in this subspace, all the observations have to be centered with the mean signature vector $\bar{x}$. After the reconstruction algorithm is applied, this mean vector is needed to be added to the final enhanced image.

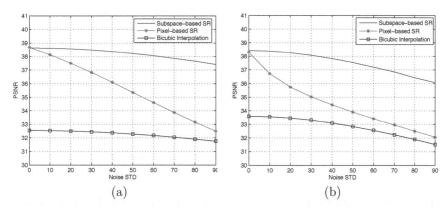

FIGURE 12.6: Robustness of subspace and pixel-based super-resolution algorithm with bicubic interpolation for (a) Moffett Field and (b) Indian Pine hyperspectral data of AVIRIS.

12.6.2   Robustness against Noise

The first set of experiments are more focused towards analyzing the behavior of the reconstruction algorithms at changing noise levels. For this, the white Gaussian noise is added to each spectral band. To simplify the analysis, the number of spectral bands (spectral resolution) of the observed and the desired high-resolution signal is assumed to be the same. This assumption is rational when the spectral resolution is satisfactory for the targeted application. Basically, in the first set of experiments only the spatial resolution enhancement is employed to hyperspectral data.

The enhancement of spatial resolution of each spectral band separately by the method of POCS with outliers of residual is called as pixel-based super-resolution. This enables one to make a fair comparison with the subspace-based super-resolution methods, which utilizes the correlation between the bands. Therefore, the main difference of two methods stems from the utilization of the correlation between the spectral bands. With the utilization

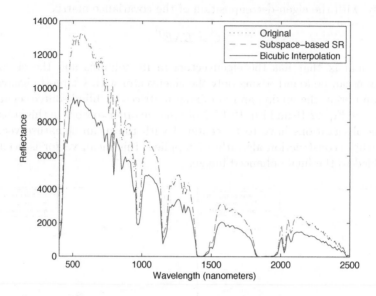

FIGURE 12.7: One-dimensional linear interpolated and super-resolved spectral signatures compared with the original one from Moffett Field data.

of spectral correlation, the subspace representation provides superior image reconstruction.

The downsampling factors at the spatial domain are denoted by f_x and f_y, and at the spectral domain by f_z. For this experiment $f_x = f_y = 2$. To account for the effects of optics, the images at each spectral band are convolved with a Gaussian kernel of size $[2f_x - 1 \times 2f_x - 1]$. In Figure 12.6 the effect of increased noise level on the PSNR of the reconstruction is shown. Note that while the standard deviation of the noise increases, the reconstruction performance of the subspace-based super-resolution (SR) stays relatively stable compared to the pixel-based super-resolution. Figures 12.8 and 12.9 show qualitative comparisons of three different approaches. To see the one-dimensional reconstruction performance at a single spectral signature, a sample selected from Moffett Field data is shown in Figure 12.7. Note that in this experiment only the spatial resolution is enhanced; yet the reconstruction result of the subspace-based SR method gets very close to the original high-resolution spectral signature.

12.6.3 Simultaneous Spatial and Spectral Super-Resolution

In the second set of experiments, the subspace-based super-resolution algorithm is tested for its spectral and spatial resolution enhancement performance. To reduce the spectral resolution, a Gaussian blur is applied before

TABLE 12.1: Reconstruction performance of a 3D interpolator and subspace-based super-resolution at various spatial and spectral scaling factor for a subimage of 224-band Moffett Field.

f_x, f_y	f_z	3D interpolator	Subspace-based SR
2	2	27.26dB	29.03dB
2	4	24.34dB	25.90dB
2	8	21.85dB	23.87dB
2	16	19.07dB	20.77dB
4	2	25.38dB	27.67dB
4	4	23.18dB	24.99dB
4	8	21.09dB	23.15dB
4	16	18.57dB	20.58dB

subsampling the spectral signature with a factor of f_z. The comparisons are done with a 3D interpolator. The 3D interpolator applies first the bicubic interpolation at the spatial domain then at each pixel location spectral vectors of size $K \times 1$ are generated by linear interpolation ($K = 224$ for this experiment). In Table 12.1, the quantitative performance of the 3D interpolator and the subspace-based super-resolution is given in terms of PSNR. Again f_x is set to be equal to f_y. Note that the subspace-based super-resolution provides 1.5dB to 2.0dB better reconstruction performance compared to the 3D interpolator. This performance can be increased if more observations are incorporated into the super-resolution reconstruction algorithm.

12.7   Conclusion

This chapter introduces an application of super-resolution techniques to multichannel data, specifically to hyperspectral images. It is shown that the requirement to design expensive equipment to capture high-resolution spectral data can be circumvented by fast, simple, and cost-efficient postprocessing methods. To achieve this, a basic spectral unmixing methods is incorporated into a super-resolution reconstruction approach, which enables calculation of accurate fractions of approximated endmembers at each pixel. These fractions are essentially the coefficients of the subspace basis vectors. The presented algorithm not only offers faster computation times compared to the pixel-based methods but also it provides a significant level of robustness against the noise due to the acquisition process. It is possible to improve the reconstruction performance of the presented method by employing better spectral mixing models and incorporating a priori information about the endmembers in the scene.

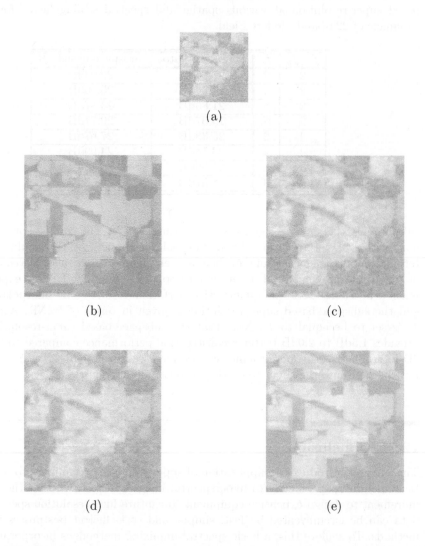

FIGURE 12.8: The results for a subimage of 220-band Indian Pines test site of the AVIRIS database. (a) Noisy low-resolution observation (downsampled in the spatial domain by a factor of two), (b) The original high-resolution data, (c) Bicubic interpolation, (d) Pixel-based super-resolution, (e) Subspace-based super-resolution.

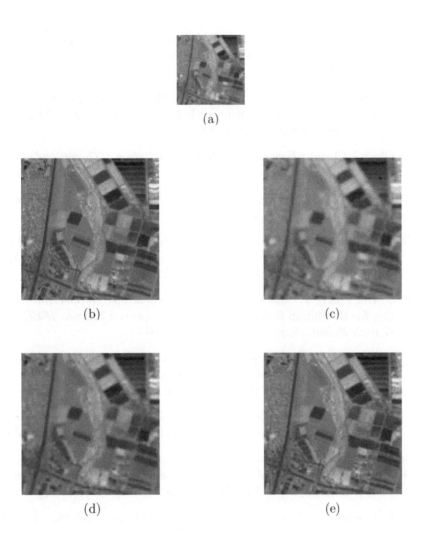

(a)

(b)

(c)

(d)

(e)

FIGURE 12.9: The results for a subimage of 224-band Moffett Field of AVIRIS database. (a) Noisy low-resolution observation (downsampled in the spatial domain by a factor of two), (b) The original high-resolution data, (c) Bicubic interpolation, (d) Pixel-based super-resolution, (e) Subspace-based super-resolution.

Bibliography

[1] T. Akgun, Y. Altunbasak, and R. M. Mersereau. Super-resolution reconstruction of hyperspectral images. *IEEE Transactions on Image Processing*, 14:1860–1875, 2005.

[2] Hardie R. C. and M. T. Eismann. MAP estimation for hyperspectral image resolution enhancement with an auxiliary sensor. *IEEE Trans. Image Proc.*, 13(9):1174–1184, 2004.

[3] Chein-I Chang. *Hyperspectral Data Exploitation : Theory and Applications*. John Wiley & Sons, 2007.

[4] AVIRIS Standard Data. *Available: http://aviris.jpl.nasa.gov/*. Nasa Jet Propulsion Laboratory, Caltech.

[5] P. E. Eren, Sezan M. I., and A. M. Tekalp. Robust, object-based high-resolution image reconstruction from low-resolution video. *IEEE Transactions on Image Processing*, 6:1446–1452, October 1997.

[6] Y. Gu, Zhang Y., and Zhang J. Integration of Spatial-Spectral Information for Resolution Enhancement in Hyperspectral Images. *IEEE Trans. Geosci. Remote Sens.*, 46(5):1347–1358, 2008.

[7] B. Gunturk, A. U. Batur, Y. Altunbasak, M. H. Hayes III, and R. M. Mersereau. Eigenface-domain super-resolution for face recognition. *IEEE Transactions on Image Processing*, 12(5):597–606, 2003.

[8] R. I. Hartley and A. Zisserman. *Multiple View Geometry in Computer Vision*. Cambridge University Press, ISBN: 0521540518, second edition, 2004.

[9] A. Hyvrinen. Fast and robust fixed-point algorithms for independent component analysis. *IEEE Transactions on Neural Networks*, 10.

[10] D. Landgrebe. Hyperspectral image data analysis. *IEEE Signal Processing Magazine*, 19:17–28, January 2002.

[11] ASTER Spectral Library. *Available: http://speclib.jpl.nasa.gov/*. Nasa Jet Propulsion Laboratory, Caltech.

[12] T.M. Lillesand, R.W. Kiefer, and J.W. Chipman. *Remote Sensing and Image Interpretation*. John Wiley & Sons, 6th edition, 2008.

[13] J. P. Nadal and N. Parga. Non-linear neurons in the low noise limit: A factorial code maximizes information transfer. *Network*, 5:565–581, 1994.

[14] J. M. P. Nascimento and J. M. B Dias. Does independent component analysis play a role in unmixing hyperspectral data? *IEEE Transactions on Geoscience and Remote Sensing*, 43:175–187, 2005.

[15] A. J. Patti and Y. Altunbasak. Artifact reduction for POCS-based super-resolution with edge adaptive regularization and higher-order inter-polants. *IEEE Transactions on Image Processing*, 10:176–186, January 2001.

[16] A. J. Patti, Sezan M. I., and A. M. Tekalp. Super-resolution video reconstruction with arbitrary sampling lattice and nonzero aperture time. *IEEE Transactions on Image Processing*, 6:1064–1076, August 1997.

[17] T. Ranchin and L. Wald. Fusion of high spatial and spectral images: The ARSIS concept and its implementation. *Photogramm Eng. Remote Sens.*, 66:49–56, 2000.

[18] G.D. Robinson, H.N. Gross, and J.R. Schott. Evaluation of two applications of spectral mixing models to image fusion. *Proc. SPIE*, 71:272–281, 2000.

[19] M. I. Sezan. An overview of convex projections theory and its application to image recovery problem. *Ultramicroscopy*, 40:55–67, 1992.

[20] O. G. Sezer, Y. Altunbasak, and A. Ercil. Face recognition with independent component based super-resolution. In *Proceedings of SPIE Visual Communications and Image Processing Conference*, volume 6077, San Jose, CA, 2006.

[21] G. Shaw and D. Manolakis. Signal processing for hyperspectral image exploitation. *IEEE Signal Processing Magazine*, 19:12–16, January 2002.

[22] H. Stark and Yang Y. *Vector Space Projections: A Numerical Approach to Signal and Image Processing*. Neural Nets, and Optics, Wiley series in Telecommunications and Signal Processing, 1998.

[23] H. J. Trussell and M. R. Civanlar. The feasible solution in signal processing. *IEEE Trans. Acous., Speech, and Signal Process.*, ASSP-32(2):201–212, 1984.

13

New Applications of Super-Resolution in Medical Imaging

M. Dirk Robinson

Ricoh Innovations Inc.

Stephanie J. Chiu, Cynthia A. Toth

Duke University

Joseph A. Izatt, Joseph Y. Lo

Duke University

Sina Farsiu

Duke University

CONTENTS

The image processing algorithms collectively known as *super-resolution* have proven effective in producing high-quality imagery from a collection of low-resolution photographic images. In this chapter, we examine some of the advantages and challenges of applying the super-resolution framework to applications in medical imaging. We describe two novel applications in detail. The first application addresses the problem of improving the quality of digital mammography imaging systems while reducing X-ray radiation exposure. The second application addresses the problem of improving the spatiotemporal resolution of spectral domain optical coherence tomography systems in the presence of uncontrollable patient motion. Experimental results on real data sets confirm the effectiveness of the proposed methodologies.

13.1   Introduction

The invention of the charge coupled device (CCD) created a new era of imaging wherein optical images could be efficiently captured by an array of solid-state detectors and stored as digital information. The resolution of the captured image depended on the size and number of these detectors. Most imaging applications critically depend on high-resolution imagery. Increasing resolution by improving detector array resolution is not always a feasible approach to improving resolution. For example, while improvements in semiconductor manufacturing have translated into higher-resolution image sensors, shrinking pixel sizes have a tendency to decrease signal-to-noise ratios (SNR) and light sensitivity. Furthermore, practical cost and physical limitations limit the ability to change detectors for most legacy imaging systems. To address this issue, the image processing community is developing a collection of algorithms known as *super-resolution* for generating high-resolution imagery from systems having lower-resolution imaging detectors. These algorithms combine a collection of low-resolution images containing aliasing artifacts and restore a high-resolution image. The ability to transcend the fundamental resolution limits of sensors using super-resolution algorithms has shown significant progress and capability in the area of photographic imaging. By far, the majority of applications using super-resolution technology have been in the area of photographic imagery for either consumer or defense-type applications, which are discussed in the other chapters of this book.

Relatively recently, researchers have begun developing methods to extend the super-resolution framework to different medical imaging applications. Medical imaging applications differ from photographic imaging in several key respects. First, unlike photographic imaging, medical imaging applications often use highly controlled illumination of the human subject during image acquisition. As with any imaging system, stronger illumination energy results in higher signal-to-noise ratios. In the case of medical imaging, however, illumination radiation is limited to prevent tissue damage, thereby limiting the SNR to well below that of photographic imaging. Second, imaging speed is more important in medical imaging applications than in photographic applications.

Short acquisition times both limit patient discomfort and minimize imaging artifacts associated with patient movement. Third, unlike photographic imaging, the goal of medical imaging is to facilitate the detection or diagnosis of disease, rather than produce visually pleasing imagery. Consequently, image processing artifacts are much less tolerable in medical images than in photographic applications. Luckily, medical imaging systems operate under highly controlled environments with highly similar objects. Algorithm developers can leverage prior knowledge about the anatomy or biology to improve image quality. Finally, the majority of medical imaging applications involve creating images from radiation propagation through three-dimensional objects. Thus, while the final images are two-dimensional, they represent some form of projection through a three-dimensional volume.

In this chapter, we describe super-resolution and its applications from the medical imaging community's point of view. In Section 13.2, we describe the general super-resolution framework and provide a brief review of the different super-resolution algorithms. In Section 13.3.1, we introduce the first of two novel applications of super-resolution in medical imaging. Namely, we describe how we tailor the super-resolution framework to improve the resolution for digital X-ray mammography. In Section 13.3.2, we describe how we apply the super-resolution framework to Optical Coherence Tomography (OCT). Finally, we conclude in Section 13.4 with some thoughts about future applications of super-resolution in medical imaging.

13.2 The Super-Resolution Framework

The goal of super-resolution image processing is to extract a high-resolution image from a collection of images containing aliasing artifacts. When a collection of aliased, low-resolution images contains sufficient variation, the high-resolution, aliased image content can be separated from the low-resolution image content thereby increasing the image resolution. This type of super-resolution is not to be confused with optical methods for transcending the optical diffraction limit (e.g., [63]). There are a number of broad reviews of super-resolution algorithms [3, 15, 35]. In this section, we describe the general super-resolution imaging framework. The section begins with a description of a generic image capture model and concludes with a general super-resolution estimation framework.

13.2.1 Image Capture Model

The image capture model describes the various physical processes involved when capturing a set of images. As with most multiframe image super-resolution algorithms, the collection of images must contain relative motion between the sets of images from which resolution is enhanced. We assume very simple translational motion for several reasons. First, even if this motion is not appropriate in a wide-field sense, the motion model is typically accurate for local regions within the images [1]. Second, the imaging acquisition system can often be controlled to induce only translational motion in the captured images. Thus, for the remainder of this chapter, bear in mind that when we refer to an image, we could also be referring to a cropped portion of a larger image. This model has been used in several previous works on super-resolution [11, 16, 23, 32].

The general image capture model, or forward model, combines the various effects of the digital image acquisition process such as point-wise blurring, motion, undersampling, and measurement noise. We represent the forward imaging model using matrix notation as

$$y_k = DHS(v_k)x + e_k, \tag{13.1}$$

where x and y are rearranged in lexicographic order. Here, the vector y_k represents $B \times B$ (assumed square without loss of generality) samples of the captured image $y_k(m_1', m_2')$, where $m_1', m_2' \in [0, B-1]$, are ordered as a $(B)^2 \times 1$ vector. The captured image is undersampled with respect to an unknown high-resolution image $x(m_1, m_2)$, where $m_i \in [0, D_iB - 1]$, by a factor of D_1 and D_2 in each of the two respective dimensions. The vector x represents samples of the unknown $D_1B \times D_2B$ high-resolution image tile $x(m_1, m_2)$ similarly ordered. The warping operator $S(v_k)$ of size $D_1D_2B^2 \times D_1D_2B^2$ represents the subpixel spatial shifts between the captured images. Without loss of generality, we assume that the image y_0 defines the coordinate system of the high-resolution image and hence we only have to estimate the unknown motion parameters for the remaining K images. Note that, here to simplify notations, instead of $v_{k,0}$, we use $v_k = [v_{k_1}, v_{k_2}]$, which is the spatial shifting between the reference frame (0th) and the kth frame. In Section 13.3.2, however, we will use the full form of $v_{k,i}$ to represent the motion between the ith and kth frames. In our model, we assume that these spatial shifts are continuous values in the range of $[-D_i, D_i]$. This corresponds to the range of subpixel motion in the captured images. The downsampling operator D of size $B^2 \times D_1D_2B^2$ captures the undersampling of the detector. The matrix H represents the blurring associated with the imaging system. This blurring could be the result of multiple processes within the imaging system. For example, this blurring could be the result of integration apertures or motion during the image capture, or scattering of radiation in the object medium as a point spread function (PSF). For the time being, we will assume that this can be reliably measured or estimated from some other process (note [39] as an example of jointly estimating the high-resolution image and the blur parameters in a Bayesian framework). Finally, e_k of size $B^2 \times 1$ represents the noise inherent in the analog-to-digital conversion. For our purposes, we assume this noise to be uncorrelated, zero-mean noise with standard deviation σ. This model is sufficiently broad as to cover a wide variety of imaging systems.

13.2.2   Super-Resolution Estimation Framework

The goal of super-resolution image processing is to estimate the high-resolution image x from the set of captured images $\{y_k\}$. The most common estimation framework begins with a cost function or penalty function relating the observed data to the unknown high-resolution image. The most common statistical framework found in super-resolution is that of the *maximum a posteriori* (MAP) penalty function of the form

$$\Omega(x, \{v_k\}) = \Omega_d(x, \{v_k\}) + \Omega_p(x). \tag{13.2}$$

The MAP functionals are based on the construction of a cost function (Ω), which is the summation of two distinct terms. One is the data penalty term Ω_d, which measures the closeness of data to the estimates. The other is the regularization term Ω_p, which applies the prior information about or constraints on the unknown high-resolution image (x).

Early MAP functionals used in super-resolution processing utilized simple quadratic data penalty and regularization terms [10, 45]. The most commonly employed regularization terms use Tikhonov type functionals despite their tendencies to reduce edge contrast. These quadratic regularization functionals penalize the amount of high spatial-frequency energy in the high-resolution image estimate. For example, using the generic imaging model of Equation (13.1), a typical quadratic MAP functional takes the form

$$\Omega(x, \{v_k\}) = \sum_{k=0}^{K} \|y_k - DHS(v_k)x\|^2 + \lambda x^T C_x^{-1} x, \tag{13.3}$$

where C_x^{-1} is often a spatial high-pass operator and λ is the weighting scalar. When C_x is the exact covariance of the unknown high-resolution image, then this cost function produces the ideal Wiener filter estimate of the unknown image. This MAP functional has the advantage of being quadratic, which means that the penalty function has an analytic solution that is a linear function of the input measurements.

Through the years, application of more advanced prior functions Ω_p such as Adaptive Kernel regression [53] and Bilateral Total-Variation (B-TV) [16] have produced higher quality estimates of the final by imposing more accurate assumptions about the image content. The tradeoff, however, is that such non-linear prior functionals are more expensive to evaluate and require more computationally-complex iterative minimization techniques. For example, the B-TV cost function is defined as

$$\Omega_{\text{B-TV}}(x, \{v_k\}) = \sum_{k=0}^{K} \|y_k - DHS(v_k)x\|_2^2 + \lambda \sum_{t_1,t_2=-L}^{L} \varrho^{|t_1|+|t_2|} \|x - S(t)x\|_1, \tag{13.4}$$

where $t = [t_1, t_2]$ is a set of integer pixel shifts and $0 < \varrho \leq 1$ is a constant

[16]. Such nonquadratic functionals can, however, preserve many important features of images such as edges. Also, MAP-based robust super-resolution techniques (e.g., [9, 16, 36]) are able to reduce the effect of outliers such as motion estimation error.

Both the quadratic and non-quadratic MAP functionals require knowledge of the relative shifts between the collection of low-resolution images. When the SNR is reasonably high and the amount of aliasing artifacts are low, then the shifting parameters $\{v_k\}$ can be reasonably estimated in an initial step from the captured images y_k. Theory as well as experimental evidence, however, suggests that using a separate shift estimation process in low-SNR cases is suboptimal. Therefore, the critical issue of joint super-resolution and motion estimation problem has been the topic of several papers (e.g., [2, 24, 43, 58, 59, 62]). Note that, additional priors on motion vector distribution may be also added to the above cost function [24].

13.3 New Medical Imaging Applications

Early, fast, and accurate detection of imaging biomarkers of the onset and progression of diseases is of great importance to the medical community since early detection and intervention often results in optimal treatment and recovery. The advent of novel imaging systems has for the first time enabled clinicians and medical researchers to visualize the anatomical substructures, pathology, and functional features *in vivo*. However, earlier biomarkers of disease onset are often critically smaller or weaker in contrast compared to their corresponding features in the advanced stages of disease. Therefore, medical imaging community strives for inventing higher-resolution/contrast imaging systems. As noted in Section 13.2, super-resolution can be beneficial in improving the image quality of many medical imaging systems without the need for significant hardware alternation.

An excellent review of previous medical imaging applications of super-resolution is given in [20]. We refer the interested reader to [20] for a broad review of applications in magnetic resonance imaging (MRI) [21, 38], functional MRI (fMRI)[37], and positron emission tomography imaging system (PET) [29, 30]. In the following two sections, we explore novel applications of the super-resolution framework to medical imaging. The first application is in the area of X-ray digital mammography. The second is in the area of Optical Coherence Tomography (OCT). Each application has its own unique properties that demand customization of the general super-resolution framework described in the previous section. In both applications, the advantage of applying the super-resolution framework is achieved by special modification of the standard image acquisition technique.

13.3.1 Super-Resolution in Low Radiation Digital X-Ray Mammography

Digital mammography provides the opportunity to efficiently control and capture digital images of the breast, while exposing the patient to the minimum amount of radiation. Today's digital detectors cannot shrink pixel sizes to increase resolution without sacrificing the SNR measurement. To maximize image resolution, we have explored digitally combining multiple low-dosage images, each containing spatial shifts. These shifts are the result of patient movement, intentional dithering of the detector, vibration in the imaging system, and small movement of the imaging gantry. In practice, the motion contained in the captured images is a combination of all such sources necessitating accurate registration of the aliased, low-resolution images.

Applying super-resolution processing to X-ray imaging requires overcoming two challenges. The first is the large amount of data associated with digital mammogram images. The captured low-resolution images could have as much as 10 megapixels worth of data. Thus, computational efficiency is extremely important during processing. Second, the total radiation exposure over the collection of images cannot exceed that of a normal X-ray image dosage. Therefore, the captured data has extremely low peak-SNR (PSNR). For example, Figure 13.1 compares a high-dosage X-ray image (computed PSNR[1] $\simeq$ 13 dB) with the very low exposure images (computed SNR $\simeq$ 3 dB) used in our multiframe scheme.

Thus, providing high-resolution imagery requires sophisticated, nonlinear reconstruction techniques to address the extremely low SNR of the captured images. To address these two challenges, we apply a divide and conquer approach to both improve efficiency while maximizing the denoising effectiveness. To achieve this, we propose a three-stage (registration, reconstruction, and restoration) algorithm. The overall algorithm procedure is shown in Figure 13.2. The entire algorithm operates on a tile-based fashion. The process begins by finding a collection of tiles with approximately equal regions-of-interest. Then, each of these tiles are registered to a subpixel precision to estimate the shifts $\{v_k\}$. Next, we apply a multiframe image restoration step with a weak quadratic prior function, resulting in a deblurred aliasing free image with reconstruction artifacts with known statistics. Next, we perform a fast estimate of wavelet coefficients, which best match the reconstruction artifacts in the previous step. Finally, we apply a nonlinear wavelet thresholding-based denoising step, to recover an efficiently denoised super-resolved image. In what follows we describe each step in detail.

---

[1]In this work, the PSNR was computed numerically as PSNR= $20log_{10}\frac{s}{n}$. In experiments on real images s is the grayscale difference between the minimum and maximum signal regions and n is the noise standard deviation estimated from flat regions. In simulated experiments, n is the RMSE error between the estimated and ground truth image.

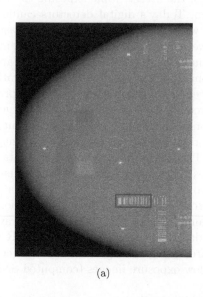

(a)

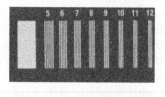 

(b)   (c)

FIGURE 13.1: Mammogram X-ray images from the phantom breast in (a). The red rectangular section in (a) is zoomed in (b) and (c). The high dosage image in (b) is captured at 226mAs (PSNR $\simeq$ 13 dB). The extremely low-dosage image in (c) is captured at 11.3 milliAmpere-second (mAs) (PSNR$\simeq$ 3 dB). Regardless of SNR, both images show aliasing artifacts.

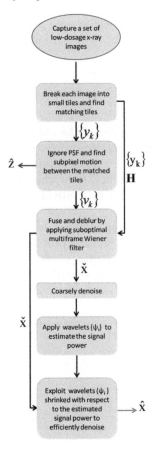

FIGURE 13.2: The block diagram shows the noniterative super-resolution algorithm we apply to digital mammogram images.

13.3.1.1 Multiframe Shift Estimation

Apart from the basic block-matching required to find a collection of approximately registered image tiles $\{y_k\}$ [41], the super-resolution algorithm begins with a multiframe subpixel shift estimation algorithm. The efficiency of this stage is improved by ignoring the optical blur. Considering the locally space-invariant PSF and shift assumptions in our models, we may reverse the order of the shifting and blur operators in Equation (13.1) [11] and reformulate the forward model without the blur operator as

$$y_k = DS(v_k)z + e_k, \qquad (13.5)$$

where $z = Hx$ is the unknown high-resolution blurry image.

The registration algorithm begins with a variant of the quadratic penalty function of Equation (13.3). The optimization process then will be formulated as

$$\Omega_1(z, \{v_k\}) = \sum_{k=0}^{K} \|y_k - DS(v_k)z\|_2^2 + \lambda z^T C_z^{-1} z, \qquad (13.6)$$

where C_z is the covariance matrix of the unknown signal z, which is typically assumed to be stationary. In other words, we can ignore the optical blur for this stage of processing. A typical solution to the above problem is the cyclic coordinate-descent method [24], in which in each iteration one unknown variable is updated based on the estimate of the other unknown variable from the previous iteration.

Noting that Equation (13.6) is known in numerical analysis literature as the Separable Nonlinear Least Squares problem [18], we momentarily assume in our Variable Projection technique [42, 58] that the non-linear parameters (motion-vectors) are known. Consequently, the estimate of the set of linear parameters (z) is computed as

$$\hat{z} = \left(Q(\{v_k\}) + \lambda C_z^{-1}\right)^{-1} g(\{v_k\}), \qquad (13.7)$$

where

$$Q(\{v_k\}) = \frac{1}{\sigma^2} \sum_{k=0}^{K} S^T(v_k) D^T DS(v_k), \qquad (13.8)$$

$$g(\{v_k\}) = \frac{1}{\sigma^2} \sum_{k=0}^{K} S^T(v_k) D^T y_k. \qquad (13.9)$$

We plug the parametric estimate of the blurry high-resolution image ($\hat{z}$) into the MAP functional (Eq. (13.6)) and after some algebraic simplifications, we get a new (maximization) cost function that only relies on the motion-vectors:

$$\Omega_1(\{v_k\}) = g(\{v_k\})^T \left(Q(\{v_k\}) + \lambda C_z^{-1}\right)^{-1} g(\{v_k\}). \qquad (13.10)$$

Note that, unlike the cyclic coordinate-descent method, we require no iterations between the sets of parameters since we do not explicitly calculate Equation (13.7). Indeed, a direct approach to maximize Equation (13.10) involves inverting a large matrix of size $D_1 D_2 B^2 \times D_1 D_2 B^2$, which is computationally challenging for even small image tiles. In [42], we described a series of numerical tricks to speed up the process. One is solving the problem in the Fourier domain and taking advantage of the spectral folding phenomenon in aliased images.

13.3.1.2   Multiframe ForWaRD Deconvolution and Denoising

The output of the previous algorithm is an estimate of the set of sampling shift offsets $\{v_k\}$ with which we can estimate the high-resolution image x. We estimate a high-quality super-resolution image using a noniterative, two-stage, linear deconvolution and nonlinear denoising algorithm. The algorithm addresses the SNR versus sharpness tradeoff inherent to quadratic-type regularization functionals without resorting to iterative, nonlinear regularization penalty functions. More information about the algorithm can be found in [44].

The first stage of the algorithm performs multiframe deconvolution using a weak quadratic penalty function. Armed with estimates of the image shifts, a sharpened, high-resolution image can be obtained using a variant of Equation (13.7) given by

$$\check{x} = B^{-1}(\hat{\underline{v}})H^T g(\hat{\underline{v}}), \qquad (13.11)$$

where

$$B(\hat{\underline{v}}) \;\; = \;\; H^T Q(\hat{\underline{v}})H + \lambda C_x^{-1}, \qquad (13.12)$$

$\hat{\underline{v}} = [v_1, ..., v_k]^T$ and Q and g were defined in Equations (13.8) and (13.9).

In this first stage of the algorithm, we use a very small value of λ so as to underregularize the estimate of the high-resolution image estimate $\hat{x}$. This creates a very sharp high-resolution image at the expense of extreme noise amplification. The second stage of the algorithm involves eliminating the noise while preserving the image signal content. We achieve this with a type of wavelet thresholding algorithm similar to the ForWaRD algorithm of [34] or the BayesShrink algorithm of [4]. The wavelet thresholding algorithm operates by first applying a wavelet transform to the noisy high-resolution image represented as

$$w = \Psi\check{x}, \qquad (13.13)$$

where the matrix Ψ represents the wavelet transform operator and w the wavelet coefficients. Then, the wavelet coefficients are scaled according to

$$w_i' = sgn(w_i)max(0, |w_i| - \gamma_i), \qquad (13.14)$$

where w_i represents the individual wavelet coefficients, sgn is the sign function, and γ_i represents the threshold value for those wavelet coefficients. After applying this threshold, the inverse wavelet transform is applied to the thresholded wavelet coefficients to get the final denoised estimate of the high-resolution image

$$\hat{x} = \Psi^{-1}w'. \qquad (13.15)$$

This type of wavelet thresholding has the ability to eliminate noise while preserving signal content. More information about how this is implemented efficiently can be found in [44].

13.3.1.3 Experimental X-Ray Results

We applied our multiframe reconstruction and restoration algorithm to real images captured on an experimental X-ray imaging system. Our experimental imaging system is based on a Mammomat NovationTOMO digital mammography prototype system (Siemens Medical Solutions, Erlangen Germany),[2] stationed at Duke University Medical Center. The system uses a stationary selenium-based detector having 85 μm pixels. Pixels with this size correspond to a Nyquist sampling rate of 5.6 line pairs per millimeter (lp/mm). We used a CIRS model 11A breast phantom (CIRS Inc., Norfolk VA) to test our super-resolution algorithms. We introduced shifts in the image by two methods. First, we allowed the X-ray tube to rotate by $\pm$ 1 degree. Second, we manually moved the breast phantom to introduce motion into the system. This manual motion was completely uncontrolled. Our dataset consisted of 15 frames at the low dosage level of 11.3 mAs at 28 kVp tube voltage. As a point of reference, we also acquired a single frame at a more typical dosage of 226 mAs at 28 kVp tube voltage (Figure 13.1), which is 25.

The breast phantom includes several testing features including a pair of resolution bar charts and small grains that mimic calcification in the breast. The results reported here are focused on the test resolution chart and the calcification grains that best represent the contrast performance and potential improved-detection abilities of the multiframe image reconstruction system.

We applied our algorithm to 100×100 pixel tiles in the captured image to estimate 400×400 pixel high-resolution images (enhancement $D = 4$). We modeled our system PSF as a heavy-tailed exponential energy distribution with $\beta = 1.5$. To get a measure of the PSNR, we calculated the standard deviation in a textureless region of the phantom. We also measured the difference in grayscale values between registration bars in the resolution chart to get an approximate PSNR value of 3 dB. We employed 2-tap Daubechies filters for the soft thresholding wavelet functions and 6-tap Daubechies filters for the coarse denoising by way of hard wavelet coefficient thresholding. We focus on the portion of the resolution chart beyond the Nyquist rate for the imaging system (5.6 lp/mm). The numbers indicate the resolution in terms of line pairs per millimeter (lp/mm). Figure 13.3 shows the images throughout the super-resolution process.

The first image (Figure 13.3a) shows one of the 15 low-dosage images. The image has very low SNR and shows some of the aliasing associated with an undersampled detector. The second image (Figure 13.3b) shows an example

---

[2]Caution: Investigational Device. Limited by US Federal law to investigational use. The information about this product is preliminary. The product is under development and is not commercially available in the US; and its future availability cannot be ensured.

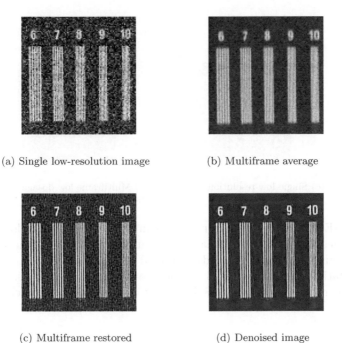

(a) Single low-resolution image   (b) Multiframe average

(c) Multiframe restored   (d) Denoised image

FIGURE 13.3: Different restoration techniques applied on the low-dosage set of images. (a) Low-dosage image, (b) Multiframe averaged image, (c) Multiframe restored $\check{x}$, (d) Denoised super-resolved image $\hat{x}$. The multiframe average image shows the aliasing present in the captured image. The super-resolved images show image contrast beyond the native sampling rate of the system. The total dosage of using 15 of these frames ($15 \times 11.3 = 170$ mAs) is still less than the high dosage image of 225 mAs in Figure 13.1b with clear aliasing artifacts on resolution bars labeled with numbers higher than 8.

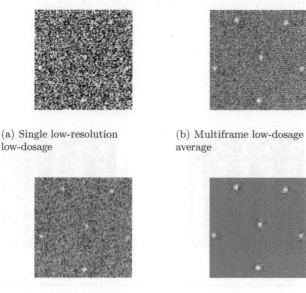

(a) Single low-resolution   (b) Multiframe low-dosage
low-dosage   average

(c) Single low-resolution   (d) Multiframe low-dosage
high-dosage   restored

FIGURE 13.4: Different restoration techniques applied on the low-dosage set of images. (a) Low-dosage image, (b) Multiframe averaged image, (c) High-dosage image at 226 mAs, (d) Denoised image $\hat{x}$ ($15 \times 11.3 = 170$ mAs). Restoration combining the 15 low-dosage frames in (d) frames, most clearly demonstrating the pentagram-shaped set of microcalcification cluster.

of averaging the 15 low-resolution frames followed by interpolation. While the SNR is improved, the aliasing contained in the low-resolution images becomes clear as the 5 bars appear as three bars above 7 lp/mm. The third image (Figure 13.3c) shows the resulting image $\check{x}$ after applying the multiframe image restoration step. This image shows contrast improvement but at the expense of significant noise amplification. The final image (Figure 13.3d) shows the final image estimate $\hat{x}$ after applying the non-linear wavelet thresholding denoising algorithm. The image shows that the contrast is preserved while eliminating most of the amplified noise.

The primary goal of mammography is detecting and diagnosing cancerous lesions in the breast. The breast phantom includes small grains of calcium for evaluating the diagnostic capability of micro-calcifications. Figure 13.4 shows another block from the same experiment demonstrating the ability of the nonlinear denoising algorithm to clearly eliminate noise while preserving the signal for a cluster of 0.275 mm-sized calcite grains. This provides confidence in the nonlinear denoising algorithm's ability to discern the signal from noise.

13.3.2 Super-Resolution in Optical Coherence Tomography

The invention and utilization of Optical Coherence Tomography (OCT), first reported in [27] in 1991, has had a profound impact on many fields of research, especially ophthalmic sciences [25, 26, 46, 49, 55]. OCT systems provide non-invasive yet high-resolution *invivo* images of retinal structures, which can be used to define imaging biomarkers of the onset and progression of many ophthalmic diseases. By employing an interferometer [8, 17], several OCT-based imaging systems have been developed throughout the years, most notably the time-domain OCT (TDOCT) and ultrahigh resolution OCT (UHROCT) [40]. The advent of the Spectral Domain Optical Coherence Tomography (SDOCT) system has further improved the image quality and acquisition speed [33, 51, 60, 61, 64, 65]. Today, several commercial SDOCT systems are available with similar capabilities and 20-30 kHz A-scan rates.

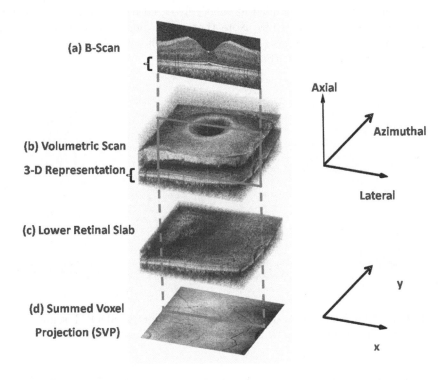

FIGURE 13.5: A volumetric SDOCT scan set is a collection of azimuthally sampled B-Scans (a), creating a 3-D view of the retina (b). The braces in (a,b) mark the lower retinal slab shown in (c) containing the shadows from the overlying larger vessels. (d) is the summed voxel projection (SVP) created by axial projection of the lower half of the B-Scans, demonstrating the vessel pattern [22, 28]. Each B-Scan corresponds to one line on the SVP.

A noninvasive, accurate characterization of retinal lesions and other pathological abnormalities is only possible with high-resolution, 3-D ocular imaging. Commonly, the lateral, axial, and azimuthal resolution of many imaging
systems, including OCT, are associated with (and calculated based on) the
illumination source characteristics (e.g., bandwidth), the optical path (e.g.,
diffraction limit due to pupil diameter, ocular aberrations, dispersion, etc.),
and other physical characteristics. Utilization of fast and efficient CCD detectors has facilitated the creation of aliasing-free 3-D images of anatomical
structures. However, for some *in vivo* imaging applications, the SDOCT acquisition time is not short enough to avoid abrupt motions such as blinking, thus
creating motion artifacts in the densely sampled volumetric measurements
(Figure 13.6). Therefore, in practice, to speed up the image acquisition process, these systems are utilized at a significantly lower than nominal resolution.
In SDOCT imaging, practical resolution in the azimuthal axis corresponds to
the number of B-scans sampled at relatively equal distances in a *volumetric*
scanning scheme (Figure 13.5). Note that, valid quantitative measurements
of retinal disease biomarkers (e.g., drusen [56] volume) are only feasible from
B-Scans with known azimuthal displacement.

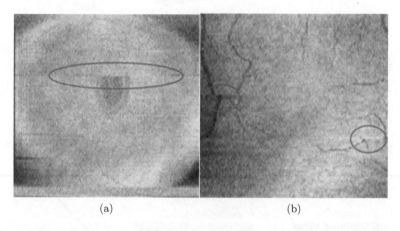

(a)   (b)

FIGURE 13.6: SVPs from densely sampled volumetric SDOCT scans of two
subjects. The ellipsoids mark motion artifact locations.

On the quest to gather useful information from SDOCT through improving the hardware design, one quickly runs into the problem of diminishing
returns. Specifically, the optical components necessary to capture very high-
quality, dense scans become prohibitively expensive or too sensitive for many
practical applications. Unlike alternative approaches that require expensive
hardware such as eye tracking systems [22], we propose a software-based image processing solution in this section based on our earlier work [12], that is

applicable to virtually any SDOCT imaging system, including the handheld SDOCT systems that are more prone to motion errors [5, 6, 47].

In this section, we introduce a novel application of the super-resolution framework for improving the azimuthal resolution of SDOCT images. We propose a method based on capturing several repeated fast sparse 3-D scans, followed by detecting and removing the ones affected by motion artifacts, and finally fusing the artifact-free scans. Our approach to reduce motion artifacts in the 3-D tomographic structure, in spirit, is close to the multi-camera time-space super-resolution [48], MRI inter-slice reconstruction [21], and video synchronization [57] problems. However, the proposed reconstruction algorithms and applications are fundamentally different and novel.

13.3.2.1 Proposed Method: Sparse Repeated Imaging

Our goal is to transcend the limitations of SDOCT imaging systems, reduce motion artifacts (Figure 13.6), and obtain densely sampled, high-quality, and accurate 3-D SDOCT images of unpredictably moving structures such as a human eye. In typical SDOCT ophthalmic imaging practice, the region of interest is swept with a relatively high number of B-Scans. For many patients, due to the multiple seconds required to capture scans, a dense scanning strategy is prone to motion artifacts such as blinking.

Alternatively, we propose to capture several (N) sparsely sampled volumetric scans with a significantly lower number of B-Scans than the target resolution. Since the number of frames in each sequence (K) is relatively small, each scan is captured very fast. Therefore, it is reasonable to assume that some of these sequences will be less affected by the abrupt patient motion. We detect such sequences, reorder and interlace their frames, and create a densely sampled, artifact-free representation of the retina. Figure 13.7 represents the main idea, where two sparsely sampled sequences are fused together creating a dense representation of the underlying pathological structures.

Putting together the frames of different scan sets (interlacing) in a correct order is a challenging task. A naive approach involves sorting via pairwise registration and computing a closeness measure (e.g., normalized cross-correlation or sum-of-squared difference) of all frames. In the case of fusing only two volumetric B-Scan sets, each frame in the first volume sequence is registered to all frames of the second sequence. Then, in the fused output sequence, this individual frame is inserted into the second sequence nearest to the frame in the second sequence with the highest cross-correlation value. This process would be repeated for all remaining frames in the first sequence. Of course, this is a simplified variation of the video synchronization problem, discussed in detail in the computer vision literature [57]. However, aside from the prohibitively heavy computational load of registering large SDOCT datasets, the SNR of the SDOCT images is significantly lower than the commercial camcorders for which the method in [57] is developed. Therefore, the commonly used closeness measures such as normalized cross-correlation may not always be sensitive enough to discriminate between very small structural changes in the neighboring SDOCT ophthalmic scans (Figure 13.7).

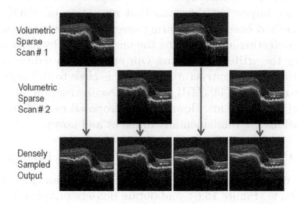

FIGURE 13.7: Fusing (interlacing) multiple sparsely sampled scan sequences to create an azimuthally higher resolution volume of B-Scans. Indeed, unlike this schematic example, in clinical applications the displacement between sequences might be noninteger as it is induced by patient motion.

To reduce the computational complexity of 3-D registration and improve accuracy, we introduce an alternative global solution based on 2-D registration. Note that, the azimuthal axis displacement is the only motion that we need to estimate to be able to interlace the 3-D volumetric scans. A quick consultation with Figure 13.5 shows that the y-axis in the 2-D SVP images corresponds to the azimuthal axis in the 3-D data volumes. Therefore, instead of dealing with full 3-D datasets, we axially project the input 3-D sequences and create corresponding SVP images (Figure 13.5). This will reduce the task of registering K sets each with B images (B-Scans) of size $[B \times L]$ pixels, to registering only K images (SVPs) each of size $[B \times B]$ pixels. In essence, we are projecting down into the SCP domain to create a collection of K images that are undersampled in only one dimension (e.g., $D_1 = 1$).

Aside from a significant reduction in data volume, the axial projection reduces the noise in the SVP images by averaging over hundreds of pixels. As SNR of the SVP images is relatively higher, outlier (motion artifact corrupted) image sets can be more accurately detected and excluded from the data pool.

We recover the order of the frames in the dense 3-D output by registering the remaining SVPs. As explained in the next subsection, we calculate the y-axis motion between different SVPs and associate this to the azimuthal motion parameters (frame number) of the 3-D volumes. For example, an estimated five pixel displacement for the SVPs of two scan sets in y-axis, indicates an offset of five frames in the corresponding B-Scan sequences. Moreover, by estimating the x-axis motion of the SVPs, we recover the lateral registration parameters needed for aligning the fused (interlaced) B-Scans in the final fused 3-D volume.

13.3.2.2 Multiframe Joint Registration

In many super-resolution applications, fast pairwise image registration is sufficient for estimating the relative shifts between the sets of low-resolution images. The basis for this approach is based on the following approximation of Equation (13.5)

$$y_k = DS(v_k)z \approx S(v_k')p + e_k + a_k \qquad (13.16)$$

where $v_k' = [v_{1,k}v_{2,k}/D_2]$ is the apparent motion in the undersampled image, p is the approximate nonaliased portion of the low-resolution image, and a_k is the aliasing artifacts that we approximately treated as noise. From the simplified model of Equation (13.16), we see that the relationship between a pair of low-resolution images y_k and y_j is approximately given by

$$y_k = S(v_j' - v_k')y_j \approx y_j + (v_j' - v_k')\nabla S(0)y_j \qquad (13.17)$$

where the second half of the equation is based on the first order Taylor approximation of the shift operator $S(v)$. In practice, the operators $\nabla S(0) = [S_x(0)S_y(0)]$ are approximately the x and y gradient operators. Equation (13.17) is known as the optical flow constraint and can be used to estimate the shift between any pair of low-resolution frames. Such an approach works as long as the energy in the aliasing artifacts a_k are minimal. We use the notation $v_{j-k}' = v_j' - v_k'$.

Due to the sub-Nyquist sampling in the azimuthal direction, the SVPs of the sparse, fast-acquired sequences are aliased in the y-axis, complicating the subpixel motion estimation task. Moreover, small estimation bias in the pairwise SVP registration is magnified to a significant misalignment error when several sequences are fused together. Therefore, to minimize the overall motion estimation error, we exploit global consistency conditions in a multiframe motion estimation framework [14, 19]. This bundle-adjusted, optical flow-based technique relies on the fact that the operator describing the motion between any pair of frames must be the composition of the operators between two other pairs of frames. In effect, by incorporating this prior information in the joint motion estimation framework of [14], we minimize the motion estimation bias while having extremely fast registration by estimating motion entirely in the low-resolution domain.

We overcome the errors associated using a global constraint enforcing the transitivity of the pairwise motion estimates. For example, if we consider three frames, then the transitivity of the motion estimates requires that

$$v_{k-j}' = v_{l-j}' + v_{k-l}' \qquad (13.18)$$

Figure 13.8 schematically describes this motion constraint. In the case of multiple translational motion vectors [14], the above conditions can be described by simple linear equations relating the motion vectors between the frames as

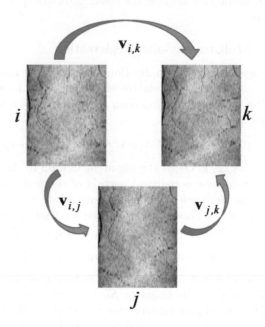

FIGURE 13.8: Global motion consistency conditions that exist for any set of images: the operator describing the motion between any pair of images is the composition of the operators between two other pairs of images: $v_{i,k} = v_{i,j} + v_{j,k}$.

$$U V = 0, \tag{13.19}$$

where $\mathbf{U}$ is a $\left[2(K{-}1)^2 \times 2K(K{-}1)\right]$ consistency condition matrix and U is a vector collecting the set of unknown motion vectors $\{v'_k\}$. Each row in the sparse matrix U has only two or three non-zero (± 1) elements. Motion vectors are estimated by minimizing a global cost function such as

$$\Omega_{\mathrm{OF}}(V) = \sum_{\substack{i \neq j}}^{\mathrm{K}} \left\| y_i - y_j - v'_{i-j} \nabla S(0) y_j \right\|_1, s.t. \ U V = 0, \tag{13.20}$$

We used nonlinear programming (*"fmincon"* function in MATLAB^{TM}) to minimize this cost function. The above conditions for the more general case of affine motion are defined in [14]. This is a simpler but faster implementation of the general framework described in Section 13.2. Since we do not estimate the high-resolution image jointly with the registration parameters our solution is suboptimal. Indeed, mathematically more rigorous solutions for registering aliased images are also possible [42], which increase computational complexity of the proposed algorithm. However, noting the extremely large SDOCT image sets (hundreds of images of size [512 × 1024] or larger), the goal of our proposed solution is to be practical for clinical implementation rather than mathematically optimal.

13.3.2.3 Experimental Results

The above registration technique recovers the order and the relative azimuthal distance of B-Scans from different scan sets, which can be exploited to reconstruct a dense 3-D view of the imaged pathological structure. Since misaligned or broken vessels are easily detectable in retinal imaging applications, the vessel pattern as seen on the SVP serves as an efficient qualitative measure of the success and accuracy of the overall algorithm. Therefore, we use the estimated motion parameters to reconstruct a fused (super-resolved) 2-D SVP map of retinal vessel structure.

Figure 13.6 shows a dense scanning of a subject, whose motion artifacts have resulted in an SVP with broken vessel structure. From the same subject, we captured 12 sparsely sampled volumetric scans (each with 50 B-Scans) and adjusted the baseline of each image using the fast registration StackReg plug-in (Biomedical Imaging Group; Swiss Federal Institute of Technology Lausanne) [54] for ImageJ (freeware; National Institutes of Health; Bethesda, MD). Following [28], by summing the lower half of the B-Scans in the axial direction, we created SVPs with distinct vessel patterns. After contrast adjustment, four of the six sequences with the highest SVP normalized cross-correlation values were manually selected to be registered. Figures 13.9a and 13.9b show two corresponding SVPs of these four sequences. Registered and sequentially ordered AVI movies of these four input sequences are available in *http://www.duke.edu/~sf59/datasets/SDOCT_SR.avi*, screenshots of which are shown in Figure 13.9f.

We used the multiframe projective bundle-adjusted motion estimation method of [14] to recover the subpixel translational motion parameters of these four SVPs (Section 13.3.2.2). We used the fast zero-order classic kernel regression-based super-resolution algorithm described in [52] to reconstruct the fused SVP. Since there are no aliasing artifacts in the x-axis (lateral direction), the SVP resolution is only enhanced in the y-axis. The SVP of the fused sequence is shown in Figure 13.9c, which has more details than any of the input SVPs.

As a point of reference, we also captured a gold-standard sequence shown in Figure 13.9d, which is the visually best of 4 densely sampled volumetric scans (100 B-Scans). The reconstruction accuracy and quality improvement is confirmed by comparing the input and output SVPs to the gold-standard.

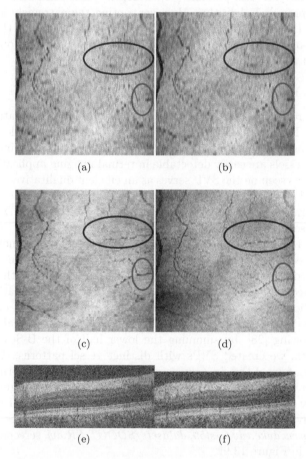

(a)   (b)

(c)   (d)

(e)   (f)

FIGURE 13.9: (a) and (b) are two representative SVPs of four input retinal SDOCT sequences (50 regularly sampled B-Scans each). (c) is the SVP of the fused sequence (200 irregularly sampled B-Scans). (d) is the Gold-Standard SVP which is the best dense (100 regularly sampled B-Scans) out of 4 such sequences (Figure 13.6b is an example of dense sampling of the same subject with motion artifacts). (e) and (f) are the screen shots of the AVI movies of registered four input B-Scan sets and the reordered and interlaced output B-Scan set, respectively (2.1 MB).

We believe the small jaggedness in reconstructed vessels of Figure 13.9c is mainly due to the dynamic structural deformation of the vessels during the cardiac cycle. Overall, the vessel pattern in Figure 13.9c shows fewer discontinuities (blue ellipsoids) compared to the input SVPs. Moreover, due to the less aggressive interpolation in the azimuthal (y-axis) direction, the vessel thicknesses are more accurate in Figure 13.9c than in any of the input frames (red ellipsoids).

13.4   Conclusion

We have provided a proof of concept for the applicability of image processing-based algorithms as an alternative to expensive hardware for creating robust high-quality X-ray mammography and SDOCT ophthalmic imaging. The proposed super-resolution-based algorithm enables ophthalmic imaging practitioners and radiologists to optimally utilize the SDOCT and X-ray systems in their highest resolution capacity.

For the SDOCT case, several implementation variations for improving the efficiency are possible. For example, rather than discarding a whole defected sequence, we may discard only those B-Scans affected by abrupt motion artifacts (e.g., blinking), and use the remaining uncorrupted B-Scans. To produce more visually appealing SVPs, more efficient super-resolution techniques such as the steering adaptive kernel [52] or robust super-resolution [16] may be also exploited. Moreover, incorporation of an advanced adaptive sparse sampling strategy (3-D extension of the method in [13]) in this imaging framework is part of our ongoing research.

While the proposed algorithm efficiently removes abrupt motion artifacts, a practical drawback is the case of imaging objects with constant deformable motion. For example, in the case of imaging pulsing blood vessels, each sparse sequence is associated with a unique position of the blood vessels compared to the background tissues. A possible remedy is synchronizing the start of image acquisition in each sparse sequence with the electrocardiogram (EKG) signal.

We note that two alternative related sparse imaging scenarios can be also considered. One is based on capturing large field of view repeated scans, dense in the azimuthal direction but sparsely sampled in the lateral direction. Then, a classic super-resolution algorithm (e.g., [16, 52]) may reconstruct the lateral resolution of individual B-Scans. Our pilot experimental results have shown moderate improvements, when imaging objects under a SDOCT microscope. However, in practical clinical trials the difficulty of capturing repeated scans from a unique azimuthal location largely voids the applicability of this strategy.

Alternatively, the authors of the published work in [66] propose to capture azimuthally dense scan sets with a small field of view in the axial-lateral plane (e.g., the *en face* view is divided into four subsections, which are imaged subsequently). A customized semi-automatic software stitches the 3-D scan volumes, creating a visually appealing, large field of view, 3-D rendition of the retina. However, unfortunately for the same practical imaging problem noted for the aforementioned strategy, as evident in the experimental results in [66], it is extremely hard (if not impossible) to recover unique, large field of view B-Scans without evident registration artifacts.

As for the mammography, we believe the design of future X-ray imaging systems would benefit from a systematic analysis of the resolution and SNR required for mammographic screening and diagnosis. In the future, we will explore the fundamental tradeoffs between radiation exposure, number of frames, and reconstruction performance. Furthermore, we will investigate more sophisticated redundant wavelet techniques such as curvelets [50] or ridgelets [7], which might show an even better performance than the proposed multiframe ForWaRD technique. In fact, recent research has shown that use of more sophisticated wavelets can improve the image quality in other medical imaging applications [31].

We believe this novel application of super-resolution can be used as a stepping stone toward many other image fusion-based medical imaging system designs, aimed especially at patients with uncontrollable motion, pediatrics, or handheld probe imaging. While this chapter was focused on X-ray and OCT image enhancement, similar strategies can be exploited for enhancing the quality of some other volumetric medical imaging devices such as ultrasound.

13.5   Acknowledgments

Our MATLAB^{TM} software implementation of the noted X-ray algorithms is in part based on the ForWaRD software, developed by Dr. Ramesh D. Neelamani of the Digital Signal Processing group at Rice University (available at *http://www.dsp.rice.edu/software/ward.shtml*). We would like to thank the Editor Prof. Peyman Milanfar for collaborating with us in the original multiframe X-ray motion estimation publications. We thank Bradley A. Bower and Yuankai K. Tao for providing invaluable SDOCT data and assistance. This work was supported in part by North Carolina Biotechnology Center Collaborative Funding Grant #2007-CFG-8005 with Bioptigen, the Duke Translational Medicine Institute Subcontract #12 of NIH Grant #5ULT-RR024128-03, Knights Templar Eye Foundation Inc. Pediatric Ophthalmology Research Grant, Hartwell Foundation, Individual Biomedical Research Award.

Bibliography

[1] Y. Altunbasak, A. Patti, and R. Mersereau. Super-resolution still and video reconstruction from MPEG-coded video. *IEEE Trans. Circuits and Syst. Video Technol.*, 12(4):217–226, Apr. 2002.

[2] L.D. Alvarez, J. Mateos, R. Molina, and A.K. Katsaggelos. High resolution images from compressed low resolution video: Motion estimation and observable pixels. *International Journal of Imaging Systems and Technology*, 14(2):58–66, October 2004.

[3] S. Borman and R.L. Stevenson. Super-resolution from image sequences - a review. In *Proc. of the 1998 Midwest Symposium on Circuits and Systems*, volume 5, Apr. 1998.

[4] S.G. Chang, B. Yu, and M. Vetterli. Adaptive wavelet thresholding for image denoising and compression. *IEEE Transactions on Image Processing*, 9(9):1532–1546, 2000.

[5] S.H. Chavala, S. Farsiu, R. Maldonado, D.K. Wallace, S.F. Freedman, and C.A. Toth. Insights into advanced retinopathy of prematurity using handheld spectral domain optical coherence tomography imaging. *in press, Ophthalmology*, 2009.

[6] G.T. Chong, S. Farsiu, S.F. Freedman, N. Sarin, A.F. Koreishi, J.A. Izatt, and C.A. Toth. Abnormal foveal morphology in ocular albinism imaged with spectral-domain optical coherence tomography. *Archives of Ophthalmology*, 127(1):37–44, 2009.

[7] M. Do and M. Vetterli. The finite ridgelet transform for image representation. *IEEE Trans. Image Process.*, 12(1):16–28, January 2003.

[8] C. Dunsby and P.M.W. French. Techniques for depth-resolved imaging through turbid media including coherence-gated imaging. *Journal of Physics D: Applied Physics*, 36:207–227, July 2003.

[9] N.A. El-Yamany and P.E. Papamichalis. Robust color image superresolution: An adaptive m-estimation framework. *EURASIP Journal on Image and Video Processing*, 8:1–12, 2008.

[10] M. Elad and A. Feuer. Restoration of a single superresolution image from several blurred, noisy, and undersampled measured images. *IEEE Transactions on Image Processing*, 6(12):1646–1658, Dec 1997.

[11] M. Elad and Y. Hel-Or. A fast super-resolution reconstruction algorithm for pure translational motion and common space invariant blur. *IEEE Transactions on Image Processing*, 10(8):1186–1193, August 2001.

[12] S. Farsiu, B.A. Bower, J.A. Izatt, and C.A. Toth. Image fusion based resolution enhancement of retinal spectral domain optical coherence tomography images. *Invest. Ophthalmol. Vis. Sci.*, 49(5):E–abstract–1845, 2008.

[13] S. Farsiu, J. Christofferson, B. Eriksson, P. Milanfar, B. Friedlander, A. Shakouri, and R. Nowak. Statistical detection and imaging of objects hidden in turbid media using ballistic photons. *Applied Optics*, 46(23):5805–5822, 2007.

[14] S. Farsiu, M. Elad, and P. Milanfar. Constrained, globally optimal, multiframe motion estimation. In *Proc. of the 2005 IEEE Workshop on Statistical Signal Processing*, pages 1396–1401, July 2005.

[15] S. Farsiu, D. Robinson, M. Elad, and P. Milanfar. Advances and challenges in super-resolution. *International Journal of Imaging Systems and Technology*, 14:47–57, 2004.

[16] S. Farsiu, M.D. Robinson, M. Elad, and P. Milanfar. Fast and robust multiframe super resolution. *IEEE Transactions on Image Processing*, 13(10):1327–1344, October 2004.

[17] A.F. Fercher, W. Drexler, C.K. Hitzenberger, and T. Lasser. Optical coherence tomography - principles and applications. *Reports on Progress in Physics*, 66(2):239–303, 2003.

[18] G. Golub and V. Pereyra. Separable nonlinear least squares: the variable projection method and its applications. *Institute of Physics, Inverse Problems*, 19:R1–R26, 2002.

[19] V.M. Govindu. Lie-algebraic averaging for globally consistent motion estimation. In *Proc. of the Int. Conf. on Computer Vision and Pattern Recognition (CVPR)*, volume 1, pages 684–691, July 2004.

[20] H. Greenspan. Super-resolution in medical imaging. *The Computer Journal*, 52:43–63, 2009.

[21] H. Greenspan, G. Oz, N. Kiryati, and S. Peled. MRI inter-slice reconstruction using super-resolution. *Magnetic Resonance Imaging*, 20(5):437–446, 2002.

[22] D. Hammer, R.D. Ferguson, N. Iftimia, T. Ustun, G. Wollstein, H. Ishikawa, M. Gabriele, W. Dilworth, L. Kagemann, and J. Schuman. Advanced scanning methods with tracking optical coherence tomography. *Optics Express*, 13(20):7937–7947, 2005.

[23] R. Hardie. A fast image super-resolution algorithm using an adaptive wiener filter. *IEEE Transactions on Image Processing*, 16(12):2953–2964, December 2007.

[24] R. Hardie, K. Barnard, and E. Armstrong. Joint MAP registration and high-resolution image estimation using a sequence of undersampled images. *IEEE Transactions on Image Processing*, 6(12):1621–1633, 1997.

[25] M.R. Hee, J.A. Izatt, E.A. Swanson, D. Huang, J.S. Schuman, C.P. Lin, C.A. Puliafito, and J.G. Fujimoto. Optical coherence tomography of the human retina. *Archives of Ophthalmology*, 113(3):325–332, 1995.

[26] D.B. Hess, S.G. Asrani, M.G. Bhide, L.B. Enyedi, S.S. Stinnett, and S.F. Freedman. Macular and retinal nerve fiber layer analysis of normal and glaucomatous eyes in children using optical coherence tomography. *American Journal of Ophthalmology*, 139(3):509–517, 2005.

[27] D. Huang, E.A. Swanson, C.P. Lin, J.S. Schuman, W.G. Stinson, W. Chang, M.R. Hee, T. Flotte, K. Gregory, C.A. Puliafito, et al. Optical coherence tomography. *Science*, 254(5035):1178–1181, 1991.

[28] S. Jiao, R. Knighton, X. Huang, G. Gregori, and C. Puliafito. Simultaneous acquisition of sectional and fundus ophthalmic images with spectral-domain optical coherence tomography. *Optics Express*, 13(2):444–452, 2005.

[29] J.A. Kennedy, O. Israel, A. Frenkel, R. Bar-Shalom, and H. Azhari. Super-resolution in PET imaging. *IEEE Transactions on Medical Imaging*, 25(2):137–147, 2006.

[30] J.A. Kennedy, O. Israel, A. Frenkel, R. Bar-Shalom, and H. Azhari. Improved image fusion in PET/CT using hybrid image reconstruction and super-resolution. *Int. J. Biomed. Imaging*, 46846, 2007.

[31] A. Khare and U.S. Tiwary. A new method for deblurring and denoising of medical images using complex wavelet transform. *Proc. IEEE Conference Engineering in Medicine and Biology*, pages 1897–1900, Sept. 2005.

[32] S. Lertrattanapanich and N.K. Bose. High resolution image formation from low resolution frames using Delaunay triangulation. *IEEE Trans. Image Processing*, 11(12):1427–1441, Dec. 2002.

[33] N. Nassif, B. Cense, B.H. Park, S.H. Yun, T.C. Chen, B.E. Bouma, G.J. Tearney, and J.F. de Boer. In vivo human retinal imaging by ultrahigh-speed spectral domain optical coherence tomography. *Optics Letters*, 29(5):480–482, 2004.

[34] R. Neelamani, H. Choi, and R. Baraniuk. Forward: Fourier-wavelet regularized deconvolution for ill-conditioned systems. *IEEE Transactions on Image Processing*, 52(2):418–433, February 2004.

[35] S.C. Park, M.K. Park, and M.G. Kang. Super-resolution image reconstruction: A technical overview. *Signal Processing Magazine*, 20(3):21–36, 2003.

[36] V. Patanavijit and S. Jitapunkul. A Lorentzian stochastic estimation for a robust iterative multiframe super-resolution reconstruction with Lorentzian-Tikhonov regularization. *EURASIP Journal on Image and Video Processing*, 2007(2), 2007.

[37] R.R. Peeters, P. Kornprobst, M. Nikolova, S. Sunaert, T. Vieville, G. Malandain, R. Deriche, O. Faugeras, M. Ng, and P. Van Hecke. The use of super-resolution techniques to reduce slice thickness in functional MRI. *International Journal of Imaging Systems and Technology*, 14(3):131–138, 2004.

[38] S. Peled and Y. Yeshurun. Superresolution in MRI: Application to human white matter fiber tract visualization by diffusion tensor imaging. *Magnetic Resonance in Medicine*, 45(1):29–35, 2001.

[39] L.C. Pickup, D.P. Capel, S.J. Roberts, and A. Zisserman. Bayesian methods for image super-resolution. *The Computer Journal*, (1):101–113, 2009.

[40] C.G. Pieroni, A.J. Witkin, T.H. Ko, J.G. Fujimoto, A. Chan, J.S. Schuman, H. Ishikawa, E. Reichel, and J.S. Duker. Ultrahigh resolution optical coherence tomography in non-exudative age related macular degeneration. *British Medical Journal*, 90(2):191–197, 2006.

[41] D. Robinson, S. Farsiu, J.Y. Lo, P. Milanfar, and C.A. Toth. Efficient multiframe registration of aliased x-ray images. *Proceedings of the 41th Asilomar Conference on Signals, Systems, and Computers*, pages 215–219, November 2007.

[42] D. Robinson, S. Farsiu, and P. Milanfar. Optimal registration of aliased images using variable projection with applications to super-resolution. *The Computer Journal*, 52(1):31–42, January 2009.

[43] D. Robinson and P. Milanfar. Statistical performance analysis of super-resolution. *IEEE Transactions on Image Processing*, 15(6):1413–1428, June 2006.

[44] M.D. Robinson, C.A. Toth, J.Y. Lo, and S. Farsiu. Efficient fourier-wavelet super-resolution with applications in low-dosage digital x-ray imaging. *Submitted to IEEE Transactions on Image Processing*, January 2009.

[45] R.R. Schultz and R.L. Stevenson. Extraction of high-resolution frames from video sequences. *IEEE Transactions on Image Processing*, 5(6):996–1011, June 1996.

[46] S.G. Schuman, A.F. Koreishi, S. Farsiu, S. Jung, J.A. Izatt, and C.A. Toth. Photoreceptor layer thinning over drusen in eyes with age-related macular degeneration imaged in vivo with spectral-domain optical coherence tomography. *Ophthalmology*, 116(3):488–496, 2009.

[47] A.W. Scott, S. Farsiu, L.B. Enyedi, D.K. Wallace, and C.A. Toth. Imaging the infant retina with a hand-held spectral-domain optical coherence tomography device. *American Journal of Ophthalmology*, 147(2):364–373, 2009.

[48] E. Shechtman, Y. Caspi, and M. Irani. Space-time super-resolution. *IEEE Transactions on Pattern Analysis and Machine Intelligence*, 27(4):531–545, 2005.

[49] V.J. Srinivasan, M. Wojtkowski, A.J. Witkin, J.S. Duker, T.H. Ko, M. Carvalho, J.S. Schuman, A. Kowalczyk, and J.G. Fujimoto. High-definition and 3-dimensional imaging of macular pathologies with high-speed ultrahigh-resolution optical coherence tomography. *Ophthalmology*, 113(11):2054–2054, 2006.

[50] J. Starck, E. Candes, and D. Donoho. The curvelet transform for image denoising. *IEEE Trans. Image Process.*, 11(6):670–684, June 2002.

[51] M. Stopa, B.A. Bower, E. Davies, J.A. Izatt, and C.A. Toth. Correlation of pathologic features in spectral domain optical coherence tomography with conventional retinal studies. *Retina*, 28(2):298–308, 2008.

[52] H. Takeda, S. Farsiu, and P. Milanfar. Kernel regression for image processing and reconstruction. *IEEE Trans. Image Process.*, 16(2):349–366, Feb. 2007.

[53] H. Takeda, S. Farsiu, and P. Milanfar. Deblurring using regularized locally adaptive kernel regression. *IEEE Transactions on Image Processing*, 17(4):550–563, April 2008.

[54] P. Thevenaz, U.E. Ruttimann, and M. Unser. A pyramid approach to subpixel registration based on intensity. *IEEE Transactions on Image Processing*, 7(1):27–41, 1998.

[55] C.A. Toth, R. Birngruber, S.A. Boppart, M.R. Hee, J.G. Fujimoto, C.D. DiCarlo, E.A. Swanson, C.P. Cain, D.G. Narayan, G.D. Noojin, et al. Argon laser retinal lesions evaluated in vivo by optical coherence tomography. *American Journal of Ophthalmology*, 123(2):188–198, 1997.

[56] C.A. Toth, S. Farsiu, A.A. Khanifar, and G.T. Chong. Optical coherence tomography in age-related macular degeneration. In Gabriel Coscas, editor, *Application of Spectral Domain OCT in AMD*, pages 15–34. Springer Medizin Verlag Heidelberg, 2009.

[57] M. Ushizaki, T. Okatani, and K. Deguchi. Video synchronization based on co-occurrence of appearance changes in video sequences. *International Conference on Pattern Recognition*, 3:71–74, 2006.

[58] P. Vandewalle, L. Sbaiz, J. Vandewalle, and M. Vetterli. Super-resolution from unregistered and totally aliased signals using subspace methods. *IEEE Transactions on Signal Processing*, 55(7):3687–3703, July 2007.

[59] P. Vandewalle, S. Susstrunk, and M. Vetterli. A frequency domain approach to registration of aliased images with application to super-resolution. *EURASIP Journal on Applied Signal Processing*, page Article ID 71459, 2006.

[60] M. Wojtkowski, R. Leitgeb, A. Kowalczyk, T. Bajraszewski, and A.F. Fercher. In vivo human retinal imaging by fourier domain optical coherence tomography. *Journal of Biomedical Optics*, 7(3):457–463, 2002.

[61] M. Wojtkowski, V. Srinivasan, T. Ko, J. Fujimoto, A. Kowalczyk, and J. Duker. Ultrahigh-resolution, high-speed, fourier domain optical coherence tomography and methods for dispersion compensation. *Optics Express*, 12(11):2404–2422, 2004.

[62] N. Woods, N. Galatsanos, and A. Katsaggelos. Stochastic methods for joint registration, restoration, and interpolation of multiple undersampled images. *IEEE Transactions on Image Processing*, 15(1):201–213, January 2006.

[63] Y. Yasuno, J. Sugisaka, Y. Sando, Y. Nakamura, S. Makita, M. Itoh, and T. Yatagai. Non-iterative numerical method for laterally superresolving Fourier domain optical coherence tomography. *Optics Express*, 14(3):1006–1020, 2006.

[64] S. Yun, G. Tearney, J. de Boer, N. Iftimia, and B. Bouma. High-speed optical frequency-domain imaging. *Optics Express*, 11(22):2953–2963, 2003.

[65] R. Zawadzki, S. Jones, S. Olivier, M. Zhao, B. Bower, J. Izatt, S. Choi, S. Laut, and J. Werner. Adaptive-optics optical coherence tomography for high-resolution and high-speed 3d retinal in vivo imaging. *Optics Express*, 13(21):8532–8546, 2005.

[66] R.J. Zawadzki, A.R. Fuller, S.S. Choi, D.F. Wiley, B.Hamann, and J.S. Werner. Improved representation of retinal data acquired with volumetric Fd-OCT: co-registration, visualization, and reconstruction of a large field of view. volume 6844, page 68440C. SPIE Ophthalmic Technologies XVIII, 2008.

14

Practicing Super-Resolution: What Have We Learned?

Nikola Bozinovic

MotionDSP Inc.

CONTENTS

14.1 Abstract

MotionDSP makes software for super-resolution (and, more generally, video enhancement) that is used in various professional and consumer applications. This chapter discusses some of the design decisions made while developing and implementing super-resolution algorithms for the real world. It also details challenges, such as robustness, automation, speed (i.e., computational complexity), and specific limitations of a video imaging pipeline (acquisition speed, noise, and compression). Based on our experience, solutions to these challenges should be incorporated early into the design of super-resolution algorithms and products.

14.2 Introduction

The past several years have clearly marked the advent of a true video age. More video is shot and published than ever before,[1] and the growing trend of using digital video for personal and professional communication seems to be accelerating. As more people are becoming involved with video, expectations of its quality are (often unrealistically) high. Casual video creators strive for a professional look in their videos; forensic video analysts hope to recreate the success of their CSI[2] counterparts. In short, everybody is looking for improvements. This chapter aims to help active SR researchers better understand the potential markets (and roadblocks) for their work and to identify applications where super-resolution can have an immediate impact.

This demand for improved video quality is partially appeased by the steady innovation around the imaging technology (such as better sensors, improved on-board image processing, and more efficient compression). But there is another, often underutilized way of improving video, and that is postprocessing. While the desired postprocessing outcome depends on the application,[3] there is a growing consensus that the objective quality improvement is possible. Postprocessing techniques most commonly used to that end generally include some variant of multiframe processing, such as super-resolution (SR), temporal denoising, or stabilization.

There are several strong arguments supporting increased interest in video post-processing (i.e., "smart" decoding). Firstly, the real-time nature of video capture sets an upper limit on the amount of processing that can be done

---

[1] At least 20 hours of video are being uploaded to YouTube every minute [37].

[2] Crime Scene Investigation, a popular TV show.

[3] For example, consumers want their videos to "look nicer" subjectively; a forensic user seeks additional information in the video.

on-board a capture device. There is not such a hard limit for "off-line" post-processing that, combined with the computing power of modern computers, makes post-processing more appealing. Secondly, many existing video cameras reside on multifeature devices (e.g., cell phones), which further limits the resources dedicated to the video module, such as processing power or battery. Finally, miniaturization of cameras has led to significant physical imaging problems related to small aperture and focal length, with an impact on output resolution and overall video quality.

14.2.1 Video Quality Trends

Let us expand our point on video quality. Contrary to popular belief, the recording quality of an average video-capable device (across all applications) has not been improving in recent years – as a matter of fact, it has been declining. This is closely following the trend in still photography of an increasing share of lower-quality cameras, such as cameraphones. For example, in June 2009, Apple's iPhone became the #1 camera used on Flickr.com,[4] surpassing DSLR (Digital Single-Lens Reflex) cameras, which had dominated the upload charts for years (Fig. 14.1(a)). Though the digital video trends typically lag several years behind trends of digital photography usage, a similar growth path is already clear. Video-enabled phones are becoming increasingly popular, as are high-speed data connections. For example, in just one week after the launch of Apple's iPhone 3GS (the first iPhone model with the video recording capability), the number of mobile uploads to YouTube, the world's largest video site, increased by a staggering 400% [20]. Even without the iPhone, YouTube has been seeing major growth across the entire mobile space – mobile uploads have gone up 1,700% just in the first six months of 2009.

This explosion of relatively low-quality mobile content has a profound impact outside consumer electronics applications and the ways people record and watch video. We have witnessed firsthand an exponential increase in the number of mobile phone videos used as evidence in criminal investigations. An increase in low-quality video is also present in the video surveillance market; the proliferation of inexpensive IP cameras and limited bandwidth and storage [7] seriously impact effectiveness [3] of many current surveillance systems.

14.2.2 The Need for Postprocessing

It is therefore obvious that a great need exists for a postprocessing solution that would alleviate the problem of poor video quality – and the field of computational super-resolution promises to do just that. Indeed, many hundreds

---

[4]Flickr.com is one of the largest photo sharing sites in the world. Updated camera statistics for the Flickr site can be found at http://www.flickr.com/cameras/.

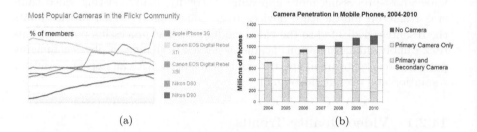

(a)   (b)

FIGURE 14.1: (a) Most popular cameras in the Flickr community for 2009 (for updated figures, please visit http://www.flickr.com/cameras/). The graphs show the number of Flickr members who have uploaded at least one photo or video with a particular camera on a given day over the last year. (b) Camera penetration in mobile phones, 2004 – 2010 [19]. It is expected that more than 1 billion cameraphones will ship in 2010, with 60% of them having video recording capabilities.

TABLE 14.1

Potential Roadblocks for Applying Current Super-Resolution Research on Existing Video Processing Problems.

	Research	**Industry**
Typical data	Clean video	Compressed video
	Synthetic sequences	Dynamic scenes
	No motion blurring	Motion blurred
	High framerates	Low framerates
	Synthetic noise	Shot noise
User interaction (forensic apps)	No research	Desired
Motion model	Simple (e.g., global)	Complex
Code speed	100 – 10,000 × slower than real-time	Real-time or near real-time

of papers on super-resolution have been published since the seminal paper of Tsai and Huang [41] in 1984, and the interest in the field seems to be picking up significantly in recent years. But, for all its promise, super-resolution failed early on to make significant inroads in the real world. At the time of writing (late 2009), there are only a handful of implementations of SR technology outside research labs. Some of the potential roadblocks for direct application of existing SR research are presented in Table 14.1 and discussed below.

14.2.3   Why is Super-Resolution Not Used More?

The first explanation for limited real-world super-resolution presence is the computational complexity that comes with it; many SR algorithms are at least an order of magnitude more computationally involved than even the most sophisticated video coders (e.g., H.264/AVC).[5] The gap between compute capability available to the average user and that required in most SR algorithms has started to close just recently, with advances in general purpose parallel hardware designs and access to better development tools. To improve their chances of being adopted in practice, new SR algorithms have to run fast. This often means that they need to be implemented using parallel computing, which should be incorporated early into algorithm designs.

There is another, often overlooked, reason for the limited deployment of super-resolution, and that is a mismatch in datasets and use cases between academic and industry applications, and the responsibility for it lies with the research community. Indeed, the majority of published SR papers deal predominantly with synthetic data, or, at best, with video recorded under carefully controlled shooting scenarios (e.g., with both camera and the scene fully controlled), but there is very little published work on the effects of video compression on super-resolution performance. Motion blur is another significant problem; while some work has been done to account for it [2] [4], its implementation is computationally prohibitive. Considering the number of cameras with the motion blurring problem, even a simple method for effective detection of blurred frames would have a significant immediate impact on SR applications. There are many more important topics that have received little or no coverage in the context of SR (e.g., modeling sensor noise, segmenting compressed and aliased video, or estimating motion for large interframe displacements).

---

[5]Mostly due to their multi-frame nature and the need for very accurate motion estimation.

14.2.4   Automation versus User Interaction

In many forensic video applications, interaction between the user (operator) and the software is not only possible but also desirable. Some very difficult video processing and computer vision problems become much easier when human assistance is available. A better understanding of a video forensic work flow would go a long way towards optimizing the time spent on different research subtasks and in focusing on those that promise to help the most. This is not to say that fully automated solutions are not relevant; for example, the enhancement of consumer video does require significant automation.[6]

14.2.5   Modeling Motion for Super-Resolution

Another important problem is the lack of focus on high-quality motion estimation. This comes as a bit of a surprise; as a matter of fact, the focus on motion has allowed us to build a robust super-resolution technology at MotionDSP. Some SR algorithms simply skip the motion estimation problem, assuming that it is known, and start from the frame fusion process; others simplify motion models to the point that they can no longer describe dynamics of real videos. Finally, many SR algorithms ignore motion altogether and assume multiple low-resolution images (LRIs) of a continuous scene are available and taken at the same time. While this assumption sets a convenient background for theoretical analysis (as it removes the relative scene displacement), it does not match reality. In practice, LRIs are always sampled at different time instances (as video frames), which leads not only to relative motion between LRIs (due to camera shift), but also to the change in the scene itself, in the form of object motion, occlusion, or illumination change.

In some cases, the proposed solution is in conflict with the research motivation. For example, the face hallucination SR methods [1] were developed with a declared goal of improving face recognition, both for human observers and for automatic face recognition systems. This is well aligned with the interests of the forensic video community, which most of the time looks for faces and persons of interest in the recorded footage. However, there is a serious roadblock with this approach: as hallucination methods admittedly do not add any new information [1], it would be hard for their results to be accepted as valid evidence (especially in courts). Furthermore, these results are dependent on the input training set and can vary significantly when different training sets are used. In addition, required normalization (i.e., pose detection and image scaling) limits algorithm robustness and presents another problem the research rarely addresses.

---

[6]However, as consumer videos are often watched in real-time, smaller resolution improvement is usually acceptable.

14.2.6   Performance Issues

Let us spend a moment to expand on the issue of speed. We have already discussed computational complexity of super-resolution algorithms in general. While it is not realistic to expect significant code optimization in the implementation of research algorithms, even a brief analysis can help assess their potential for practical (industry-strength) implementation. For example, it is safe to assume that at least a tenfold increase in speed from the research-level code can be done by using code optimizations and faster hardware. But if the algorithm is slower than required by 3 or 4 orders of magnitude (e.g., it takes minutes to output a video frame for the application that aims at real-time performance), its chances of contributing to the solution of any real problem are seriously hampered. What is encouraging is that many leading companies are making a focused effort [24] to evangelize new development platforms among research institutions. This has led to improved awareness of the importance of fast algorithm implementation.

14.2.7   Relationship to Existing Standards

Finally, for any new video technology to have its chance of being adopted it usually has to follow important industry canons and, preferably, piggyback on existing standards or infrastructure. As a case study, a new paradigm of distributing the load (and complexity) between the encoding and decoding side of video channel was introduced in the early 2000s, with several independent efforts [15] [28] jump-starting significant interest in distributed video coding. Such a scheme (or, more generally, the model of shifting the computational load to the decoding end of the video communication channel that we strongly support) makes even more sense these days with the ascent of high-power personal computers or super-computers [23] [36]. However, similar to some earlier initiatives aimed at re-designing video encoding standards from scratch,[7] there was no real impact on practical applications. It is safe to claim that for any new video coding technology to be successful it will have to be aligned with the large investment into infrastructure based on existing video-coding standards and transport mechanisms, like H.264/AVC and MPEG-TS.

In the following part of this chapter, we do not attempt to provide ready-made recipes on how to do super-resolution.[8] Instead, we offer our observations of the needs for super-resolution in practice, which are based on our interaction with dozens of companies and thousands of individual users. We hope that

---

[7]Technically sound but commercially unsuccessful JPEG2000 and Wavelet-based Scalable Video Coding come to mind.

[8]Not only because we do not have all the answers (our algorithms are constantly being tweaked and reworked), but also because publishing all details of our algorithm would be detrimental to our business.

our experience will benefit both academic researchers and those working on commercialization of super-resolution technologies.

14.3   MotionDSP: History and Concepts

MotionDSP was founded in 2005, with its roots in super-resolution research done at the University of California at Santa Cruz in the 2000s.[9] MotionDSP started by licensing super-resolution patents from UCSC, but has since then built a strong internal SR portfolio since 2006. The joint motion estimation (partially based on [5]) and segmentation as well as a novel frame fusion technology form its core intellectual property. Following the academic roots, all of our SR algorithms fall under a reconstruction-based category, as they follow the classic framework for image formation and solving the inverse problem [12], [13], [14]. As our understanding of practical super-resolution problems and users' needs have evolved over the years, most of the recent research has been driven by a direct demand from the users.[10] Working directly with our users, we have witnessed firsthand many real problems requiring immediate attention, which has led us to several design decisions presented below.

MotionDSP: A Brief History

The first engineering goal of MotionDSP in 2006 was to produce a practical super-resolution software that runs on generic hardware (i.e., common PC). Combining classical reconstruction-based super-resolution framework with the advanced spatial motion modeling, we were able to deliver the first working prototype to the market in late 2006. As a result, MotionDSP attracted new investors (In-Q-Tel) and early customers (US government agencies) for its "Ikena" forensic video products. Ikena was designed as a multi-threaded x86 Windows application, running efficiently on Intel- or AMD-based PCs.

2007 saw us refining our core super-resolution technology and expanding our customer base in the forensic and intelligence video market. In parallel, we developed and launched a consumer web

---

[9]Disclosure: the editor of this book is one of MotionDSP's founders.

[10]Although we found another (somewhat unexpected) driver for SR research in the hardware industry. At this moment, super-resolution is one of a handful of compute-thirsty multimedia applications having the potential to grow into a mainstream business. Many industry leaders (Intel, NVIDIA, AMD) have shown interest in informing their future hardware designs with critical SR computational operations.

application[11] that was designed as a cloud-based video enhancement service, offering automated video enhancement (including super-resolution) to the wide consumer audience. The site attracted tens of thousands of users and received excellent reviews (such as winning CNET's WebWare 100 award [38]). It also provided us with direct user feedback and gave us access to a large number of training videos for algorithm development.

In the first half of 2008 we successfully ported our parallel algorithms to NVIDIA's CUDA programming platform [22], which enabled a further three-to five fold improvement in performance over already highly optimized CPU software. As a result, NVIDIA became an investor and partner [30]. Our web service was transformed into a consumer-grade software application named vReveal. vReveal was launched in March 2009 to great reviews [10] and it reached hundreds of thousands of users within the first 9 months. In the meanwhile, the customer-base for Ikena grew to include many video forensic labs, government agencies, and the U.S. Navy.

14.3.1 Design Decisions

One of our earliest design decisions was to operate as a purely post-processing technology, without requiring any modifications of the existing encoder/channel/decoder processing pipeline. That way, we could reach as many users as possible.

Another early (and more controversial) choice was not to use any motion/encoding information already available in the video stream. While one can argue that re-using some of this precomputed data could, in theory, improve the quality of the final reconstruction, we have observed at least three separate problems with this approach. First, there is too much customization involved in supporting various codecs and their different profiles. In addition, this motion information is external to our core algorithm (which operates in an uncompressed pixel domain), and it might not be always available (this depends on the decoder over which we often have no control). Second, the confidence in motion fields required to drive the super-resolution algorithm is qualitatively different from those used to encode video.[12] Naturally, motion for video encoding is intentionally designed to be "flawed," as there is always

---

[11]www.fixmymovie.com

[12]Not only because motion accuracy has to be real-valued (non-quantized), but also because a simplistic motion model such as block-matching cannot support the SR algorithm.

TABLE 14.2
Improvement in Maximum Performance of Ikena over the
past 4 years; Input: VGA Frames (640 × 480),
Super-Resolution 2×, 7 Matching Frames. Measured on a
High-End PC.

	2006	2007	2008	2009
CPU (cores)	2	4	4	8
GPU (cores)	-	-	240	240
Max. processing speed (fps)	0.2	1.5	10	30

a prediction error layer to correct for it. Finally, even if used for the simple
initialization of the SR problem, time saved is minimal at best.

Another important design decision was to rely exclusively on off-the-shelf
hardware PC components. Although some competitive solutions (such as
Sarnoff's Acadia Video Processors [33]) have relied on custom ASICs to maximize processing speed,[13] we believed that only general purpose computing
platforms could provide sufficient flexibility and allow for quick algorithm
tune-ups. All our software is x86 compatible, with implementations of the
core video-processing engine available on Windows and Linux. In addition, a
highly parallel version of our core technology that runs on NVIDIA's CUDA-
capable GPUs is also available on both operating systems. (Our front-facing
applications, Ikena and vReveal, are Windows-only.) We have spent significant time working directly with NVIDIA and Intel on the optimization of our
software. As a result of this work, we have achieved dramatic performance improvement over the last three years (Table 14.2), and also enabled larger input
resolutions (up to 1080p) and new product features (real-time processing) in
more recent versions of our products.

14.4   Markets and Applications

In the past few years, we have seen the emergence of three clearly distinct markets for super-resolution technology: consumer, forensic, and so-called "real-
time" professional market. Their overview is presented in Table 14.3. As we
will shortly see, the applications in these markets are very different, but the
same core technology can be applied to improve video resolution in all of them.

---

[13]Application-specific integrated circuit.

TABLE 14.3 Markets for Super-Resolution in 2009.

	Consumer	**Forensic**	**Real-Time Pro**
Domain	full frame	details (ROI)	both
Resolution	$\leq$ HD	$\leq$ SD	$\leq$ SD
Interest	general quality	details	both
Typical input	digicam	CCTV	live feed (UAV)
User interaction	minimal	high	variable
High-speed processing	desired	not critical	critical
Hardware platforms	diverse	diverse	very high-end
Price tolerance	$	$$$	$$$$
Market size	tens of millions	tens of thousands	thousands

14.4.1   Forensic and Real-Time Markets: MotionDSP's Ikena

The forensic (sometimes referred to as the "law enforcement") market consists of thousands of forensic video labs, both public (police departments and government agencies) and private. Revealing more details is critical as any new information recovered from video evidence can potentially be very valuable. Subjective enhancements that do not contribute new information are of minimal value (e.g., improved white balance or better upscaling). Good forensic results are sometimes even visually incorrect (e.g., some analysts use negatives for better legibility of white-on-black text). Finally, the tolerance towards longer core processing times is much higher,[14] and, consequently, more complex algorithms can be used. Also, the users are typically willing to interact more with the software. As a matter of fact, user-interaction is desirable and can help get the best possible results.

Most of the installed base of cameras and imaging sensors in 2009 are still on the low end in terms of quality as the transition to new, digital surveillance systems is still in progress. Typical video in this market suffers from significant compression artifacts, as DVR[15] systems used for video storage try to fit days, and sometimes weeks, of footage onto limited disk space [7]. Many scenes are dark (as more incidents happen at night-time), and the dynamic range of these sequences is very low. Especially critical is the low frame rate, which makes any kind of multiframe analysis difficult. On the application level, the user interface should be rich, but also intuitive, in order to allow novice users to

---

[14]The duration of end-to-end processing, which includes tasks such as video import or setup actions, should still remain reasonably short, in order to process the high volume of cases typically handled by modern-days video forensic labs. Any software aiming to become a standard tool in this market should try to streamline various pre- and postprocessing tasks and closely follow established workflows and procedures.

[15]Digital Video Recording.

(a) Ikena Workspace. Main panels: Project List (left), Player (middle), Enhancements (right), Assets (bottom).

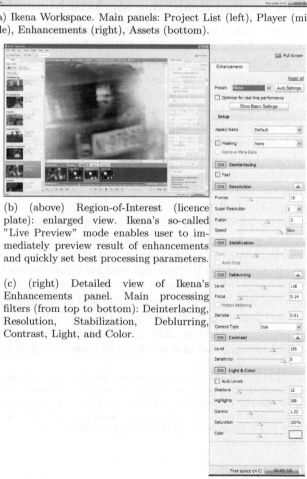

(b) (above) Region-of-Interest (licence plate): enlarged view. Ikena's so-called "Live Preview" mode enables user to immediately preview result of enhancements and quickly set best processing parameters.

(c) (right) Detailed view of Ikena's Enhancements panel. Main processing filters (from top to bottom): Deinterlacing, Resolution, Stabilization, Deblurring, Contrast, Light, and Color.

FIGURE 14.2: Screenshots of MotionDSP's Ikena 1.5.

use the software. In addition, the enhancement process needs to be relatively simple, straightforward, and reproducible.[16]

Currently, most video forensic labs do not use sophisticated multiframe postprocessing. Many police departments still rely on consumer-level video players, like Window Media Player, to review and visually inspect video. Adobe Photoshop is a popular tool for applying advanced spatial transforms and filters, but it lacks an automated solution for the multiframe enhancement. This is usually replaced with a series of long and tedious manual steps. Firstly, multiple video frames are aligned, so that Areas-of-Interest are matched in exactly the same spatial position in every frame. Secondly, temporal averaging of all available frames (sometimes called "frame averaging") is performed to reduce noise and increase effective signal-to-noise ratio. Ikena, MotionDSP's flagship forensic product, is designed to help operators quickly produce much higher quality results (due to superior motion estimation and SR processing) using a process that is significantly less labor-intensive (as many steps of the standard workflow are combined and performed automatically). Ikena's screenshots are presented in Figure 14.2.

The real-time professional SR market is dominated by the high-end surveillance or defense applications, that require live stream processing. Compared to forensic markets, there is a much smaller variety of video sources (e.g., most areal platform use some flavor of DV video). Due to strict specifications for the software's performance (processing speed, quality, and latency), the product is typically delivered on a certified PC workstation (such as R5400 by Dell) and on high-end video capture platform (such as Xena by AJA). Much of the video in this market is still analog; major improvements can be obtained by high-quality SR deinterlacing.

14.4.2 Consumers: MotionDSP's vReveal

Our first observation about the consumer market for super-resolution is that there is a phenomenal variety of video cameras (from cameraphones to prosumer[17] devices) and video resolutions (from QCIF to HD). Handling different formats and decoders is a daunting challenge in its own right, even before any enhancement is involved, but it is necessary. Regarding enhancement, the resolution improvement per se is less critical to the end-user than overall video quality perception. For example, using a multiframe SR approach simply to remove noise often produces excellent results (i.e., upscaling is not always required), especially after applying contrast or gamma corrections.

Additionally, consumer videos are almost always viewed in real-time, and the important enhancement metric is related not to better resolution or PSNR, but rather to the satisfaction level of the videographer and audience. In other

---

[16]The process has to produce the same results each time for given video and processing settings. Reliance on training data is not desirable.

[17]Market segment typically describing high-end consumer devices, often on a par with equipment used by professionals.

(a)

(b)

FIGURE 14.3: Screenshots of MotionDSP's vReveal 1.2. (a) Gallery view:
users can quickly navigate through their local folders containing video files
using built-in video browser (left panel). vReveal's internal video player (right
panel) can also be used for quick editing tasks (trim and rotate). (b) Enhancement view: Available basic filters (from top to bottom): Clean, 2× Resolution,
Sharpen, Auto Contrast, Stabilize, and Fill Light. vReveal also offers fully automated ("One Click Fix") video enhancement solution.

words, this is a very subjective market, and better video quality assessment tools (discussed in more details in Section 14.7) would go a long way in improving subjective video enhancement.

vReveal offers the market-leading video postprocessing solution for consumers (see Figure 14.3 for screenshots). vReveal is based on the same fundamental principles and technologies as Ikena.[18] In vReveal, we adjusted our core algorithms to match consumer's requirements and maximize subjective video quality. Despite using state-of-the-art multiframe post-processing technologies, vReveal is still very easy to use (for example, users can enhance their videos using "One Click Fix" option) and fast (it is optimized for NVIDIA's CUDA). In order to keep vReveal's focus on its core strength (video enhancement), we designed and positioned it, both in terms of features and the price, as a complementary tool to popular nonlinear editing (NLE) software tools, such as Adobe Premiere or Sony Vegas, rather than as their direct competitor.

14.5 Technology

In Section 14.3 we briefly discussed the origins and basic design principles of our technology. In order to classify our SR approach, we briefly review three major categories of SR algorithms: frequency-based, learning-based, and reconstruction-based algorithms.

Frequency-based SR. Frequency-based SR methods were first proposed in the mid-1980s [41]. Tsai and Huang tried to de-alias LRIs by exploiting phase differences among them. This first algorithm operated on observations that are free from degradation and noise. Though computationally attractive, frequency-based methods have significant disadvantages as the assumption of ideal sampling is unrealistic and observation noise is not considered. Most importantly, their global translation model is, for many applications, inappropriate.

Learning-based SR. Learning-based SR algorithms have gained significant popularity since their introduction around 2000 [1]. They incorporate strong application-dependent priors in order to reconstruct HRI. One of the most frequently used approaches is so-called "hallucination" reconstruction. This is the process of obtaining a high-resolution image (usually of a human face) from an input low-resolution image (or images, although an SR result can be obtained even from a single image), with the help of a large collection of other high-resolution images (a training set). While the result of this process often looks sharp and eye-pleasing, it is indeed "hallucinated" and is strongly dependent on the training set used. As such, its usefulness in forensic applications is limited. General consumer applications of learning-based SR methods

---

[18]Naturally, with fewer options and lesser capabilities.

are possible (e.g., improving quality of thumbnail images), but they require a large amount of training. In addition, it is often necessary to normalize video input (i.e., resize it to standard size, or compensate for a difference in pose) before applying a learning-based SR algorithm, which introduces complexity and limits overall robustness of the system.

Reconstruction-based SR. Classic SR framework, which we use for our work, is also known as reconstruction-based super-resolution. In it, the high-resolution image is computed as the MAP estimate from multiple LRIs, using explicit temporal correlation between them and various regularization priors. In the conventional approach to reconstruction-based super-resolution, temporal data correlation plays a critical role. There is a wide consensus that the accuracy of motion estimation is crucial for the success of the super-resolution process (although a couple of promising methods that use probabilistic motion estimation have been proposed recently [27], [26], [40]). While we cannot share details of our motion estimation and frame fusion process, we here provide the broad outline of our reconstruction process. Our algorithms combine the parametric global camera model[19] with the parametric motion over segmented frame patches and local, optical-flow based motion. This mixture of motion models is computed over several levels of Gaussian pyramid, which facilitates faster implementation and improves robustness. For the frame fusion step, we depart from the conventional "shift-and-add" algorithm, which in most cases ends up being computationally expensive (due to the unpredictable cost of non-uniform interpolation). Instead, inspired by motion-compensated temporal filtering [5] [17] [18] [35], which is effectively used in scalable video coding, we developed a proprietary fusion method that is very robust and efficient.

In terms of a desired outcome of the SR process, most algorithms (including ours) attempt to improve subjective visual quality. Depending on the application, some SR algorithms [16] use, as a reconstruction criterion, a more specific, application-driven metric, such as improving the success ratio of a face recognition system when SR is used as a preprocessing step.

14.5.1   Robust Parametric Motion Estimation

Classic implementation of super-resolution consists of four phases: initial motion estimation, motion refinement, frame fusion, and deblurring. As an illustration, we here outline the first two steps for computing the parametric global camera motion.

Initial motion estimation. Let I_1 and I_2 represent two successive images corresponding to two time instances t_1 and $t_2 = t_1 + \Delta t$. Also, let a parametric model M represent a mapping between the two image index spaces:

$$I_2 = M(I_1; a), \tag{14.1}$$

---

[19] An illustration of which is presented in Section 14.5.1.

where $a = (a_1, a_2, \ldots, a_n)$ is the vector of parametric motion.

The goal of parametric motion estimation is to determine the parameter vector a in such a way that:

$$\hat{a} = \underset{a}{\operatorname{argmin}} \|I_2 - M(I_1; a)\|_2^2, \tag{14.2}$$

which represents the standard minimization in the least square sense.

As a first step, the motion between successive video frames is calculated using the iterative Newton-Raphson algorithm for non-linear least square minimization [42], where the following iterative minimization procedure can be used to determine $\hat{a}$:

$$\hat{a}_{k+1} = \hat{a}_k - [H(\hat{a}_k)]^{-1} \cdot \frac{\partial J(\hat{a}_k)}{\partial a}, \tag{14.3}$$

where $H(\hat{a}_k)$ is an approximation of Hessian matrix:

$$H(\hat{a}_k) = \sum \left(\frac{\partial J(\hat{a}_k)}{\partial a}\right) \cdot \left(\frac{\partial J(\hat{a}_k)}{\partial a}\right)^T, \tag{14.4}$$

and $J(\hat{a}_k) = \|I_1 - M(I_1; \hat{a}_k)\|_2^2 = \|e(\hat{a}_k)\|_2^2$ is the squared motion prediction error in iteration k.

In order to improve the chance of finding a global minimum, this minimization algorithm is implemented on a Gaussian pyramid built on original video frames.[20] The number of levels in the pyramid is determined heuristically, based on the spatial resolution of the frame and temporal video dynamics. Typically, the model used for global camera motion at this stage consists of four motion parameters corresponding to camera zoom (z), rotation around principal camera axis (r), and camera's horizontal and vertical translation (t_x and t_y). This model describes the index space mapping that can be represented by homogeneous transform matrix T_{TRZ} (translation-rotation-zoom) of the following form:

$$T_{TRZ} = \begin{bmatrix} z & r & t_x \\ r & z & t_y \\ 0 & 0 & 1 \end{bmatrix}. \tag{14.5}$$

Using this notation, we can express mappings as matrix multiplication, which facilitates fast implementation. For example:

$$\begin{bmatrix} x_2 \\ y_2 \\ 1 \end{bmatrix} = T_{TRZ} \begin{bmatrix} x_1 \\ y_1 \\ 1 \end{bmatrix}, \tag{14.6}$$

---

[20] Additional techniques for improving robustness of the solution, such as smart initialization of motion parameter vector a through temporal motion modeling, can be deployed to avoid local minima.

is transforming pixel coordinate (x_1, y_1) from frame I_1 to coordinate (x_2, y_2) in frame I_2. As the initial motion estimation is completed for each pair of consecutive frames, frames from the user-definable temporal window are stored in the frame cache and remain available to next stages of SR algorithm.

Motion refinement. Before frame fusion, another motion estimation phase, called motion refinement, is performed over all frames present in the frame cache. The algorithm starts by concatenating initially available index-space mappings. These concatenated mappings, describing index mapping between each of the reference frames and the anchor frame (the frame that is being enhanced), are then used as the initial solution for the motion refinement phase.

An effective solution for the motion refinement in the case of global camera motion is a robust version of non linear iterative Newton-Raphson's least squares algorithm.[21] The robustness against outliers is achieved by substituting L_2 norm (14.2) with Geman-McClure M-estimator [34]:

$$\rho_{GM}(z) = \frac{z^2}{1+z^2}, \qquad (14.7)$$

where $z = \frac{e(a_k)}{\sigma}$ is the motion prediction error normalized with scale factor σ (approximating its standard deviation). Derivative of $\rho_{GM}(z)$, denoted $\Psi_{GM}(z) = \frac{2z}{(1+z^2)^2}$ is called the *influence function*, and, as we will show below, can be used to compute weights for a reformulated least-squares problem. Our robust motion estimation problem now becomes:

$$\hat{a} = \underset{a}{\operatorname{argmin}}\big[\rho_{GM}(\sigma_k, \ I_1 - M(I_2; a))\big]. \qquad (14.8)$$

In the context of Newton-Rapshon iterative solution, this means that Hessian approximation (14.4) needs to be modified in the following way:

$$H(\hat{a}_k) = \sum \big(\frac{\partial E(\hat{a}_k)}{\partial a}\big) \ diag(\Psi'_{GM}(\sigma_k, E(\hat{a}_k)) \ \big(\frac{\partial E(\hat{a}_k)}{\partial a}\big)^T, \qquad (14.9)$$

which means that each element of the Hessian matrix is weighted by the gradient of Geman-McClure potential:

$$\Psi'(z) \approx \frac{\Psi(z)}{2z}, \qquad (14.10)$$

calculated at each pixel. The standard variation σ is evaluated at each iteration using approximation formula for standard deviation of the signal with Gaussian distribution with zero mean [25]:

$$\sigma_k = 1.4826 \ median(|e(\hat{a}_k)|).$$

---

[21]Other higher-order parametric motion models, such as affine or projective, can replace the original TRZ model here.

Obviously, this simplified motion estimation will successfully describe the entire scene only in the case of global camera motion, without any objects exhibiting local motion. Although the robustness term in the refinement stage will prevent the parametric solution from exploding with relatively small outliers in the scene, foreground objects that occupy 25% or more of the frame will cause parametric camera motion estimation to fail. This is a soft (empirical) threshold, and it depends greatly on the quality and strength of textures in both background and foreground areas. However, if global motion estimation is sufficient for a particular application (e.g., stabilization), these motion parameters may be useful and are also very quick to compute. To guarantee a good performance in general case, this global camera motion model is extended to include both parametric local motion and local pixel-based motion (i.e., optical flow).

14.6 Results

In this section we present some super-resolution results from typical consumer and forensic sequences. All results are generated directly from either Ikena or vReveal (versions available to the public, for details visit `www.motiondsp.com`), and all processing parameters are accessible from application's graphical user interface (GUI). Processing time used to generate these results was minimal; even at maximum quality settings (e.g., using a frame cache of 50 frames, processing speed set to slow), processing time was under 200ms per frame for mobile and digicam clips and under 400 ms per frame for DV and HD content. Results are generated on an off-the-shelf PC workstation with Intel i7 Core CPU (2.66GHz) and G200 NVIDIA GPU (GTX280). The cost of this system is under $1,500 at the time of writing.

One of the most challenging super-resolution problems is a lack of a viable metric for assessing the quality of the reconstruction. While it is possible to design a scheme that will allow us to measure PSNR in a prepared experiment (when the ground truth is known), this means little in real-world scenarios (where ground truth is never known). That is why we have decided to avoid PSNR altogether; we present all our results in visual form (as frames from the output sequence). Even then, it is impossible to demonstrate video results accurately – the only solution is to see the full motion video. We, therefore, encourage the reader to try our software itself (available online at `www.motiondsp.com`), or to visit `www.motiondsp.com/SRbook` for input video sequences and results presented in this chapter.

FIGURE 14.4: Sample super-resolution results for cellphone (top) and digital still camera video (bottom). (a) Original frame, upscaled 2× using bicubic interpolation. Source: H.263 video at 170 kbps, 176×144 at 5 fps, (b) Super-resolution results (2×) using 21 matching frames. (c) Detail from original frame, upscaled 2× using bicubic interpolation. Source: MotionJPEG video at 2.7 Mbps, 320×240 at 15 fps, (d) Super-resolution results (2×) using 31 matching frames.

14.6.1   Mobile and Digital Still Camera Video

As mentioned earlier, cameraphones are quickly becoming the most popular video sources among consumers. These videos usually suffer from high compression, low dynamic range, and are very susceptible to motion blur as there are generally no advanced recording controls. In Figure 14.4 we show some typical results for super-resolution of consumer-level videos.

Users who shot video on mobile phones and digital still cameras (as well as the most casual video creators) care not so much about pure resolution improvements as about the noise removal. For these users, applying high-quality long-term motion-compensated temporal filtering (MCTF) leads to good results most of the time, with significantly reduced compression artifacts. Resolution gains are often obtained implicitly (i.e., without additional upscaling) as the result of MCTF. multiframe super-resolution technology produces especially good results in high-noise clips, where sensor noise (shot noise due to high ISO settings) is significant.

14.6.2 DV and HD Video

DV was one of the most popular formats for recording and playing back digital video from 1995 until the mid-2000s. DV video suffers from relatively low chroma resolution, interlacing, and unsophisticated compression (though at a very high bit-rate). Super-resolution can address all of these very successfully, and our software produces excellent results on most DV material. In addition to consumer applications, many CCTV[22] and surveillance systems deploy DV cameras. A large number of these systems is still in use and many are in need of postprocessing enhancement. In addition, as DV was extensively used by independent filmmakers and journalists, large collections of DV material exist and often need to be processed (cleaned and upscaled) before they are edited and combined with newer, higher-resolution content.

Consumer-grade high-definition (HD) cameras started to gain popularity in the late 2000s. They produce excellent video and are usually equipped with higher-quality lenses ensuring image sharpness. But despite a very high pixel count, residual aliasing is often present in HD footage. This high spatial-frequency content can be used to obtain excellent SR results (such as in Figure14.5) even at HD input resolutions, especially when video is shot in low light. While there are currently no commercially available video displays with resolutions above HD[23] to truly benefit from full-frame HD upscaling, high-quality denoising (obtained as a result of SR process with no upscaling), can significantly improve visual quality of HD content.

14.6.3 Handling Complex Motion

As we have discussed earlier, the key to successful super-resolution of full-frame video lies in a robust mix of different spatial motion models and their effective temporal recombination. Many spatial-domain SR algorithms use only translation or other parametric global motion that can produce good results for some scenes, but are incapable of handling more complex motion present in nearly all videos. To obtain high-quality reconstruction results without visually annoying artifacts (e.g., ghosting), local motion needs to be employed

---

[22]Closed-circuit television.

[23]And first experimental 4K displays (4096×2160) are just starting to emerge, like Panasonic's 4K HDTV [8].

(a)   (b)

(c)   (d)

(e)   (f)

FIGURE 14.5: Super-resolution (2×) of DV and HDV video. Left: Detail from
DV video (shot on Canon ZR200 DV camcorder, 720×480, interlaced, 29.97
fps). (a) Original frame (two fields weaved together), bicubic upscaling, (c)
Motion-compensated deinterlacing using two fields. (e) SR deinterlacing result
with 15 matching frames. Right: High Definition video shot on Canon HV30
HDV camera (1080p @ 24 fps) (b) original detail at 2× (bicubic upscaling),
(d) original detail with improved contrast, (f) SR result with 21 matching
frames.

for foreground handling, in addition to parametric global camera motion used to process background pixels.

Figure 14.6 illustrates the use of such local motion. The clip shown here (baby crawling) is representative of many videos shot indoors when additional lighting is not available. Original frame (upscaled 2× using bicubic interpolation) from the video sequence is shown in (a). The first obvious problem is that the video is underexposed (only about 30% of a full dynamic range is used). Frame (b) shows the same frame with contrast correction applied. While more details are visible due to improved contrast, noise is also much more visible. Furthermore, its random nature makes it even more perceptible when video is played at full frame rate.

The result of SR reconstruction using only global camera motion and 15 matching frames[24] is presented in (c). Notice much improved reconstruction of details on the floor and in the background (e.g., the character in the middle of blue flower – blanket, middle right) and overall greatly reduced noise levels. However, the use of global motion results in a blurry foreground object (baby) and unacceptable ghosting artifacts. Many details on both baby's head (hair) and body (pajamas) are lost, and baby's left hand has almost completely disappeared from the frame.

For best results, we perform joint motion-based segmentation and motion estimation that includes a combination of local parametric motion (e.g., baby's body), local pixel-based motion (e.g., baby's hair), and global camera motion (background). The result is a sharp output displayed in frame (d). Notice the preserved details in the background and denoised and sharper view of the baby (details on the back, right hand, and in the hair).

14.6.4   Practical Limits of Super-Resolution

Several efforts to determine the theoretical limit of resolution enhancement were carried out in the past. For example, Lin and Shum suggest that the limit of resolution enhancement on real-world video is 1.6×, though up to 5.7× improvement is possible under synthetic conditions [21]. As we have seen in previous sections, more frames with consistent content (and sufficient high-frequency information) will result in better reconstruction. Contrary to the result of Lin and Shum, and as demonstrated theoretically in [31], it is clear that there is no single answer to the question "how many frames is enough?" as this mostly depends on the amount of useful high-frequency information available in neighboring frames. Other factors that play a role in determining sufficient size of processing cache are video noise (compression or imaging), the geometry of the scene (in how many frames is the object at least partially visible), and corruption of frames, due to transmission errors or motion blur. We here present several heuristic rules based on our experience (as with any practical observation, exceptions are possible):

---

[24]7 frames before and 7 frames after the anchor frame.

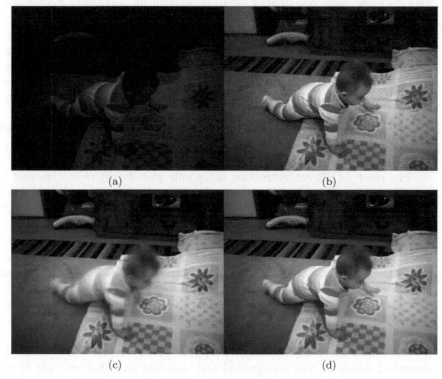

FIGURE 14.6: Example of efficient global and local motion mixture: (a) Original video frame (1.5Mbps AVC/H264 at 640×480 at 15fps), upscaled 2× using bicubic interpolation ; (b) Contrast correction applied (notice the increased noise visibility), 2× using bicubic interpolation; (c) SR result (2×, 15 matching frames) using only global parametric motion model. Notice the visible ghosting due to inability of this motion model to account for local motion. (d) SR (2×, 15 matching frames) using mixture motion model.

- The minimal number of frames required for useful reconstruction of aliased information is not smaller than 5;

- For better visual quality, as many as 50 frames can be effectively used;

- In analyzing the results of SR, human perception is what matters.

We illustrate these points through a real-world example. In Fig.14.7 (a)–(d), we show how the license plate number is recovered from a noisy video. Typically, a frame cache containing between 5 and 15 frames should provide enough aliased data[25] for a good reconstruction. Notice the improved legibility of the result using 5 matching frames compared to the original; while visually better,

---

[25]If such high-frequency spatial information is present in the video.

the result is still not definitely clear (especially the first digit). This illustrates a real-world problem of subjectivity of visual enhancement. For improved confidence, more frames should be used, when available. We have seen meaningful improvements in reconstruction quality even with an increase in frame cache size beyond 40 frames.

However, that much useful data is not always available. Much of surveillance video is captured at a very low frame-rate, sometimes in the 1–3 FPS[26] range. In such systems, the chance of capturing many frames containing different observations of objects of interest (e.g., people, vehicles) is significantly diminished. In order fully to exploit benefits of multiframe processing (such as super-resolution), any major decisions to upgrade a video surveillance system should be accompanied with a clear analysis of the content's nature (especially its dynamics).

Even when the goal of reconstruction is not forensic, using more frames (beyond 15) can help produce a better result. Fig. 14.7 (e)–(h) demonstrates how Moiré patterns can be remove using super-resolution principles.

14.7   Lessons Learned

Based on our experience, we have identified ten critical super-resolution issues that we discuss in more detail below.

1. **Not all algorithms are equal in front of the law.** As already discussed, the need for forensic super-resolution analysis is expected to continue to grow exponentially in the foreseeable future. Over the course of several years working with law enforcement, we became aware of the role that video evidence plays in the justice system and of the critical impact of legal procedure on the way video is handled and used as valid evidence. For example, great attention is given to preserving the integrity of any video material used, typically by employing various hashing techniques.

   In addition, legal proceedings can often take months or years to finish. For the result of any video processing algorithm to be accepted in court as evidence, the algorithm itself has to be fully deterministic and its results reproducible at any time during this process. Based on our experience, if the output video depends on a training set (as in many learning-based SR methods), its chance of being treated as a valid evidence is small.

   With this in mind, certification of current and future SR algorithms for their use in the judicial system presents a critical task in which

---
[26]Frames Per Second.

FIGURE 14.7: Demonstrating visual improvements in reconstruction when increasing the number of matching frames. (a)–(d) Improving resolution. a) Original frame upscaled 2× using bicubic interpolation, b) Original frame after contrast adjustment (2× bicubic upscaling), c) 2× SR result using 5 matching frames, d) 2× SR result using 15 matching frames. The license plate number is clearly readable only in the last image (NI-949 29). (e)-(h) Removing compression noise and Moiré patterns. (a) no SR (2× bicubic upscaling), (b) 2× SR, 5 matching frames, (c) 2× SR, 15 matching frames, (d) 2× SR, 50 matching frames. Notice how in this example 15 frames are not sufficient to completely remove Moiré patterns. MotionJPEG at 3.4 Mbps, 320 × 240 at 20 fps.

FIGURE 14.8: Three frames from the sequence affected by motion blur. Automatic detection of blurred frames could improve SR results in general by excluding these frames from processing.

super-resolution research community should play an important role.[27] In addition, all SR researchers who aim to advance practical super-resolution should be aware of this issue before starting their work.

2. **Motion blur is a significant practical problem.** Motion blurring presents a very significant reconstruction problem, especially when doing fully automated or semi-automated super-resolution. In many modern cameras, used for both consumer and professional applications, the user has no explicit access to camera gain and shutter speed. Even when shutter priority options exist on a camera, many users are unaware of it, and they record video using either default (Automatic) mode or "Night" mode, which usually makes the motion blurring problem even worse.

When video is captured in low light,[28] shutter speed in many consumer cameras usually drops too low and causes motion blurring. Small aperture and small sensor size of these devices just aggravate this problem as they limit the amount of collected light. What is even more problematic is that for all videos shot from handheld devices, direction and intensity of motion blurring varies greatly over time, and often from frame to frame. Any technique of estimating motion blur and using the time-varying point-spread function (PSF) for reconstruction is prohibitively computationally expensive (in addition to this problem being very hard in the general case). In the context of super-resolution, control over the minimum shutter speed is desirable but rarely available. New research on robust detection of motion blur would help automate the process of selecting frames that maximize the output quality.

Also, most camera manufacturers decide to use lower shutter speeds ag-

---

[27]In collaboration with makers of SR products, professional associations (such as LEVA [11]), and the justice system itself.

[28]A large majority of the clips shot indoors without additional lighting fall under this category.

gressively when total illumination of the scene decreases.[29] It is obvious that limiting the minimum shutter speed will result in noisier video, but this, at least, can be easier to fix in the postprocessing phase than motion blurring.

3. **Speed matters.** For SR algorithms to be of practical use to a larger number of users, they need be reasonably fast (i.e., run in real-time or close to it). We found that most users are motivated to wait to get results,[30] especially if their videos are of high personal value, but that the interest in doing SR postprocessing drops dramatically if significantly higher processing times are required. With high-definition video cameras becoming ubiquitous (e.g., Flip HD, Kodak HD) and providing SR algorithms with large amounts of data to process, this "need-for-speed" is even more pronounced. It is also important to keep in mind that the majority of consumers have hardware that is typically 2-3 years behind the consumer state of the art. This translates into 3–5× performance drop,[31] which should be considered when projecting the impact of any new technology, including SR.

Making the problem even more challenging, real-time SR applications (e.g., enhancement of UAV[32] video [9]), require not only highly efficient implementation but also a minimal latency. Algorithm latency becomes absolutely critical when super-resolution software is part of a closed-loop system. In that case, no forward-looking (i.e., future) frames may be used for reconstruction. In closed-loop systems,[33] maximal tolerable latency is typically around 100 ms (about three frames at 30 frames per second), which also includes the time needed for video acquisition and other tasks not directly related to core SR processing. This hard limit on the processing time should be included into any algorithm being developed for this market.

4. **User interaction with the algorithm is often desirable.** The subjective nature of the super-resolution reconstruction emphasizes the importance of human perception. In professional applications, users/operators are willing (and often motivated) to assist in the reconstruction. Examples of what the user may do to assist processing include manual object segmentation and feature identification (e.g., marking vertices of the licence plate). It is, therefore, beneficial to keep in mind a possibility for user's input when designing new SR algorithms as this can dramatically reduce the complexity of some image processing problems.

---

[29]Therefore, opting for more motion blur over more shot noise, which is likely driven by consumer preference for brighter video.

[30]But not for too long; typically up to five times of the video duration.

[31]Based on Moore's Law.

[32]Unmanned Areal Vehicle.

[33]For example, those including image trackers, where the output of processing is used to change the camera direction.

(a)   (b)

FIGURE 14.9: User interaction: In forensic applications, trained users are willing to interact with the software (in this case, by visually inspecting original frame (a) and marking a region-of-interest (b)) in order to quickly obtain a super-resolution result.

At the same time, increased user interaction does require more effort in designing the user interface and a larger investment in user training – the benefits of which are that optimal or near-optimal results can be obtained by a large number of users.

5. **Beauty is in the eyes of the beholder.** As many researchers before us have concluded, it is often hard to draw direct correlation between PSNR and visual quality (even worse, results are sometimes deceiving, with higher PSNR video of obviously lower visual/forensic quality). In addition, the nature of super-resolution makes the problem of objectively estimating resolution/quality gains inherently impossible, as in real applications there is no ground truth that we can use as a reference. In this context, an effort on standardizing new subjective enhancement metrics when ground truth is not available is very much needed [43], [44], [6].

6. **Some sources are better then others (in SR terms).** User generated content (UGC) is, in general, a great source for super-resolution.[34] Most of the time, UGC represents the first generation of video. UGC is typically shot using handheld cameras, which introduces necessary subpixel relative frame displacement and allows for full-frame resolution

---

[34]Since large quantities of user-generated content are freely available online, the easiest way to get access to it is by using one of the video downloaders, such as [29]. (Be sure to understand all legal implications before using any such tool.) Also, different sites store their videos in different formats for video streaming (typically through Adobe's Flash Player), but it is often possible to obtain original content, which was not additionally transcoded on the server. At MotionDSP, we used UGC uploaded to our own service (www.fixmymovie.com) to test and develop our algorithms.

(a)   (b)   (c)

FIGURE 14.10: Examples of user-generated and preproduced content. (a) Good video source, often found in user-generated content; (b) Poor video content for super-resolution, usually professional footage mixing natural video with synthetic elements; (c) Explicit masking of synthetic content can significantly improve super-resolution and stabilization results but it requires user intervention (screenshot from Ikena's built-in mask editor).

improvement even when there is no intrinsic motion in the scene. Also, UGC rarely includes text overlay or other static content mixed with natural video. One example of UGC video frame is shown in Figure 14.10a.

On the other hand, pre-produced video containing both natural and synthetic content (Figure 14.10b) can often be difficult to segment and process automatically. In general, this problem is very hard; one solution to limiting the impact of burned-in text is to ask users for an explicit segmentation of natural and synthetic content. We implemented this as a standard feature in Ikena UAV, where the user can easily create a new "mask" in the integrated mask editor (illustrated in Figure 14.10c). Similarly to our previous point on user interaction with the algorithm, even a small input from the operator can go a long way in reducing complexity of the problem.

7. **Super-resolution is a great denoiser.** Even without explicit upscaling, SR can improve perceivable resolution by reducing noise. This is especially effective at larger frame resolutions (VGA and higher), where it also makes sense from the computation point of view (processing larger frames require more memory and time, so skipping upscaling is sometimes even necessary). More importantly, the multiframe nature of SR means that it can implicitly help with many common problems with video, including shot noise and compression artifacts. With more and more people being involved with the process of video creation – and often shooting in subideal conditions (resulting in noisy videos) – SR denoising, indeed, has a very large addressable market and a very promising future.

8. **It is generally good to avoid core design customization.** Our ap-

proach to core SR algorithm design was to limit the level of customization in order to expand the range of applications that could benefit from it. For mass-market applications, it is rarely beneficial to tie the algorithm's performance to a particular class of videos.[35] The finite research and development resources are usually better spent on making core algorithmic improvements that will result in improved general case performance.

9. **Custom imaging solutions can benefit from incorporating SR into their designs.** If the imaging system is envisioned to work with a super-resolution postprocessing system [32], [39], it is a good idea to include some requirements of the SR algorithm into the very design of the imaging pipeline (for example, the amount of aliasing in the camera output might be controlled to optimize SR processing). As a rule of thumb, when used in any forensic application, sharper, aliased video is always preferred to smoother, non-aliased video, regardless of subjective appearance.

10. **Super-resolution is not the end of it.** As we have illustrated in this chapter, super-resolution can produce great results and can make a significant impact in both consumer and professional applications. Nevertheless, super-resolution is not the only way to improve video quality, nor is it always the most computationally effective solution. This is particulary true when video is processed and viewed in real-time. In that case, the most effective way for maximum quality improvement is to combine super-resolution with other video enhancement technologies, especially those stemming from in-depth understanding of the scene motion (like robust stabilization). Other high-impact enhancements may include adaptive contrast enhancement, histogram equalization, out-of-focus deblurring, high-quality deinterlacing, rolling shutter correction, or frame-rate upscaling. Joint optimization of visual impact and time spent on each processing block presents an interesting research opportunity. Also, sharing and reusing already computed data between various processing blocks is crucial for an efficient implementation; ideally, this should be incorporated early in the design process.

---

[35]For example, we tried to include the information about key-frames into the SR reconstruction process (by assigning larger weights to key-frames during reconstruction). While we saw certain improvements in reconstruction quality when this method was applied on videos compressed with some older codecs, this approach, in general, was not very successful (as newer codecs, like H.264/AVC, have much better control over temporal quality variation).

14.8 Conclusions

Improving video resolution is a very important task and there are numerous real-world applications where super-resolution is adding the real value. multiframe video processing has a great potential to become a mainstream "killer" application, especially with the increase in compute power available in the vicinity of the video decoder. We hope that our experience will help in bringing more exciting technologies from labs to the market.

Bibliography

[1] S. Baker and T. Kanade. Hallucinating faces. In *Fourth International Conference on Automatic Face and Gesture Recognition*, pages 83–88, 2000.

[2] B. Bascle, A. Blake, and A. Zisserman. Motion deblurring and super-resolution from an image sequence. In *ECCV '96: Proceedings of the 4th European Conference on Computer Vision-Volume II*, pages II:571–582, 1996.

[3] BBC News. 1,000 cameras solve one crime. http://news.bbc.co.uk (Web site), August 2009. http://news.bbc.co.uk/2/hi/uk_news/england/london/8219022.stm.

[4] M. Ben-Ezra, A. Zomet, and S. K. Nayar. Video super-resolution using controlled subpixel detector shifts. *IEEE Transactions on Pattern Analysis and Machine Intelligence*, 27(6):977–987, 2005.

[5] N. Bozinovic. *Advanced motion modeling for 3D video coding*. PhD thesis, Boston University, ECE Deptartment., December 2006.

[6] T. Brand and M. P. Queluz. No-reference image quality assessment based on DCT domain statistics. *Signal Processing*, 88(4):822–833, 2008.

[7] D. Bulwa. San Francisco security cameras' choppy video. http://www.sfgate.com, January 2008. http://www.sfgate.com/cgi-bin/article.cgi?f=/c/a/2008/01/28/MN37TKH6O.DTL.

[8] B. Drawbaugh. Panasonic does it again, introduces 152-inch 3D 4K HDTV. http://www.engadget.com (Web site), 2009. http://www.engadget.com/2010/01/06/panasonic-does-it-again-introduces-152-inch-3d-4k-hdtv/.

[9] C. Drew. Military is awash in data from drones. *The New York Times*, page A1, January 10, 2010. http://www.nytimes.com/2010/01/11/business/11drone.html.

[10] A. Eisenberg. Those big bright eyes may soon be brighter. *The New York Times*, page BU3, June 20, 2009. http://www.nytimes.com/2009/06/21/business/21novel.html.

[11] Law Enforcement and Emergency Services Video Association. www.leva.org (Web site), 2010. http://www.leva.org.

[12] S. Farsiu, M. Elad, and P. Milanfar. Video-to-video dynamic super-resolution for grayscale and color sequences. *EURASIP Journal of Applied Signal Processing, Special Issue on Superresolution Imaging*, 2006:1–15, 2006.

[13] S. Farsiu, D. Robinson, M. Elad, and P. Milanfar. Advances and challenges in super-resolution. *International Journal of Imaging Systems and Technology*, 14(2):47–57, 2004.

[14] S. Farsiu, D. Robinson, M. Elad, and P. Milanfar. Fast and robust multi-frame super-resolution. *IEEE Transactions on Image Processing*, 13(10):1327–1344, Oct. 2004.

[15] B. Girod, A.M. Aaron, S. Rane, and D. Rebollo Monedero. Distributed video coding. *IEEE Transactions on Image Processing*, 93(1):71–83, January 2005.

[16] B. K. Gunturk, A. U. Batur, Y. Altunbasak, M. H. III Hayes, and R. M. Mersereau. Eigenface-domain super-resolution for face recognition. *IEEE Transactions on Image Processing*, 12(5):597–606, 2003.

[17] S.-T. Hsiang and J.W. Woods. Invertible three-dimensional analysis/synthesis system for video coding with half-pixel-accurate motion compensation. In *SPIE Visual Communications and Image Processing*, volume 3653, pages 537–546, January 1999.

[18] S.-T. Hsiang, J.W. Woods, and J.-R. Ohm. Invertible temporal subband/wavelet filter banks with half-pixel-accurate motion compensation. *IEEE Transactions on Image Processing*, 13:1018–1028, August 2004.

[19] iSuppli Corporation. Camera penetration in mobile phones, 2004-2010. http://www.isuppli.com (Web site).

[20] J. Kincaid. Youtube mobile uploads up 400 percent since iPhone 3GS launch. http://www.techcrunch.com (Web site), June 2009. http://www.techcrunch.com/2009/06/25/youtube-mobile-uploads-up-400-since-iphone-3gs-launch/.

[21] Z.C. Lin and H.Y. Shum. Fundamental limits of reconstruction-based superresolution algorithms under local translation. *IEEE Transactions on Pattern Analysis and Machine Intelligence*, 26(1):83–97, January 2004.

[22] NVIDIA Corporation. NVIDIA CUDA. http://www.nvidia.com (Web site), 2009. http://www.nvidia.com/cuda.

[23] NVIDIA Corporation. NVIDIA Tesla personal supercomputer. www.nvidia.com (Web site), 2009. http://www.nvidia.com/object/personal_supercomputing.html.

[24] NVIDIA Corporation. Universities teaching CUDA. www.nvidia.com (Web site), 2009. http://www.nvidia.com/object/cuda_university_courses.html.

[25] P.J. Rousseeuw and A.M. Leroy. *Robust Regression and Outlier Detection*. John Wiley & Sons, Inc. New York, NY, USA, 1987.

[26] M. Protter and M. Elad. Super resolution with probabilistic motion estimation. *IEEE Transactions on Image Processing*, 18(8):1899–1904, August 2009.

[27] M. Protter, M. Elad, H. Takeda, and P. Milanfar. Generalizing the non-local-means to super-resolution reconstruction. *IEEE Transactions on Image Processing*, 18(1):36–51, Jan 2009.

[28] R. Puri, A. Majumdar, and K. Ramchandran. Prism: A video coding paradigm with motion estimation at the decoder. *IEEE Transactions on Image Processing*, 16(10):2436–2448, October 2007.

[29] RealNetworks, Inc. Video downloader. http://www.real.com (Web site), 2010. http://www.real.com/realplayer/video-downloader.

[30] B. D. Rizzo. NVIDIA announces strategic partnership with MotionDSP. http://www.nvidia.com (Web site), 2008. http://www.nvidia.com/object/io_1222345067642.html?_templateId=320.

[31] D. Robinson and P. Milanfar. Statistical performance analysis of super-resolution. *IEEE Transactions on Image Processing*, 15(10):1413–1428, June 2006.

[32] D. Robinson and D. Stork. Joint digital-optical design of imaging systems for gray-scale objects. In *SPIE European Optical Design Conference*, volume 7100, pages 710011–710011.9, 2008.

[33] Sarnoff Corporation. Acadia video processors. http://www.sarnoff.com (Web site), 2010. http://www.sarnoff.com/products/acadia-video-processors.

[34] H. S. Sawhney and S. Ayer. Compact representations of videos through dominant and multiple motion estimation. *IEEE Transactions on Pattern Analysis and Machine Intelligence*, 18:814–830, 1996.

[35] H. Schwarz and D. Marpe and. T. Wiegand. MCTF and scalability extension of H.264/AVC. In *Procedings Picture Coding Symposium*, December 2004.

[36] S. Seguin. NVIDIA launches Tesla personal supercomputer. http://www.tomshardware.com (Web site), 2008. `http://www.tomshardware.com/news/Nvidia-Tesla-Supercomputer,6616.html`.

[37] MG Siegler. Every minute, just about a days worth of video is now uploaded to YouTube. http://www.techcrunch.com (Web site), 2009. `http://www.techcrunch.com/2009/05/20/every-minute-just-about-a-days-worth-of-video-is-uploaded-to-youtube/`.

[38] Webware staff. Webware 100 winner: FixMyMovie. http://news.cnet.com (Web site), 2008. `http://news.cnet.com/8301-13546_109-9914732-29.html`.

[39] D. Stork and D. Robinson. Theoretical foundations for joint digital-optical analysis of electro-optical imaging systems. *Applied Optics*, 47:B64–B75, April 2008.

[40] H. Takeda, P. Milanfar, M. Protter, and M. Elad. Superresolution without explicit subpixel motion estimation. *IEEE Transactions on Image Processing*, 18(9):1958–1975, Sep. 2009.

[41] R. Y. Tsai and T. S. Huang. Multiframe image restoration and registration. *Advances in Computer Vision and Image Processing*, 1:317–339, 1984.

[42] Y. Wang, Y-Q. Zhang, and J. Ostermann. *Video Processing and Communications*. Prentice Hall PTR, Upper Saddle River, NJ, USA, 2001.

[43] Z. Wang and A. C. Bovik. *Modern Image Quality Assessment*. Morgan and Claypool Publishers, 2006.

[44] Z. Wang, H.R. Sheikh, and A.C. Bovik. No-reference perceptual quality assessment of JPEG compressed images. In *IEEE International Conference on Image Processing*, pages I: 477–480, 2002.

Index

T - #0065 - 101024 - C0 - 234/156/27 [29] - CB - 9781439819302 - Gloss Lamination